Ancient Seismites

Edited by

Frank R. Ettensohn
Department of Geological Sciences
University of Kentucky
Lexington, Kentucky 40506-0053
USA

and

Nicholas Rast
Department of Geological Sciences
University of Kentucky
Lexington, Kentucky 40506-0053
USA

and

Carlton E. Brett
Department of Geology
University of Cincinnati
Cincinnati, Ohio 45221-0013
USA

THE
GEOLOGICAL
SOCIETY
OF AMERICA

Special Paper 359

3300 Penrose Place, P.O. Box 9140 ▪ Boulder, CO 80301-9140 USA

2002

Copyright © 2002, The Geological Society of America, Inc. (GSA). All rights reserved. GSA grants permission to individual scientists to make unlimited photocopies of one or more items from this volume for noncommercial purposes advancing science or education, including classroom use. For permission to make photocopies of any item in this volume for other noncommercial, nonprofit purposes, contact the Geological Society of America. Written permission is required from GSA for all other forms of capture or reproduction of any item in the volume including, but not limited to, all types of electronic or digital scanning or other digital or manual transformation of articles or any portion thereof, such as abstracts, into computer-readable and/or transmittable form for personal or corporate use, either noncommercial or commercial, for-profit or otherwise. Send permission requests to GSA Copyright Permissions, 3300 Penrose Place, P.O. Box 9140, Boulder, Colorado, 80301-9140, USA.

Copyright is not claimed on any material prepared wholly by government employees within the scope of their employment.

Published by The Geological Society of America, Inc.
3300 Penrose Place, P.O. Box 9140, Boulder, Colorado 80301
www.geosociety.org

Printed in U.S.A.

GSA Books Science Editor Abhijit Basu
Cover design by Heather L. Sutphin

Library of Congress Cataloging-in-Publication Data

Ancient Seismites / edited by Frank R. Ettensohn, Nicholas Rast, and Carlton E. Brett.
    p. cm. — (Special papers ; 359)
  Includes bibliographical references.
  ISBN 0-8137-2359-0
    1. Seismites—Congresses. I. Ettensohn, Frank R. II. Rast, Nicholas, 1927- III. Brett, Carlton E. (Carlton Elliot) IV. Special papers (Geological Society of America) ; 359.

QE472 .A53 2002
551.22—dc21

2002020336

Cover: Two Upper Ordovician (lowermost Edenian), probable seismite horizons in the upper tongue of the Tanglewood Limestone Member, Lexington Limestone, sharply overlie poorly fissile, medium-gray, calcareous shales in a tongue of the Clays Ferry Formation. The locale is situated on the southwestern margin of the Tanglewood buildup where calcareous, Tanglewood, shoal-water sands pinched out into the deeper water Clays Ferry muds (see Chapter 13, Fig. 2). The lower horizon is part of a channel-like thickening that consists of a jumble of blocks and flow rolls up to 1.9 m (6.2 ft) thick, which were deformed and foundered into underlying Clays Ferry shales during an earthquake. This horizon was originally a unit of thin-bedded, fine- to medium-grained, crossbedded calcarenite interbedded with shales and capped by a hardground. Features not visible in the photo indicate deformation by liquefaction. Lower parts of the second deformed horizon sharply truncate the lower horizon and consist of randomly oriented clots and clasts of sediment floating in a massive, argillaceous calcisiltite with poorly defined, secondary flow lamination. Except for flow rolls reworked from the seismite below, the unit has been completely homogenized and represents very intense deformation by fluidization during an earthquake. View is to the east along U.S. Highway Bypass 127, 2.0 mi northwest of Lawrenceburg, Kentucky. Map case for scale is 32 cm (1.1 ft) high. Photo by F.R. Ettensohn.

10 9 8 7 6 5 4 3 2 1

Contents

Dedication v
    Nicholas Rast, 1927–2001

Preface vii

Acknowledgments ix

1. *Distinguishing seismic from nonseismic soft-sediment structures: Criteria from seismic-hazard analysis* 1
    Russell L. Wheeler

2. *Paleoliquefaction studies in continental settings* 13
    Stephen F. Obermeier, Eric C. Pond, Scott M. Olson, and Russell A. Green

3. *Late Quaternary paleoseismites: Syndepositional features and section restoration used to indicate paleoseismicity and stress-field orientations during faulting along the main Lima Reservoir fault, southwestern Montana* 29
    Mervin J. Bartholomew, Michael C. Stickney, Edith M. Wilde, and Robert G. Dundas

4. *Stratigraphic evidence of coseismic faulting and aseismic fault creep from exploratory trenches at Mt. Edna Volcano (Sicily, Italy)* 49
    Luca Ferreli, Alessandro Maria Michetti, Leonello Serva, and Eutizio Vittori

5. *Mid-Tertiary paleoseismites: Syndepositional features and section restoration used to indicate paleoseismicity, Atlantic Coastal Plain, South Carolina and Georgia* 63
    Mervin J. Bartholomew, Brendan M. Brodie, Ralph H. Willoughby, Sharon E. Lewis, and Frank H. Syms

6. *Late Pleistocene soft-sediment deformation structures interpreted as seismites in paralic deposits in the city of Bari (Apulian foreland, southern Italy)* 75
    Massimo Moretti, Piero Pieri, and Marcello Tropeano

7. *Indicators of paleoseismicity in the lower to middle Miocene Guadagnolo Formation, central Apennines, Italy* 87
    Goffredo Mariotti, Laura Corda, Marco Brandano, and Giacomo Civitelli

8. *Stratigraphic and sedimentological evidence for late Paleozoic earthquakes and recurrent structural movement in the U.S. Midcontinent* 99
    Daniel F. Merriam and Andrea Förster

9. *Critical evaluation of possible seismites: Examples from the Carboniferous of the Appalachian basin* 109
   Stephen F. Greb and Garland R. Dever Jr.

10. *Late Mississippian paleoseismites from southeastern West Virginia and southwestern Virginia* ... 127
    Kevin G. Stewart, John M. Dennison, and Mervin J. Bartholomew

11. *Anomalous paleoflow orientations: A potential methodology for determining recurrence rates and magnitudes in paleoseismic studies* 145
    Gerald J. Smith and Robert D. Jacobi

12. *Seismically induced soft-sediment deformation in some Silurian carbonates, eastern U.S. Midcontinent* 165
    Charles M. Onasch and Charles F. Kahle

13. *Interpreting ancient marine seismites and apparent epicentral areas for paleo-earthquakes, Middle Ordovician Lexington Limestone, central Kentucky* 177
    Frank R. Ettensohn, Mark A. Kulp, and Nicholas Rast

Dedication

**Nicholas Rast
(1927–2001)**

The death of Nicholas Rast on August 28, 2001, due to complications arising from earlier surgery, ended a distinguished career in research, teaching, and service, which encompassed nearly all aspects of geology.

Nick's career in geology spanned 53 years and ranged from mid-eastern Asia and Europe to North America. His ability to integrate diverse aspects of geology and his wide experience throughout the Caledonian-Appalachian mountain belt made him a major contributor in understanding the belt at a time when mountain belts were first being explained in terms of plate tectonics. In addition to Appalachian tectonics, his breadth of knowledge extended to the fields of structural geology, seismology, igneous and metamorphic petrology, and regional geology, all of which he saw as parts of a larger tectonic milieu.

Not long after arriving at the University of Kentucky as Hudnall Professor of Geology in 1979, Nick was quick to recognize that most of the unusual convoluted beds that were present in the Middle-Upper Ordovician Lexington Limestone were probably "seismites," an interpretation that did not resound well with many U.S. geologists at the time. As papers in this volume will point out, Nick was almost certainly correct, and Nick never tired in his determination to demonstrate the importance of seismicity in epicontinental sedimentation and its relationship to larger tectonic events at continental margins.

Nick was the moving spirit behind the symposium "Interpreting Fossil Earthquakes from the Stratigraphic Record" in Toronto in 1998, from which this volume evolved. As a volume editor, he was instrumental in seeking appropriate papers and in seeing that many of those papers made it into the volume, a task that he continued until shortly before his death. His keen eye, sharp wit, sage advice, and eagerness to help others understand Earth will be greatly missed. However, several of Nick's later aspirations are reflected in this Special Paper, and we believe that it is a suitable memorial to Nick and his work. Hence, with admiration and affection, we dedicate this volume to his memory.

Preface

This volume is the result of a symposium held at the October 1998 Annual Meeting of the Geological Society of America in Toronto, Canada. It represents the record of the first session dedicated wholly to ancient seismites at a national GSA meeting. The timing of the volume is propitious, because with the recent advent of major earthquakes in the northwestern United States, El Salvador, India, and Turkey, earthquakes are once again in the public and geologic eye. The damage and loss of life are of paramount concern to all, so preventing this devastation and predicting earthquakes are becoming increasingly common topics of discussion. We believe that there is much important information to be gleaned on both topics from the sedimentary record, especially relating to number, location, intensity, and recurrence interval of ancient earthquakes. As the seismograph was not invented until the 1880s, most of our knowledge of previous earthquakes and their effects is essentially pre-instrumental, because historically, relatively few earthquakes have been instrumentally recorded. Thus, searches for the indicators and effects of older earthquakes as predictive examples were begun by, and continue to be largely the concern of, engineers, who have been attempting to identify late Quaternary, earthquake-altered sequences on the basis of sometimes debatable geomorphic and depositional features. Although geologists have made suggestions that some of the penecontemporaneous, soft-sediment deformation commonly observed in the sedimentary rocks might have had seismogenic origins, geologists were relative latecomers in interpreting likely seismogenic effects from the geologic record. This is surprising, in view of the important role of earthquakes in the plate-tectonics paradigm. Nonetheless, the interpretive term "seismite" for seismogenically produced or altered structures and patterns in the sedimentary record was not introduced into the geological vocabulary until 1969, by A. Seilacher.

Since that time, there has been a plethora of geologic papers about seismites, but nearly all have dealt with deformed clastic sediments of late Tertiary or Quaternary age on active margins, where location and age were significant, corroborative lines of evidence for seismogenic origin. There are, however, just as many, if not more, examples of soft-sediment deformation in intraplate settings within older Phanerozoic rocks that display characteristics consistent with a seismogenic origin. Given the significance and frequency of historic intraplate seismicity in regions like the New Madrid–Wabash Valley area of the east-central United States, and most recently, the Kutch area of western India, the presence of ancient intraplate seismicity is predictable. In fact, the continued occurrence of intraplate seismicity merely confirms the fact that continental amalgamations retain a "memory" of their accretion, and that this memory can be "jogged" later by superimposed stresses. The result is the repetitive regeneration of motion on old fault planes. An important point here is that even in plate interiors, many old, basement fault traces are present and that distal stressors as seemingly benign as glacial loading or as inimical as coeval, plate-margin orogenies, though hundreds of kilometers removed, may reactivate them. Yet, despite this reality, the possibility of seismites in older Phanerozoic rocks has only rarely been considered, and evidence has commonly been overlooked or misinterpreted. Part of the interpretation problem in many plate-interior settings is the fact that potential seismites are present in ancient marine rocks, especially carbonates, for which modern analogues are absent. Clearly, few, if any, soft-sediment deformation features are wholly unique to seismites, and there seems to be a greater number of alternative deformational processes, such as storm- and wave-induced loading, that must be excluded in marine settings before a seismogenic interpretation can be assumed.

In this volume we have tried to assemble papers that deal with the problems of interpreting seismites in a variety of terrestrial and marine settings. Overall, these papers help us to answer two questions: (1) On the basis of more recent examples, what are the preeminent characteristics of seismites? (2) How can we use these characteristics to interpret possible seismites in older Phanerozoic rocks from more distal, commonly marine, intraplate settings? The papers are arranged in nearly reverse chronological order (younger to older) relative to the ages of the rocks involved, but the order also approximates relative proximity to active margins, the younger examples being more proximal and the older ones more distal.

The initial two chapters, by Wheeler and by Obermeier et al., respectively, define basic terms and discuss the criteria that have been used and, perhaps more important, should be used, in determining the likelihood of seismogenic origin for soft-sediment deformation. Although both papers work with criteria that are most applicable to Quaternary terrestrial settings, the reasoning employed and some criteria are equally valuable in analyzing pre-Quaternary marine or terrestrial settings.

The next three chapters, by Bartholomew et al. (Chapter 3), Ferreli et al. (Chapter 4), and Bartholomew et al. (Chapter 5), deal with seismites associated with fault scarps in volcanic and intraplate settings. The connection between fault scarps and seismites might seem to be an unequivocal one. However, aseismic creep and alteration by subsequent colluvial processes can make discrimination of ultimate seismic causes difficult. Criteria for differentiation of origins are presented, and the two papers by Bartholomew and his coauthors illustrate the use of fossils in support of interpretations.

The papers by Moretti et al. (Chapter 6), and by Mariotti et al. (Chapter 7) deal with probable Quaternary and Tertiary seismites on currently and formerly active margins in Italy. In dealing with marginal-marine and marine sediments respectively, both chapters emphasize the importance of excluding other possible nonseismic causes. The chapter by Mariotti et al. is one of the first studies to utilize petrographic and compositional analyses in support of seismite interpretations.

The last six chapters deal with probable seismites from Paleozoic marine and marginal-marine sediments in various intraplate settings. Pennsylvanian seismites are discussed in chapters by Merriam and Förster (Chapter 8) and by Greb and Dever Jr. (Chapter 9); Mississippian seismites are discussed by Greb and Dever Jr., (Chapter 9) and by Stewart et al. (Chapter 10); Devonian seismicity is discussed by Smith and Jacobi (Chapter 11); Silurian seismites are discussed by Onasch and Kahle (Chapter 12); and Ettensohn et al. present work on Ordovician seismites in Chapter 13. In nearly every one of these chapters, the significance of syndepositionally active faults in producing seismites is stressed. Thus, the coincidence of soft-sediment deformation with potentially active basement structures, commonly indicated at the surface by more recent, extant structures, can be an important line of evidence supporting seismogenic origins. Moreover, the coeval nature of this intraplate, syndepositional fault activity and resulting deformation with craton-margin orogenies, including the Ouachita, Alleghanian, Acadian, Salinic, or Taconian orogenies, indicated or inferred in the above chapters, points out the critical role of the far-field or flexural forces generated by these orogenies in reactivating basement structures hundreds of kilometers inboard of the orogeny.

Three of the chapters dealing with Paleozoic deformation provide other insights. Chapter 9 by Greb and Dever Jr. shows the full development of arguments used in discriminating between likely seismic and nonseismic causes. Chapter 11 by Smith and Jacobi, on the other hand, does not specifically deal with seismites, but rather uses the stratigraphic record of redirected paleoflow in time and space to suggest the magnitude and recurrence intervals of earthquakes along faults that were responsible for the redirection. Finally, in Chapter 13, Ettensohn et al. suggest ways in which epicentral areas may be located and ways in which the issues of very ancient earthquake intensity, directivity, recurrence, clustering, and site effects may be approached from the stratigraphic record of marine rocks.

In the time since the GSA symposium was convened, the interest and attention given to seismites by the geologic community has continued to grow. We hope that criteria developed in this volume will help in discriminating seismic from nonseismic causes in the older geologic record so that the true informational and predictive value of these deposits can be recognized, not only on the active margins where they were first recognized, but also across the expansive continental interiors, where they appear to be just as common in the epicontinental sedimentary cover. It is our additional hope that this volume will introduce an even broader group of geologists to the recognition, study, and application of seismites in their studies of Earth's history.

Frank R. Ettensohn
Nicholas Rast
Carlton E. Brett
March 2001

Acknowledgments

We appreciate very much the work of the authors in preparing their contributions for this volume. Mary Sue Johnson was especially helpful in keeping track of papers, authors, and reviewers. Our special thanks goes to the reviewers for their constructive criticism of the papers. Reviews for this volume were provided by M. Bennett, C.E. Brett, M.C. Chapman, K.A. DeJong, R.V. Demicco, D.I. Doser, S. Greb, P.H. Heckel, D.L. Kidder, D.R. Kolata, R.L. Martino, W. McGuire, W.J. Nelson, J. Pashin, R.K. Pickerill, M.C. Pope, B.R. Pratt, D. Rust, J. Schieber, G. Schumacher, J.W. Sears, J.P. Smoot, R. Smosna, P.A. Thayer, W.A. Thomas, C.H. Trupe, M.P. Tuttle, R.B. Van Arsdale, G. Wiezorek, D. Woodrow, and two anonymous readers. Finally, we thank GSA Books Science Editor A. Basu for helping to keep our project alive and moving forward.

Distinguishing seismic from nonseismic soft-sediment structures: Criteria from seismic-hazard analysis

Russell L. Wheeler
U.S. Geological Survey, P.O. Box 25046, MS 966, Denver, Colorado 80225, USA

ABSTRACT

Most studies of the geologic records of prehistoric earthquakes are driven by the need of seismic-hazard analyses for estimates of the locations, ages, and magnitudes of individual large earthquakes. In North America, most such studies analyze surface ruptures and earthquake-induced sand blows, dikes, and sills. In contrast, small, soft-sediment structures in one or a few beds, possibly induced by seismic shaking, are little used in North American hazard analyses. The reason is that present methods for studying such structures often cannot demonstrate an earthquake origin, rule out alternatives, or estimate location, age, and magnitude as well as can methods developed for the study of surface ruptures and sand blows, dikes, and sills. This chapter proposes six tests to distinguish seismic from nonseismic origins of small, soft-sediment structures in one or a few beds. The tests utilize evidence for sudden formation, synchroneity and zoned map distribution over many exposures, size of the structures, and tectonic and depositional settings. The main barriers to more rigorous utilization of these soft-sediment structures in hazard analysis are (1) few observations of their formation during historical earthquakes, and (2) few or no measurements of the threshold earthquake magnitudes and threshold horizontal accelerations that are required to form the structures.

INTRODUCTION

The information needs of seismic-hazard analysis drive most studies of the geologic records of prehistoric earthquakes. Hazard calculations need estimates of the locations, ages, and magnitudes of large individual prehistoric earthquakes. The need arises because the historical records (catalogs) of earthquakes in most areas are short compared to the recurrence intervals of damaging earthquakes on individual faults or at given localities. Sole reliance on short historical catalogs can preclude the detection of centuries-long fluctuations in rates of earthquake recurrence. One example is the 300 yr periodicity detected by McGuire (1979) in the then 3126-year-long historical catalog for northern China. In addition, short historical catalogs can preclude recognition of prehistoric earthquakes larger than any observed historically in the same area. Examples include the eight large Holocene and latest Pleistocene earthquakes that struck southern and central Illinois and southwestern Indiana (Obermeier, 1998), and the great Holocene plate-boundary earthquakes of the Cascadia subduction zone (Atwater et al., 1995; Clague, 1997). Therefore, the main reason for studying geologic records of prehistoric earthquakes is to extend the short historical record of large earthquakes (e.g., Wheeler and Frankel, 1999, 2000).

The geologic study of large prehistoric earthquakes is the subject of the two-decades-old specialty of paleoseismology. McCalpin (1996) and his collaborators summarized, integrated, and critiqued the scattered, but rapidly growing, paleoseismological literature. Accordingly, in many cases, I cite parts of McCalpin's book in preference to individual older reports. Paleoseismologists study many kinds of geomorphic, stratigraphic, and sedimentological features, including but not restricted to those mentioned here (McCalpin and Nelson, 1996). In North America,

Wheeler, R.L., 2002, Distinguishing seismic from nonseismic soft-sediment structures: Criteria from seismic-hazard analysis, *in* Ettensohn, F.R., Rast, N., and Brett, C.E., eds., Ancient seismites: Boulder, Colorado, Geological Society of America Special Paper 359, p. 1–11.

the most common features studied are (1) surface ruptures, the outcrops of large seismic ruptures that nucleated in the upper crust and propagated to the ground surface to produce scarps and related landforms, and (2) sand blows, dikes, and sills that can be shown to have formed by prehistoric earthquake shaking. Studies of surface ruptures and sand blows, dikes, and sills have evolved beyond the simple recognition of one or more large prehistoric earthquakes, to the quantitative estimation of the location, magnitude, and age of individual earthquakes (e.g., McCalpin, 1996; Obermeier, 1998).

In contrast, the recognition and study of individual beds, or thin intervals of a few beds, that contain small, soft-sediment structures possibly induced by earthquake shaking have received less attention in North American paleoseismology (although note, e.g., Sims, 1973, 1975). Such beds or intervals are among the seismites recognized by Seilacher (1969), and include beds containing load casts, ball-and-pillow structures, pseudonodules, and other load features; fault-graded beds and other evidence of incipient, but arrested, slope failures (Seilacher, 1984); water-escape features; and the like (e.g., McCalpin and Nelson, 1996, Table 1-1; Obermeier, 1996b, Fig. 7.23). These generally small, soft-sediment structures (SSSSs), which are confined to one or a few beds, and can form seismically and nonseismically, are the focus of this chapter. This focus excludes larger features such as tsunami deposits, landslides, and turbidites. I will not use the term "seismite" further, because some recent authors have expanded its definition to include all geologic records of earthquakes (e.g., Vittori et al., 1991), thereby rendering the term ambiguous.

Although this chapter concentrates on SSSSs, most of the examples used later are from studies of sand blows, dikes, and sills that form as seismic liquefaction features (SLFs), or from studies of surface ruptures. SSSSs and SLFs share some characteristics. Like SSSSs, some sand blows and sills are restricted to a single or a few beds, some sand dikes and sills are small, and all form in soft sediments. However, studies of SLFs benefit crucially from established criteria for distinguishing sand blows, dikes, and sills of seismic origin from those formed nonseismically (Obermeier, 1996a, 1996b). In contrast, paleoseismologists observe and describe SSSSs, and in some cases evidence allows authors to conclude that most of their SSSSs are most likely of nonseismic origin (e.g., Tuttle, 1994; Tuttle et al., 1996, 1999). However, I am unaware of generally accepted criteria that can distinguish seismic from nonseismic SSSSs, unless their formation was observed. Accordingly, I distinguish SSSSs from SLFs in order to concentrate here on SSSSs, but it is unavoidable that most features of known origin that I will cite as examples are of SLFs, not SSSSs.

The purpose of this chapter is to suggest criteria for testing the seismic origin of SSSSs. The criteria could be used in two ways. First, they could be applied to SSSSs that are exposed in pre-Quaternary rocks, where results may aid studies of depositional and lithification processes or of paleoenvironments. However, as explained next, information about pre-Quaternary earthquakes is unlikely to benefit seismic-hazard assessments. Second, the criteria could be applied to SSSSs in Quaternary sediments, where results could improve hazard assessments.

SMALL, SOFT-SEDIMENT STRUCTURES IN THE CONTEXT OF SEISMIC-HAZARD ANALYSIS

Hazard analysis concentrates on the comparatively few earthquakes that are large enough to cause damage. Hazard analysis uses the far more numerous small earthquakes mainly to infer the likely properties of damaging earthquakes. The geologic record of a large pre-Quaternary earthquake will probably have little impact on hazard calculations unless a similar earthquake is known to have occurred recently at or very near the same place. The reason is that a pre-Quaternary earthquake that has not recurred in the late Quaternary is likely to have an annual probability of occurrence that is too small to affect the results of the hazard calculations. For example, the calculations that are performed in support of building codes for noncritical structures, such as residences and low-rise buildings, are relatively insensitive to earthquakes with annual probabilities smaller than 10^{-4}.

TABLE 1. SUMMARY OF SUGGESTED TESTS OF SMALL, SOFT-SEDIMENT STRUCTURES OF POSSIBLE SEISMIC ORIGIN

Test name	Critical observation	Limitations
1. Sudden formation	Structure formed more suddenly, and perhaps more violently, than by any nonseismic alternative	May be unable to rule out some nonseismic origins without additional evidence
2. Synchroneity	Nearby structures of same type formed at times indistinguishable from each other	May be unable to rule out some nonseismic origins; dating and correlation lack resolution to distinguish synchroneity from near-synchroneity
3. Zoned distribution	Indicators of strength of shaking decrease outward from a central area	Cannot rule out an earthquake origin
4. Size	Structure cannot be larger than all similar structures formed by historical earthquakes	Maximum sizes unknown; cannot rule out an earthquake origin for small structures
5. Tectonic setting structure	Seismic shaking strong enough to form the structure occurs more often than does any nonseismic alternative in modern analog settings	Threshold magnitudes and accelerations for formation are largely unknown
6. Depositional setting	Seismic shaking by itself forms the structure in modern analog sediments	Difficulty in recognizing some newly formed structures in the field

Accordingly, the calculations are not very sensitive to pre-Holocene earthquakes. However, some hazard calculations for special purposes may have to consider less likely earthquakes with smaller annual probabilities. The times between consecutive earthquakes at or very near the same place can vary, so it is possible that a pre-Holocene earthquake might have a long-term average annual probability of, say, 10^{-3}, even though it has not recurred during the past 10^4 yr. However, a pre-Quaternary earthquake that has not recurred yet would be extremely unlikely to have a long-term annual probability as large as 10^{-3}, and would be unlikely to impact even special-purpose hazard calculations.

Pre-Quaternary SSSSs of demonstrated seismic origin might be useful in determining whether an area has had pre-Quaternary earthquakes larger than any known from the local historical record (regardless of the annual probability of occurrence). However, we already know that even the least active parts of continents can have infrequent earthquakes larger than would be indicated by the identification of earthquake-induced SSSSs. Obermeier (1996a, p. 2–3), Obermeier and Pond (1999), and Obermeier et al. (this volume), have noted that SSSSs can form at shaking levels too low to cause damage to buildings. The smallest earthquakes that typically damage buildings have moment magnitudes M_w within approximately half a unit of 5. In contrast, the global historical record of continental earthquakes shows that land areas at all levels of seismic activity can have earthquakes much larger than M_w 5. In particular, a global survey that included the least seismically active parts of continents, the unrifted stable cratons, found that these areas have had infrequent historical earthquakes as large as M_w 6.5–7.1 (Johnston, 1994). Therefore, use of SSSSs to determine whether pre-Quaternary earthquakes were larger than any known historically must await methods for estimating an earthquake's magnitude from the SSSSs it caused.

Thus, as explained earlier, SSSSs in sediments of Holocene to late Pleistocene age are little used in hazard assessment in North America, because the paleoseismological study of surface ruptures and SLFs of those ages is so much more advanced. However, in some large continental areas, surface ruptures tend not to form and be preserved, and liquefiable deposits are not widespread. Hazard analyses of these areas would be aided by the development of additional indicators (which some SSSSs might eventually provide) of Holocene or late Pleistocene seismicity. In addition, early Pleistocene and older SSSSs, most of which are in rocks or sediments that are too lithified to liquefy, might be useful in other types of studies. Such usefulness must await the development of widely accepted criteria to (1) determine whether SSSSs formed seismically or nonseismically, (2) recognize the SSSSs that formed by a single earthquake and distinguish them from those formed by other nearly coeval earthquakes, and (3) use SSSSs to estimate the epicenter and magnitude of the single earthquake. The rest of this chapter suggests criteria that may lead to distinguishing seismic from nonseismic SSSSs of any age.

TESTS OF CHARACTERISTICS OF SMALL, SOFT-SEDIMENT STRUCTURES

This section concentrates on tests that are based on characteristics of the SSSS itself, whereas the next deals with tests based on the setting of the SSSS. For brevity, all tests are described in a generalized form, from which they could be made more specific to apply to a particular type of SSSS. Each test will have one of three outcomes: (1) the SSSS passes the test, if the available evidence demonstrates or strongly favors an earthquake origin, (2) the SSSS fails, if the evidence rules out or strongly disfavors an earthquake origin, or (3) the test cannot be applied or is inconclusive, probably for lack of sufficient evidence. Obermeier (1996a, 1996b, 1998) proposed equivalents of tests 1–3. The tests can be applied in any order. Tests 2 and 3 cannot be performed on a single SSSS, but instead require information from a group of features of the same type. Few, if any, individual tests can show or disprove a seismic origin alone. The clearest result is likely to come from the passing or failing of several tests, and preferably of all six.

The strategy of assessing SSSSs by performing tests one at a time differs from the more common strategy of assessing nonseismic alternative causes one at a time—for example, large storms. Nonetheless, the two strategies utilize some of the same information. Thus, while an investigator performs the tests on an SSSS, it should also be possible to assess several alternative origins. The following descriptions of individual tests contain a few examples.

Test 1: Sudden formation

A fundamental characteristic of most of the features formed by surface rupture or seismic shaking during moderate to large earthquakes is that the features formed suddenly, even violently or turbulently. Therefore, any characteristics of an SSSS that indicate sudden or violent formation would strengthen the case for an earthquake origin. It is not necessary to show that an SSSS formed in the seconds or tens of seconds that a large fault rupture and its earthquake shaking would last, but only that the SSSS formed in a time interval significantly shorter than that required for the formation of a similar-appearing feature by any nonseismic mechanism (Table 1).

A prehistoric normal-fault scarp provides a useful example of evidence and reasoning that can be used in a test for sudden formation. The scarp formed suddenly if (1) it is steeper than the angle of repose, (2) it is buried by a colluvial wedge formed by debris eroded from the scarp, and (3) the base of the wedge contains intact blocks of soil or bedded units from the top of the footwall, and the rest of the wedge fines upward and outward (McCalpin, 1996, chap. 3). Scarp steepness and the colluvial wedge that buries the scarp demonstrate that the scarp formed faster than it could have been carved by subaerial erosion. The intact blocks at the base of the wedge demonstrate that the scarp formed too fast to have been cut by a stream or created by aseismic fault creep. In addition, landsliding and catastrophic flooding might form a scarp suddenly, but these nonseismic alternatives can be assessed by consideration of the continuity, geometry, and geomorphological setting of the scarp. Finally, the steepness, colluvial wedge, intact blocks, and internal sedimentary facies and stratigraphy of the wedge match those of historical earthquake scarps and their erosional products.

Some sedimentary features such as load or water-release structures might form nonseismically and rapidly enough to make a test for sudden formation inconclusive. In the Discussion section, I suggest a line of argument whereby the fault-graded beds of Seilacher (1969, 1984) probably would pass this test.

A search for evidence of a sudden or violent origin may help support or rule out an earthquake origin. For example, Schafer et al. (1987) ruled out seismic shaking as the cause of some filled vertical fissures in Wisconsinan meltwater deposits in Connecticut, in part because thin strata in the fissure walls can be traced downward into the fissure fillings. Schafer et al. (1987) concluded that the delicate strata could not have survived the violent process of shaking-induced liquefaction and upward venting of liquefied sand through the fissures. Thorson et al. (1986) suggested that the stratified fissure fillings represent fissure sidewalls that collapsed downward into fissures that were formed by lateral spreading. However, no evidence appears to require that the collapse was sudden.

Test 2: Synchroneity

Much of the geologic record of a large earthquake forms synchronously—for example, a sand blow and its feeder dike, or most of the surface rupture formed by a single earthquake. Other parts of the record form within days to a few years after the earthquake—for example, the basal parts of most colluvial wedges, or redeposited lake sediment that Doig (1991) concluded had been resuspended or caused to slump by shaking. Thus, arguments for synchroneity or near synchroneity of all nearby SSSSs of the same type would favor an earthquake origin for them all (Obermeier, 1996b, p. 385–386). An example might be numerous ball-and-pillow structures that are seen in many scattered exposures of a single correlatable bed. A more complex example could involve load structures in two or more adjacent beds. In such an example, load structures, diapirs, water-release features, or clastic dikes might link beds and indicate that all beds were liquefiable during the same time span, thereby supporting an inference of synchroneity. Synchroneity alone may not be able to rule out nonseismic origins that act quickly, such as a single powerful storm, without additional evidence. Synchroneity can be evidence against formation by local sediment loading or local slope failures triggered by gravity or rapid deposition, because these would likely occur at different times in different places. In contrast, demonstrated nonsynchroneity can rule out formation by a single earthquake. For example, suppose that load structures in two stratigraphically close lacustrine beds are separated by an interval of mud-cracked sediment. The mud cracks would indicate the passage of sufficient time to demonstrate nonsynchroneity.

Synchroneity can only be demonstrated to within the time resolution of the dating or correlation methods used. Even within a historical earthquake sequence, SSSSs that might have formed during foreshocks, mainshocks, or aftershocks might not be distinguishable from each other without eyewitness accounts (Table 1; e.g., Saucier, 1989; Sims and Garvin, 1995; Johnston and Schweig, 1996). The datable geologic records of late Pleistocene and Holocene earthquakes have age uncertainties of decades at best and centuries or millennia more commonly. Nonetheless, despite an uncertain date, correlation can demonstrate synchroneity. Correlation of pre-Quaternary SSSSs to test for synchroneity might be done most precisely by comparing the bed-by-bed stratigraphy of nearby exposures. Examinations of modern depositional analogs may allow one to estimate whether the time interval represented by the SSSSs is years, decades, centuries, or millennia. The next section explains how the map distribution of SSSSs of the same type, but of various sizes, can reinforce evidence of synchroneity.

Test 3: Zoned map distribution

SLFs that form synchronously over a large area can allow estimation of the location and magnitude of the single, large, causal earthquake, and the same might prove to be true for some kinds of SSSSs. For example, Holocene and latest Pleistocene sand blows and sand dikes of seismic origin are widely distributed across southern Illinois and southwestern Indiana. Obermeier (1998) reported that dikes of ages so similar that they cannot be distinguished tend to be distributed zonally over their area of occurrence, the widest dikes being in or close to the center of the area and increasingly narrow dikes outward. The zoned distribution of dikes of approximately the same age supports the interpretation that the dikes all formed in a single event centered near the widest dikes. If the dikes are all attributed to the same event, the best age determination on any of them could be applied to all of them. The zoned distribution, its areal extent, and other evidence confirm a seismic origin (Obermeier, 1998). The center of the area, where the widest dikes are most numerous, is taken as the epicentral area of the causal earthquake. (The earthquake location determined in this way approximates the area of greatest energy release. This area is not necessarily the place where the earthquake rupture nucleated, nor the epicenter that would have been calculated seismologically. All three locations are the same or similar for small to moderate earthquakes, but they can be many kilometers apart for large earthquakes with rupture zones tens to hundreds of kilometers long.) Finally, a global compilation of the distances between the epicenters of historical earthquakes of known magnitudes and the most distant observed seismic liquefaction allowed estimation of the magnitudes of the prehistoric earthquakes in Illinois and Indiana.

A similar approach might be taken with some types of SSSSs (Obermeier, 1996b, p. 385–386), for example, with ball-and-pillow structures that might be found in correlatable beds exposed at numerous places over a large area. If the ball-and-pillow structures were observed to be larger, more numerous, or of greater vertical extent in the center of the area and smaller, fewer, or shorter outward, then synchroneity and a common origin could be inferred. In contrast, if the largest, most numerous, or tallest features were found at several places that were separated by smaller, fewer, or shorter features of the same type, and if all the features formed in host materials with the same susceptibility to shaking and the same geomorphic setting, then the cause could be either nonseismic or several small or moderate, nearly coeval earthquakes instead of one large one. Thus, it does not appear possible to rule out an earthquake origin with a test for zoned distribution alone (Table 1).

Test 4: Size

An SSSS is unlikely to be of seismic origin if it is significantly larger than any SSSS of the same type that has been observed to form during a historical earthquake. SLFs provide two possible examples of this test. First, the liquefied source beds of most historical SLFs are within "several tens of meters" of the ground surface (Obermeier and Pond, 1999, p. 38; Obermeier et al., this volume). Therefore, a sand dike markedly taller than several tens of meters probably would not have formed by earthquake-induced liquefaction. Second, SLFs are restricted to the vicinities of the epicenters of medium-sized historical earthquakes (Table 2). A field of coeval SLFs that spans an area significantly wider than the radius indicated in Table 2 for the inferred magnitude probably did not form in a single earthquake.

A size test for an SSSS could depend on the width, depth, or volume of a load cast or ball-and-pillow structure, or the thickness of a convolutely laminated bed or a fault-graded bed. Nevertheless, size tests of SSSSs lie in the future. Sizes of some SSSSs may depend on sediment thickness, slope, and other depositional and geographic factors as much as on earthquake shaking. The maximum dimensions and areal distributions of most types of seismically induced SSSSs do not appear to be well determined, because methods for distinguishing features of seismic origin remain to be developed and field tested (Table 1). Furthermore, even if maximum sizes become known for some types of seismic SSSSs, an SSSS much smaller than the maximum might still have formed during a small or distant earthquake. Therefore, a size test alone can rule out an earthquake origin for large features, but probably not for small ones.

This test does not involve estimating earthquake magnitude from some measure of SSSS size. SSSS size may be controlled complexly by many aspects of the sediment, pore waters, burial depth, earthquake size and distance, and shaking at the site. At present, few if any of these potential controls are likely to be sufficiently understood and well enough verified by field testing to support magnitude estimation.

TESTS OF SETTINGS OF SMALL, SOFT-SEDIMENT STRUCTURES

Test 5: Tectonic setting

This test focuses on establishing the relative frequencies of seismic shaking and of alternative causes, at the time and place in which an SSSS formed. The strategy is (1) to identify a modern analog of the tectonic setting in which the SSSS formed, and (2) to demonstrate that earthquakes sufficiently large to create SSSSs like those being tested are more common in the modern analog setting than the occurrences of nonseismic alternatives that form similar-looking structures. Note that a test of tectonic setting by itself can neither demonstrate nor disprove any origin. It can only show which possible origin is the most likely.

The first step is to determine the tectonic setting and a modern analog. The tectonic and analog settings should cover large areas to smooth out the local fluctuations in seismicity rates that can occur on time scales of centuries or millennia and that affect some individual faults and seismic zones (Coppersmith, 1988; Crone et al., 1997). The level of seismicity is influenced strongly by the type and degree of modern deformation, plate boundaries being more seismically active than plate interiors. In addition, the geographic extent of strong shaking from a given earthquake is governed by the attenuation of seismic energy as it propagates away from the rupture zone. Attenuation is greater in active plate-boundary regions than in more stable plate interiors. Therefore, greater seismicity in plate-boundary regions is partly balanced by lower attenuation in plate interiors. Seismic hazard calculations use seismicity, type and degree of modern deformation, and attenuation to estimate the strength and likelihood of seismic shaking at a point. Accordingly, maps of probabilistic seismic hazard, considerations of regional structure and modern tectonics, and boundaries of extensive and compressive stress provinces allow a crude division of North America into two broad tectonic settings (Fig. 1).

The tectonic environment of an SSSS at the time it formed can identify the tectonic setting as either plate boundary or stable continental region. For example, an SSSS in the early Mesozoic rift basins of eastern North America would have formed in a plate-boundary setting, although now the SSSS is within the North American stable continental region. In Cambrian time, the

TABLE 2. LARGEST OBSERVED DISTANCE OF EARTHQUAKE-INDUCED SAND BLOWS, DIKES, AND SILLS FROM HISTORICAL EPICENTERS

M_W*	Distance (km)
5	2
6	20
7	110
8	210
9	600

Note: From Obermeier and Pond (1999, Fig. 9A) and Obermeier et al. (this volume).
*Moment magnitude.

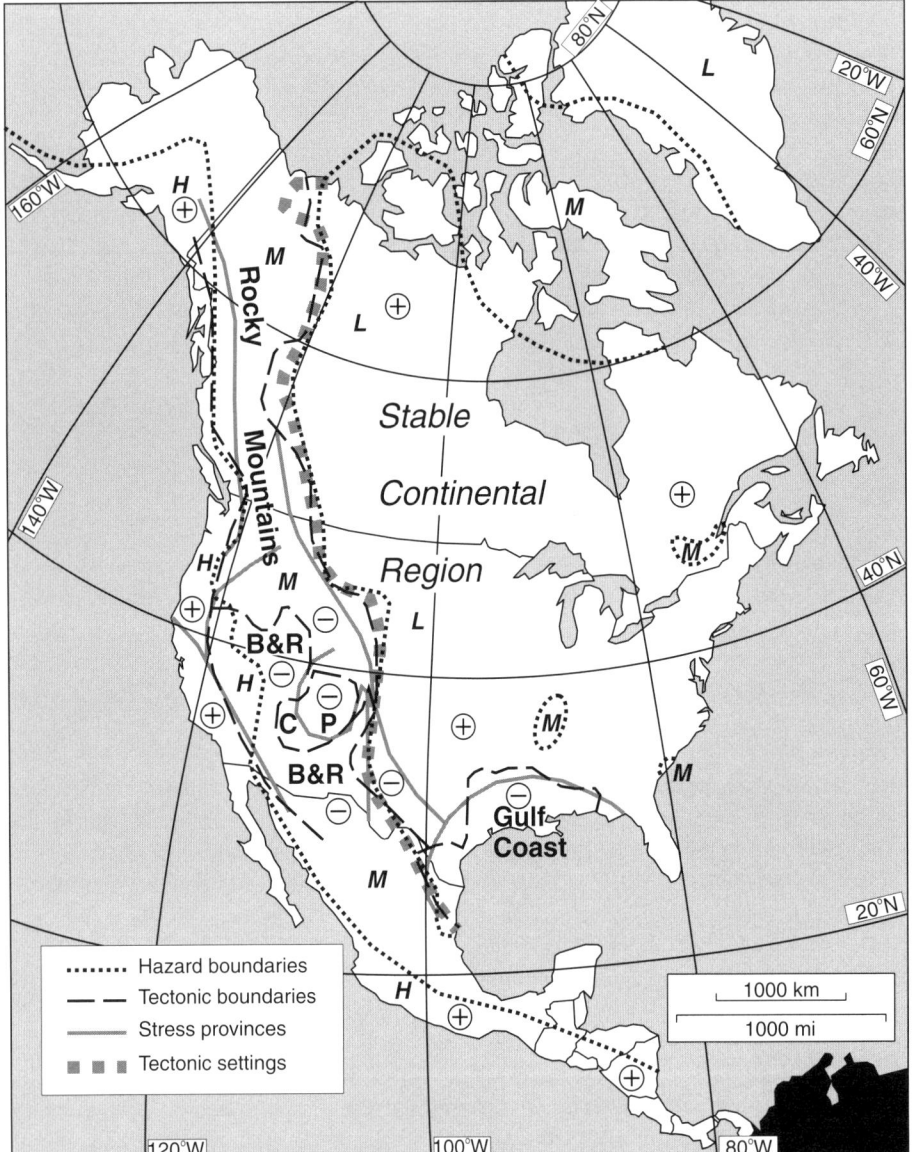

Figure 1. Generalized tectonic settings of North America (white), as defined for use in test of tectonic setting (see text). *Hazard boundaries* are generalized from Basham et al. (1997), Frankel et al. (1997), and Giardini et al. (1999). Seismic hazard is generally high (H) in westernmost North America, moderate (M) in rest of western one-third of continent, and low (L) in eastern two thirds. *Tectonic boundaries* are generalized from Ewing and Lopez (1991) and Muehlberger (1996). Tectonically distinct regions include, from west to east, subduction and transform plate boundaries (unlabeled); Basin and Range province (B&R), Colorado Plateau (C P), Rocky Mountains, and Gulf Coast; and central and eastern North America (unlabeled). *Stress provinces* are simplified from Zoback and Zoback (1991). Some province boundaries are not closed because of insufficient data. Circled plus signs identify compressive provinces of western plate boundaries and of most of central and eastern North America. Circled minus signs identify extensive provinces between compressive provinces. Adjacent extensive provinces have different extension directions. *Tectonic settings* are (1) stable continental region in the east, mostly under compressive stress, and (2) plate boundaries in the west and active continental crust of Rocky Mountains, Basin and Range province, and Colorado Plateau between plate boundaries and stable continental region. Boundary between the two tectonic settings is generalized from Johnston (1989) and Kanter (1994).

boundary between plate-boundary and stable continental region settings in eastern North America was hundreds of kilometers farther northwest than during the Mesozoic Era (Wheeler, 1995). The present-day plate-boundary setting of western North America includes rifting, transform, and ocean-under-continent subduction regimes, but no active collisions of North America with other continental lithosphere or with island arcs. Accordingly, any SSSS of Paleozoic age in what is now the Appalachian or Ouachita orogens would have formed in a tectonic setting now found only outside North America, perhaps in the Alpine, Zagros, Himalayan, or southeast Asian orogens.

Some tectonic settings can produce larger earthquakes than others. Larger earthquakes shake larger areas and increase the likelihood of equaling or exceeding any specific level of acceleration and, therefore, the likelihood of forming seismic SSSSs.

Johnston (1994) summarized worldwide observations of maximum magnitudes in various tectonic settings (Table 3). The table shows that active subduction zones can produce earthquakes approximately three magnitude units larger than have been observed in Paleozoic contractional orogens and cratons. Magnitude, M_w, and seismic moment, M_o, are related by $M_w = (2/3)(\log_{10} M_o) - 10.7$ (Hanks and Kanamori, 1979). Thus, the largest earthquake observed in active subduction zones released $10^4 - 10^5$ times as much energy (seismic moment) as did the largest earthquake observed in Paleozoic orogens and cratons.

The second step in a test of tectonic setting would be to compare the expected frequencies of occurrence of earthquakes and of nonseismic alternative causes. The comparison would be performed for the modern analog of the tectonic setting in which a SSSS formed. The comparison would require four items of information about the earthquakes of the modern analog setting: (1) the threshold or smallest magnitude of earthquake that is typically required to form features like the SSSS, (2) the threshold horizontal acceleration that is typically required to form such features, (3) a recurrence equation of the form $\log_{10} N(M_w) = a - bM_w$, in which $N(M_w)$ is the annual number of earthquakes of moment magnitude M_w or larger, per unit area, in the modern analog setting, and a and b are constants (Gutenberg and Richter, 1954, p. 17), and (4) an attenuation equation, which allows estimation of the area over which an earthquake of magnitude M_w in the modern analog setting typically produces horizontal acceleration at least equal to the threshold value (e.g., Abrahamson and Shedlock, 1997). The recurrence equation would determine N(thresh), the annual number of earthquakes at least as large as the threshold magnitude, per unit area. The attenuation equation would determine A(thresh), the area shaken at or above the threshold acceleration by each of these earthquakes. For each magnitude interval above the threshold magnitude, $[N(\text{thresh}) \times A(\text{thresh})]^{-1}$ would give the average number of years between shakings that are strong enough to form SSSSs. Combining results for all magnitude intervals would give the average number of years between SSSS formation at a typical place in the modern analog setting.

Recurrence and attenuation equations are available in the seismological literature. Unfortunately, values of threshold magnitudes and threshold accelerations are largely unknown for SSSSs (S.F. Obermeier, 1999, oral commun.; but see Sims, 1973), and presumably differ between different types of SSSSs. SLFs typically require earthquakes of M_w 5.5 or larger (Obermeier and Pond, 1999; Obermeier et al., this volume), and horizontal accelerations of roughly 0.1 g or greater (Obermeier, 1996b, p. 331). In contrast, soft-sediment deformation can occur at much lower magnitudes and accelerations (Obermeier and Pond, 1999; Obermeier et al., this volume). In fact, some soft-sediment structures can form in the absence of seismic shaking (e.g., Obermeier, 1996b, p. 382). Accordingly, a quantitative test of tectonic setting will remain unfeasible until field and laboratory investigations can specify the threshold magnitude and acceleration for the formation of a particular type of SSSS (Table 1).

Test 6: Depositional setting

This test focuses on the likelihood that the sediment in which an SSSS formed would have been sufficiently susceptible to seismic shaking during an earthquake (Table 1). The test involves two steps. The first is to determine the material properties of the sediment at the time the SSSS formed. Properties that can influence susceptibility to shaking include grain size, sorting, and fabric; clay mineralogy and pore-water chemistry at deposition and afterward; sediment composition and bedding characteristics; degree of compaction, dewatering, and incipient cementation; density inversions and their magnitudes; water depth and depth below the sediment-water interface if the setting was subaqueous; and depth of water table and depth below ground level if the setting was subaerial (e.g., Seed, 1979).

TABLE 3. MAXIMUM MOMENT MAGNITUDE OBSERVED IN VARIOUS TECTONIC SETTINGS

M_W	Setting
Plate boundaries	
9.5	Subduction zones
8.5	Active continental crust* (e.g., North American Cordillera)
Stable continental regions†	
8.1 ± 0.3	Intracontinental rift, particularly those formed or reactivated in extension less than 250 m.y. ago (e.g., Reelfoot rift, central United States)
7.7 ± 0.2	Passive margins (e.g., Atlantic and Gulf of Mexico margins)
6.8 ± 0.3	Cratons
6.4 ± 0.2	Orogenic crust 100–550 m.y. old (e.g., Appalachian and Ouachita orogens)

Note: Modified from Johnston (1994, 1996) and Kanter (1994).
*Continental crust that has undergone deformation, metamorphic or igneous heating, or abundant active faulting since the Early Cretaceous. Has had less than 100 m.y. to reach thermal equilibrium.
†Has had longer than 100 m.y. to reach thermal equilibrium, except passive margins, which have had longer than 25 m.y.

The second step is to demonstrate that modern sediment with the same pertinent material properties, when shaken, forms features like the SSSS being tested. The demonstration may require laboratory experiments in addition to field observations, because SSSSs that include only a single bed or a few beds and do not erupt onto the ground surface or deform it might be difficult to detect in the field and to assign to an individual historical earthquake. An added complication is that a given level of acceleration can come from either a small, nearby earthquake or a large, distant one. However, weak shaking from a large, distant earthquake may last longer or be dominated by different frequencies, and perhaps duration and frequency of shaking could affect the formation of some types of SSSSs.

DISCUSSION

Proposed classification based on test results

The strength of these tests is that there are six of them. None is likely to apply to all, or perhaps even most, SSSSs. At present none can demonstrate an earthquake origin by itself; perhaps the test for sudden formation comes closest. Most of the tests have limitations that stem either from the natures of geologic and seismologic data (Table 1; Tests 2, 3, 4, and 6) or from insufficient quantitative knowledge of SSSS formation (Tests 4 and 5). Despite these limitations, the six tests are multiple tools with which to build a case for or against a seismic origin.

It will likely remain difficult to show or disprove an earthquake origin of a pre-Quaternary SSSS, because the geologic details on which most of the tests depend are progressively erased through time. Weathering, erosion, compaction, cementation, fluid alterations, lithification, and postlithification physical and chemical processes can destroy much of the primary sedimentary record of the conditions under which an SSSS formed. Accordingly, I suggest a more modest goal than proof. The more tests an SSSS passes, the more likely is its seismic origin. Similarly, the more tests are failed, the more likely is its nonseismic origin. Each test that cannot be performed or which has an inconclusive outcome will increase the uncertainty of the inferred origin, be it seismic or nonseismic. The result of applying all the tests can be a classification of SSSSs according to the numbers of tests considered, passed, and failed (Fig. 2).

Examples

Examples illustrate use of the classification in Figure 2. Obermeier (1996b, 1998) evaluated Holocene and latest Pleistocene SLFs in southern Illinois and southwestern Indiana, using criteria specifically designed for earthquake-induced SLFs there and elsewhere. The descriptions of Obermeier (1996b, 1998) also demonstrate that the dikes and sand blows would pass all six of the tests proposed here for SSSSs (Table 1). (1) Sand blows consist of sequences dominated by sand and silt. The sequences fine upward, thereby demonstrating that a mixture of sediment and water was erupted faster than the finer silt grains could settle out of the mixture. In addition, some vented material contains gravel larger than 4 cm in diameter, which indicates a rapid, perhaps violent upward flow. The only likely nonseismic origins, artesian flow and landsliding, would be inconsistent with the low-relief geomorphic setting. Accordingly, these SLFs pass the test for sudden formation. (2) Archeological and radiocarbon dates imply that the dikes and sand blows formed within a few short time intervals that were separated by much longer intervals (Obermeier, 1996b, 1998). SLFs that formed within a given time interval formed at times so close as to be indistinguishable at the available dating resolution, which implies synchroneity or near synchroneity. Thus, a synchroneity test is passed. (3) Within a given time interval, the widest dikes clustered in a core region a few tens of kilometers or less in diameter, successively narrower dikes being exposed successively farther outward from the core region. Therefore, the SLFs pass a test for zoned distribution. (4) A size test is passed because dikes and sand blows similar in size to those in Illinois and Indiana were seen to form during large earthquakes near New Madrid, Missouri, in 1811–1812 and Charleston, South Carolina, in 1886. (5) The tectonic setting during the Holocene and latest Pleistocene was in the middle of the North American stable continental region, as it is now. Accordingly, no modern analog setting is needed. Within the stable continental region, the New Madrid and Charleston earthquakes produced shaking strong enough to form SLFs like those in Illinois and Indiana. As explained earlier, nonseismic origins would be inconsistent with the geomorphic setting. Therefore, the test of tectonic setting is successful. (6) The sediments in which the Illinois-Indiana SLF formed were, and still are, saturated sands and silts with finer grained capping soils and water tables at depths of a few meters. This depositional setting resembles those in which historical SLFs formed abundantly near New Madrid and Charleston. Thus, the SLFs of Illinois and Indiana pass all six tests. The features plot together in Figure 2 as example A.

The origins of pre-Quaternary SSSSs are harder to determine because of poorly preserved details, uncertain tectonic or depositional settings, or insufficient information on threshold magnitudes and accelerations. For example, consider the fault-graded beds of the finely laminated Miocene Monterey Shale of coastal California (Seilacher, 1969, 1984). Downward from the top of a fault-graded bed, a homogenized layer of mud, interpreted as resuspended, grades into a rubble layer of laminated clasts, which in turn grades into a faulted layer in which the faults become larger and fewer downward. The homogenized upper layer and the rubble layer indicate sudden, probably violent, formation. The gradations between layers suggest that resuspended mud, rubble, and faults all formed at the same time (see photographs in Seilacher, 1969, 1984). The fault-graded beds formed on a submarine slope (Seilacher, 1984). The small sizes of the fault offsets and the failure of the faults to integrate their slips into a single large detachment surface indicate that probably the causal forces were short-lived. Sudden onset and short duration characterize cyclic shaking from an earthquake. Sudden onset and short duration would not be expected from the prolonged, cyclic stresses of repeated storm waves, or from the permanent, constant stresses caused by gravity on a slump-prone submarine slope (Seilacher, 1969). If this reasoning is valid, then the fault-graded beds would pass a test of sudden formation. In contrast, the single exposure described by Seilacher spanned only

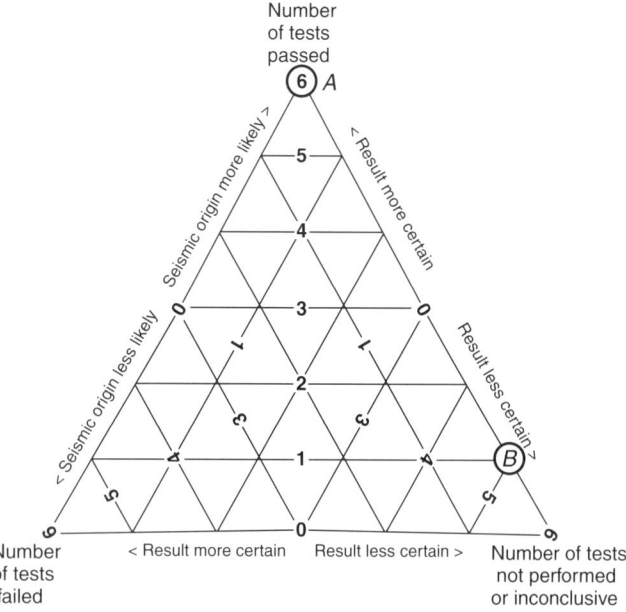

Figure 2. Suggested classification of small, soft-sediment structures in one or a few beds, according to outcomes of six tests described in text. Numbers decreasing away from each vertex give the numbers of tests that produced the result written at the vertex. Circles labeled A and B refer to examples discussed in text.

10 m by 50 m, so either tests of synchroneity and zoned distribution would require additional exposures, or they would be inconclusive because of insufficient data. Similarly, tests of size and tectonic and depositional settings require additional examples, observations of the formation of fault-graded bedding during laboratory or earthquake shaking, and values of threshold magnitude and acceleration, otherwise these tests, too, would be inconclusive. Accordingly, the fault-graded beds plot in Figure 2 as example B. Test results hint at an earthquake origin, but the hint is highly uncertain, as cautioned by Seilacher (1984). Additional field examination with the tests in mind might produce evidence with which to strengthen or weaken the hint of a seismic origin.

A classification like that of Figure 2 provides a ranking of SSSSs according to the likelihood of a seismic origin and the uncertainty of the result. For example, a seismic origin might be equally likely for examples A and B, because all the performed tests favor seismic origins in both cases. However, this result is far more certain for example A than for example B, because more tests could be performed for example A.

Note that Figure 2 contains no sharp boundaries between features that are and are not of seismic origin. Incomplete geologic evidence could cause many SSSSs to cluster in the middle of Figure 2, or toward its lower right vertex. The implications of incomplete evidence and inconclusive tests will vary with the use to which Figure 2 will be put. For basic geologic research, additional study of example B might advance our understanding of earthquakes nearly as much as example A. The two examples could be seen as equally likely to have a seismic origin. In contrast, for hazard analysis with its potentially large social, economic, and regulatory implications, the most useful test results will be the most certain ones. If the six tests proposed in this paper are effective in distinguishing seismic from nonseismic SSSSs, the most certain test results would plot at or very near the upper or lower left vertices. Perhaps hazard analysts would accept as seismically induced only SSSSs that plot at or very near the upper vertex.

Closeness of small, soft-sediment structures to known faults

Note that no test described in this report involves closeness of an SSSS to an individual known fault. Usually, the individual causal faults of historical earthquakes are difficult to identify unless the earthquake ruptures the ground surface along the fault trace. For prehistoric earthquakes the difficulty is commonly much greater, depending on the degree to which any surface rupture remains recognizable. The two main reasons for this difficulty are (1) the uncertain locations of earthquakes and faults at depth and (2) the abundance of faults in continental crust.

Earthquakes in the North American stable continental region nucleate at a median depth near 10 km, approximately half the depths being between 5 and 15 km, well within metamorphic and igneous basement (Wheeler and Johnston, 1992). Depths in other North American tectonic settings are generally similar, except in subduction zones, where the zone of earthquake occurrence can extend to greater depths. At depths of 5–15 km, earthquake locations and fault locations usually are uncertain by at least several kilometers. The uncertainty of the fault location at depth is likely to increase if the fault dips shallowly or moderately, or if it is curved. These uncertainties alone preclude confident association of most earthquake foci with the few faults that might be traceable to focal depths.

In addition, faults large enough for their larger earthquakes to create abundant SSSSs are probably present more or less everywhere in continental crust. The following reasoning supports this assertion. As mentioned earlier, SLFs typically require earthquakes of M_w 5.5 or larger and accelerations of 0.1 g or higher, but at least some SSSSs are likely to form at lower accelerations and thus at smaller magnitudes. Table 4 summarizes estimates of the sizes of typical rupture zones of earthquakes of various magnitudes. An earthquake cannot fit on a fault smaller than its rupture zone, but few historical earthquakes large enough to have formed surface ruptures appear to have ruptured the full lengths of their host faults (McCalpin, 1996, p. 467). For discussion only, let us postulate that a typical SSSS-forming earthquake occurs on a fault twice as long and wide as the earthquake rupture zone. The same earthquake also could have occurred on a smaller fraction of a larger fault. However, large faults are much less numerous than small ones. Accordingly, an SSSS-forming earthquake that ruptures only half the length and half the width of its causative fault may still be a reasonable postulate. In addition, note that earthquakes of any M_w are very roughly ten times as numerous as earthquakes of $M_w + 1$. Finally, suppose that most seismic SSSSs are caused by earthquakes of M_w 4 to 6. Smaller earthquakes are more numerous, but they shake smaller areas. Larger earthquakes shake larger areas, but they are rarer. Then, from the values in Table 4, we might expect that most seismic SSSSs should be caused by earthquakes on faults 1.2–25.2 km long and 1.2–16.2 km in down-dip width.

Many detailed geologic maps of exposed basement rocks show numerous faults of these sizes. Accordingly, it is reasonable to suppose that faults of the proper size to cause SSSSs are likely

TABLE 4. ESTIMATED DIMENSIONS OF TYPICAL RUPTURE ZONES

M_W*	Length (km)†	Down-dip width (km)†
2.0	0.06, —	0.06, —
3.0	0.2, —	0.2, —
4.0	0.6, —	0.6, —
5.0	1.8, 3.2	1.8, 3.9
5.5	3.3, 6.4	3.3, 5.6
6.0	8.0, 12.6	4.1, 8.1
7.0	30, 49	11, 17
8.0	150, 190	22, 35

* Moment magnitude.
† Estimates given in pairs. First estimate in each pair is for stable continental regions (Johnston, 1993). Second estimate is for all tectonic settings taken together and earthquakes of M_W 4.8–8.1 (Wells and Coppersmith, 1994).

to be widely and densely distributed in continental crust. In addition, some of the faults will have moderate to shallow dips, and many will exist at earthquake focal depths of, say, 5–15 km. Therefore, an SSSS-causing fault could be too small or too deep to be exposed for mapping or to be detected geophysically. Thus, any SSSS is likely to be close to one or more faults, whether or not the faults are known and whether or not the faults caused the SSSS. In fact, a fault that is both shallow enough to be mapped and deep enough to host SSSS-causing earthquakes is likely to be large, and therefore comparatively rare. Smaller faults are more numerous and, therefore, one or more of them are likely to be closer to the SSSS. Thus, in the absence of independent evidence that a known fault was active frequently during the geologically short time when the SSSS probably formed, the known fault might be among those least likely to have caused the SSSS.

For example, the M_w 5.9 Saguenay, Quebec, earthquake of 1988 produced liquefaction at five sites 25–30 km northeast of the epicenter (Tuttle et al., 1990). The largest fault mapped nearby is the Lac Kenogami fault, which strikes west-northwest, has a trace approximately 140 km long, and is between the mapped liquefaction and the epicenter (Du Berger et al., 1991). At its closest approach to the epicenter, the Lac Kenogami fault is 16 km to the northeast. However, the two nodal planes of the mainshock focal mechanism project upward from the earthquake focus to intersect ground level 15 km southwest and 34 km southeast of the epicenter (Du Berger et al., 1991). The upward projections of the nodal planes reach ground level tens of kilometers from the largest fault mapped nearby, and on the opposite side of the epicenter from the fault. The earthquake is unlikely to be related to the largest nearby fault.

CONCLUSIONS

Small, seismically induced, soft-sediment structures in one or a few beds older than Quaternary, and commonly older than Holocene, represent earthquakes with annual probabilities of occurrence too low to impact seismic-hazard assessments. Even if structures of this type are of Holocene age, they are little used in North American paleoseismology because the study of surface ruptures and earthquake-induced sand blows, dikes, and sills is more advanced, often being able to provide estimates of the location, magnitude, and age of individual prehistoric earthquakes. If methods can be developed to extract the same estimates from SSSSs, they could contribute to hazard assessments where they are of Quaternary age, and to sedimentological, tectonic, or historical geologic studies where they are older.

The relative likelihoods of seismic and nonseismic origins of one or a group of these structures could be indicated by six tests for sudden formation, synchroneity and zoned distribution over many exposures, size, and tectonic and depositional settings. Each test is either passed, failed, or not performed or inconclusive. A triangular graph of the test results could allow ranking of different structures or groups of structures according to their overall likelihoods and uncertainties of seismic origins.

There are two main barriers to the use of Quaternary small, soft-sediment structures in hazard assessment. The first is the apparent lack of numerous published field descriptions of structures that might have been observed to form during historical earthquakes. The second barrier is a lack of data on the threshold earthquake magnitudes and threshold horizontal accelerations that are required to form specific types of these structures.

ACKNOWLEDGMENTS

Familiarity with the geologic requirements of hazard analysis allowed me to recognize the need for tests like those suggested here, but negligible experience in paleoseismology and sedimentology forces me to leave the testing of these ideas to readers. If any bear fruit, credit the testers; errors are mine alone. Discussions with S. Greb, A.R. Nelson, S.F. Obermeier, D.M. Perkins, and J.P. Smoot sharpened ideas. Suggestions from Greb, Nelson, Obermeier, N. Rast, Smoot, and especially M.P. Tuttle improved the manuscript.

REFERENCES CITED

Abrahamson, N.A., and Shedlock, K.M., 1997, Overview: Seismological Research Letters, v. 68, no. 1, p. 9–23.

Atwater, B.F., et al., 1995, Summary of coastal geologic evidence for past great earthquakes at the Cascadia subduction zone: Earthquake Spectra, v. 11, p. 1–18.

Basham, P., Halchuk, S., Weichert, D., and Adams, J., 1997, New seismic hazard assessment for Canada: Seismological Research Letters, v. 68, no. 5 p. 722–726.

Clague, J.J., 1997, Evidence for large earthquakes at the Cascadia subduction zone: Reviews of Geophysics, v. 35, p. 439–460.

Coppersmith, K.J., 1988, Temporal and spatial clustering of earthquake activity in the central and eastern United States: Seismological Research Letters, v. 59, no. 4, p. 299–304.

Crone, A.J., Machette, M.N., and Bowman, J.R., 1997, Episodic nature of earthquake activity in stable continental regions revealed by palaeoseismology studies of Australian and North American Quaternary faults: Australian Journal of Earth Sciences, v. 44, p. 203–214.

Doig, R., 1991, Effects of strong seismic shaking in lake sediments, and earthquake recurrence interval, Temiscaming, Quebec: Canadian Journal of Earth Sciences, v. 28, p. 1349–1352.

Du Berger, R., Roy, D.W., Lamontagne, M., Woussen, G., North, R.G., and Wetmiller, R.J., 1991, The Saguenay (Quebec) earthquake of November 25, 1988: Seismologic data and geologic setting: Tectonophysics, v. 186, p. 59–74.

Ewing, T.E., and Lopez, R.F., 1991, Principal structural features, Gulf of Mexico basin, in Salvador, A., ed., The Gulf of Mexico Basin: Boulder, Colorado, Geological Society of America, Geology of North America, v. J, plate 2, scale 1:2 500 000, 1 sheet.

Frankel, A., Mueller, C., Barnhard, T., Perkins, D., Leyendecker, E.V., Dickman, N., Hanson, S., and Hopper, M., 1997, Seismic-hazard maps for the conterminous United States: U.S. Geological Survey Open-File Report 97-131, scale 1:7 000 000, 12 color sheets.

Giardini, D., Grunthal, G., Shedlock, K., and Zhang, P., 1999, Global seismic hazard map: Rome, Italy, Instituto Nazionale di Geofisica, Swiss Seismological Service, and U.S. Geological Survey, scale 1:35 000 000, 1 sheet.

Gutenberg, B., and Richter, C.F., 1954, Seismicity of the Earth: Princeton, New Jersey, Princeton University Press, 310 p.

Hanks, T.C., and Kanamori, H., 1979, A moment magnitude scale: Journal of Geophysical Research, v. 84, p. 2348–2350.

Johnston, A.C., 1989, The seismicity of "stable continental interiors," *in* Gregersen, S., and Basham, P.W., eds., Earthquakes at North-Atlantic passive margins: Neotectonics and postglacial rebound: Dordrecht, Netherlands, Kluwer Academic Publishers, p. 299–327.

Johnston, A.C., 1993, Average stable continental earthquake source parameters based on constant stress drop scaling [abs.]: Seismological Research Letters, v. 64, no. 3–4, p. 261.

Johnston, A.C., 1994, Seismotectonic interpretations and conclusions from the stable continental region seismicity database, *in* Schneider, J.F., ed., The earthquakes of stable continental regions, Volume 1, Assessment of large earthquake potential: Palo Alto, California, Electric Power Research Institute, p. 4-1–4-103.

Johnston, A.C., 1996, Seismic moment assessment of earthquakes in stable continental regions. 3. New Madrid 1811–1812, Charleston 1886 and Lisbon 1755: Geophysical Journal International, v. 126, p. 314–344.

Johnston, A.C., and Schweig, E.S., 1996, The enigma of the New Madrid earthquakes of 1811–1812: Annual Review of Earth and Planetary Sciences, v. 24, p. 339–384.

Kanter, L.R., 1994, Tectonic interpretation of stable continental crust, *in* Schneider, J.F., ed., The earthquakes of stable continental regions, Volume 1, Assessment of large earthquake potential: Palo Alto, California, Electric Power Research Institute, p. 2-1–2-98.

McCalpin, J., ed., 1996, Paleoseismology: San Diego, Academic Press, 588 p.

McCalpin, J.P., and Nelson, A.R., 1996, Introduction to paleoseismology, *in* McCalpin, J.P., ed., Paleoseismology: San Diego, Academic Press, p. 1–32.

McGuire, R.K., 1979, Adequacy of simple probability models for calculating felt-shaking hazard, using the Chinese earthquake catalog: Bulletin of the Seismological Society of America, v. 69, p. 877–892.

Muehlberger, W.L., compiler, 1996, Tectonic map of North America: Tulsa, Oklahoma, American Association of Petroleum Geologists, scale 1:5 000 000, 4 sheets.

Obermeier, S.F., 1996a, Use of liquefaction features for paleoseismic analysis: An overview of how seismic liquefaction features can be distinguished from other features and how their regional distribution and properties of source sediment can be used to infer the location and strength of Holocene paleo-earthquakes: Engineering Geology, v. 44, p. 1–76.

Obermeier, S.F., 1996b, Using liquefaction-induced features for paleoseismic analysis, *in* McCalpin, J., ed., Paleoseismology: San Diego, Academic Press, p. 331–396.

Obermeier, S.F., 1998, Liquefaction evidence for strong earthquakes of Holocene and latest Pleistocene ages in the states of Indiana and Illinois, USA: Engineering Geology, v. 50, p. 227–254.

Obermeier, S.F., and Pond, E.C., 1999, Issues in using liquefaction features for paleoseismic analysis: Seismological Research Letters, v. 70, p. 34–58.

Saucier, R.T., 1989, Evidence for episodic sand-blow activity during the 1811–1812 New Madrid (Missouri) earthquake series: Geology, v. 17, p. 103–106.

Schafer, J.P., Obermeier, S.F., and Stone, J.R., 1987, On the origin of wedge structures in southern New England: Geological Society of America Abstracts with Programs, v. 19, no. 1, p. 55.

Seed, H.B., 1979, Soil liquefaction and cyclic mobility evaluation for level ground during earthquakes: Journal of Geotechnical Engineering, v. 105, p. 201–255.

Seilacher, A., 1969, Fault-graded beds interpreted as seismites: Sedimentology, v. 13, p. 155–159.

Seilacher, A., 1984, Sedimentary structures tentatively attributed to seismic events: Marine Geology, v. 55, p. 1–12.

Sims, J.D., 1973, Earthquake-induced structures in sediments of Van Norman Lake, San Fernando, California: Science, v. 182, p. 161–163.

Sims, J.D., 1975, Determining earthquake recurrence intervals from deformational structures in young lacustrine sediments: Tectonophysics, v. 29, p. 141–152.

Sims, J.D., and Garvin, C.D., 1995, Recurrent liquefaction induced by the 1989 Loma Prieta earthquake and 1990 and 1991 aftershocks: Implications for paleoseismicity studies: Bulletin of the Seismological Society of America, v. 85, p. 51–65.

Thorson, R.M., Clayton, W.S., and Seeber, L., 1986, Geologic evidence for a large prehistoric earthquake in eastern Connecticut: Geology, v. 14, p. 463–467.

Tuttle, M., Law, K.T., Seeber, L., and Jacob, K., 1990, Liquefaction and ground failure induced by the 1988 Saguenay, Quebec, earthquake: Canadian Geotechnical Journal, v. 27, p. 580–589.

Tuttle, M., Chester, J., Lafferty, R., Dyer-Williams, K., and Cande, R., 1999, Paleoseismology study northwest of the New Madrid seismic zone: U.S. Nuclear Regulatory Commission Technical Report No. NUREG/CR-5730, 155 p.

Tuttle, M.P., 1994, The liquefaction method for assessing paleoseismicity: U.S. Nuclear Regulatory Commission Technical Report No. NUREG/CR-6258, 38 p.

Tuttle, M.P., Dyer-Williams, K., and Barstow, N., 1996, Seismic hazard implications of a paleoliquefaction study along the Clarendon-Linden fault system in western New York State: Geological Society of America Abstracts with Programs, v. 28, no. 3, p. 106.

Vittori, E., Labini, S.S., and Serva, L., 1991, Palaeoseismology: Review of the state-of-the-art: Tectonophysics, v. 193, p. 9–32.

Wells, D.L., and Coppersmith, K.J., 1994, New empirical relationships among magnitude, rupture length, rupture width, rupture area, and surface displacement: Bulletin of the Seismological Society of America, v. 84, p. 974–1002.

Wheeler, R.L., 1995, Earthquakes and the cratonward limit of Iapetan faulting in eastern North America: Geology, v. 23, p. 105–108.

Wheeler, R.L., and Frankel, A., 1999, USGS probabilistic seismic-hazard maps of the United States. 2. Use of geology: Geological Society of America Abstracts with Programs, v. 31, no. 7, p. A194.

Wheeler, R.L., and Frankel, A., 2000, Geology in the 1996 USGS seismic-hazard maps, central and eastern United States: Seismological Research Letters, v. 71, no. 2, p. 273–282.

Wheeler, R.L., and Johnston, A.C., 1992, Geologic implications of earthquake source parameters in central and eastern North America: Seismological Research Letters, v. 63, no. 4, p. 491–514.

Zoback, M.D., and Zoback, M.L., 1991, Tectonic stress field of North America and relative plate motions, *in* Slemmons, D.B., Engdahl, E.R., Zoback, M.D., and Blackwell, D.D., eds., Neotectonics of North America: Boulder, Colorado, Geological Society of America, Decade Map Volume 1, p. 339–366.

MANUSCRIPT ACCEPTED BY THE SOCIETY MAY 11, 2001

Paleoliquefaction studies in continental settings

Stephen F. Obermeier
Emeritus, U.S. Geological Survey, Reston, Virginia, USA, and EqLiq Consulting, Rockport, Indiana, USA

Eric C. Pond
Kleinfelder Inc., Consulting Engineers, Albuquerque, New Mexico, USA

Scott M. Olson
URS Corporation, St. Louis, Missouri, USA

Russell A. Green
Department of Civil and Environmental Engineering, University of Michigan, Ann Arbor, Michigan, USA

ABSTRACT

Research of the past 15 years has reported the manifestations of seismically induced liquefaction that occur in the sedimentary conditions commonly found in continental settings. And, criteria have been developed and published that can demonstrate a seismic origin for features of liquefaction origin. We present guidelines for conducting a paleoliquefaction search by means of geologic and geotechnical parameters. We also address the interpretation of results of a paleoliquefaction study in terms of locating the source region of a paleo-earthquake and back-calculating its strength of shaking and magnitude. Our critique of the geotechnical methods for these back-calculations points out uncertainties in the techniques that are most commonly used.

The guidelines that we present for a paleoliquefaction search are in terms of both geologic and geotechnical parameters because it is the combination that is critical. Neither suffices alone, and the relations between the geologic setting and geotechnical properties must be appreciated in order to understand why seismically induced liquefaction features are to be found in some locales and not in others.

INTRODUCTION

The study of paleoliquefaction features for seismic analysis is a new technique, developed over the past 15 years. The systematic search for paleoliquefaction features throughout large geographic areas is being used to interpret the paleoseismic record through much of the Holocene into latest Pleistocene time. Searches have been conducted chiefly in the southeastern, central, and northwestern United States, through different settings and seismotectonic conditions. Despite extensive studies on paleoliquefaction, their scope is not widely appreciated even by paleoseismic researchers. It is, thus, not well known that geologic field studies can yield clues about the severity of earthquake shaking and, in many settings, the probable location of the tectonic source zone. Realistic estimates can be made in many settings, even though some of the procedures are very recent and not fully developed and applicable in all situations. The techniques used in the continental United States to verify a seismic origin for suspected features are well developed (Obermeier, 1996). In contrast, some uncertainty is usually inherent

Obermeier, S.F., Pond, E.C., and Olson, S.M., 2002, Paleoliquefaction studies in continental settings, *in* Ettensohn, F.R., Rast, N., and Brett, C.E., eds., Ancient seismites: Boulder, Colorado, Geological Society of America Special Paper 359, p. 13–27.

in using a single procedure to back-calculate the strengths of shaking and magnitude (Olson et al., 2001).

This chapter critiques issues concerning field searches and the interpretation and back-analysis of strength of shaking and earthquake magnitude by using geologic and geotechnical-seismological procedures. The chapter is restricted to features developed on ground that was less than a few degrees in slope when an earthquake struck, and excludes slumps and flow failures induced by liquefaction, which may occur on steeper slopes. We focus on the following aspects of level-ground liquefaction: (1) mechanisms that form seismic liquefaction in the field; (2) field settings where liquefaction should be present if strong seismic shaking had occurred; (3) settings where the absence of paleoliquefation features indicates an absence of strong paleoseismic shaking; (4) how liquefaction features should be used to interpret the tectonic source region of a paleo-earthquake; and (5) how effects of liquefaction can be used to interpret the strength of prehistoric paleomagnitude and shaking. We include material relevant to geologists, seismologists, and engineers.

The summary here of the process of liquefaction is restricted to features that characteristically form in a clay- or silt-rich layer (i.e., host) lying above a liquefied sand-rich deposit. The discussion is based on the premise that collecting adequate data for the analyses described herein requires a search over an area at least tens of kilometers in radius. Exposures must be examined in scattered places often by searching banks of ditches or streams.

The paleoliquefaction record extends through much of Holocene time into the latest Pleistocene in many locales. A typical setting for these ages is in a river valley on the modern flood plain or a terrace a few meters higher, where the depth to the water table is less than several meters and where there are thick, sandy deposits. The liquefaction-induced features most commonly found here are steeply dipping, tabular, clastic dikes that cut a fine-grained host. Intrusions more horizontally inclined, such as sills, also abound in places.

Deformation of soft sediments that involves mud and freshly deposited cohesionless sediment (Allen, 1982; Obermeier, 1996) are not included in this discussion. Furthermore, not only is a seismic origin difficult to verify for plastically deformed soft sediments, but they may form without seismic shaking (e.g., Sims, 1975). Conversely, the origin of clastic dikes can usually be determined easily. Seismically induced dikes and sills typically involve a significantly elevated pore-water pressure. Their formation requires significant strength or duration of strong shaking. The minimum earthquake magnitude to form liquefaction features in most settings is moment magnitude M_w ~5.5 (Ambraseys, 1988), which is about the same as the threshold for damage to human-made structures. The minimum value of peak accelerations to form liquefaction features decreases with increasing magnitude; reported values are as low as 0.025 g for M_w 8.25 and 0.12 g for M_w 5.5 (Carter and Seed, 1988). It is commonly accepted that the vibration frequencies of interest are less than ~10 Hz, because higher frequencies do not induce shear strains large enough to break down the grain-to-grain contacts in granular sediments.

Herein we generally cite articles that contain expanded discussions and comprehensive references. The following section concerning the process of liquefaction and its manifestations is largely from Obermeier (1998a, 1998b); see also articles by Seed (1979), Ishihara (1985), Castro (1987, 1995) and Dobry (1989). Some recent critiques of geotechnical and seismological techniques are by Trifunac (1995, 1999), Pond (1996), Obermeier (1998a), and Olson et al. (2001). The Obermeier (1998b) paper includes numerous photographs showing features, with and without a fine-grained cap, in various types of field settings.

The purpose of this chapter is to serve as a state-of-the-art discussion of the seminal concepts being used for paleoliquefaction studies, as well as to discuss the application of those concepts and to properly credit, by references, from whence the concepts came. This chapter is not intended to serve as a listing of the numerous ensuing paleoliquefaction studies that use those concepts.

THE PROCESS OF LIQUEFACTION AND ITS MANIFESTATIONS

Following Seed and Idriss (1971) and Youd (1973), we define liquefaction as the transformation of a saturated granular material from a solid to a liquefied state due to increased pore-water pressure. Liquefaction is caused by the application of shear stresses and accumulation of shear strain, resulting in a breakdown of the soil skeleton and buildup of interstitial pore-water pressure. The process is typical of cohesionless, or nearly so, sediments, and most readily for fine- to coarse-grained sands, especially where uniform in size. The liquefied mixture of sand and water reacts as a viscous liquid with a greatly reduced shear strength. Liquefaction can be induced by seismic shaking, by nonseismic vibration, and by wave-induced shear stresses. In some loose sediments located on slopes, liquefaction can be triggered by static forces; static triggering mechanisms include an increased shear stress caused by toe erosion or an increased seepage force due to a changing water table.

The shear stresses that induce seismic liquefaction are primarily due to cyclic shear waves propagating upward from bedrock and through the soil column, although waves traveling along the ground surface can be important locally. Sediment on level ground undergoes loading, the shear stresses typically being somewhat random but nonetheless cyclic. Loosely packed, cohesionless sediments tend to become more compact when sheared. When subjected to earthquake shaking, the pore water does not have time to escape from the soil voids and allow the sediment to compact as the grain-to-grain skeleton is collapsing. Complete level-ground liquefaction occurs when the pore-water pressure increases to carry the static confining (overburden) pressure—i.e., the grain-to-grain stress equals zero, which permits large strains, flow of water, and suspension of sediment. Partial liquefaction occurs when the increase in pore-water pressure is not enough to fully carry the static overburden pressure. While, by definition, this is not true level-ground liquefaction, ground failure can occur under a condition of partial liquefaction.

A large increase in pore-water pressure commonly occurs during the transition to the liquefied state. The pore-water pressure carries the weight of overlying sediment and water. In many field situations, the pore-water pressure can increase several-fold within a few seconds, thereby hydrofracturing a fine-grained cap lying above the liquefied zone.

Subsequent densification occurs throughout the column of liquefied sediment during dissipation of excess pore-water pressure, and large quantities of water can be expelled. The water flows upward, carrying along sediment. This process is referred to as "fluidization" by some geologists (e.g., Lowe, 1975). The flowing water causes sediment to be carried or dragged along by other grains. The process of fluidization transports the sediment that fills clastic dikes and sills observed in paleoliquefaction studies.

Liquefaction can result from only a few cycles from many cycles of shaking. For a very loose packing of sediment grains, the breakdown of grain structure can be abrupt and liquefaction is virtually simultaneous with the onset of shaking (National Research Council, 1985, Fig. 2.26). Such loose packing is relatively common in delta and eolian dune deposits as well as in very young (less than 500 yr) river channel deposits (Youd and Perkins, 1978). However, some very young fluvial sands have such a dense initial packing that any pore-water pressure increase during seismic shaking is insignificant, and liquefaction does not occur (Seed et al., 1983). For older Holocene-age river deposits, the buildup of pore-water pressure generally tends to be more gradual and requires more cycles of shearing than for younger deposits. Deposits of Pleistocene age are often very resistant to liquefaction owing to effects of aging and weathering (Youd and Perkins, 1978), but deposits hundreds of thousands of years old and still highly susceptible to liquefaction have been encountered (e.g., Obermeier et al., 1993; Martin and Clough, 1994). A broad range of susceptibilities to liquefaction is commonly encountered in a local field setting.

Liquefaction Susceptibility

Liquefaction susceptibility refers to a sediment property and takes into account the depth to the water table and other factors (e.g., static stress conditions, density aging) that affect the ability of a deposit to liquefy. The relative state of packing of sand deposits (called the "relative density" by geotechnical engineers) is a principal determinant of liquefaction susceptibility in most Holocene deposits; the relative density is, in turn, related to standard penetration test (SPT) blow counts measured in situ (Table 1). Relative density is by definition a measure of how densely the sand grains are packed in comparison to the laboratory determined loosest and densest reference states (Terzaghi and Peck, 1967). Correlations of relative density with SPT blow counts are listed in Table 1. For practical purposes, sediments having blow counts in excess of 30 will not liquefy, even if other factors in the field are very favorable. Loose and very loose sands are generally highly susceptible to liquefaction. The moderately compact sands are generally moderately susceptible.

Liquefaction susceptibility is nearly always measured using an in situ test because of the extreme difficulty and expense of collecting samples that are sufficiently undisturbed for laboratory testing. In recent years, there has been a tendency to use the cone penetration test (CPT) to measure liquefaction susceptability in situ (e.g., Stark and Olson 1995; Robertson and Wride, 1997). The CPT permits more detailed measurements of sediment properties and stratigraphy than does the SPT, and thereby is likely to provide a more accurate evaluation of in-situ liquefaction susceptability. Cone penetration testing is also relatively inexpensive compared to standard penetration testing. Because of the larger database where SPT results are available, however, particularly in the central and eastern United States, the following discussion focuses mainly on the SPT.

SPT data commonly provide a reasonable measure of relative density in Holocene clean sands (sediment that is composed almost entirely of the sand-sized fraction). Exceptions occur where the sediment has been cemented with chemical precipitates, where the fines content (silt, and clay fraction less than 0.075 mm) is greater than about 15%–20%, and where the mean grain size (50% of the material by weight) is greater than ~2 mm (Seed et al., 1983). Sites where stress conditions in the sediment are unusually high in the horizontal plane (as by prior glacial loading) can also cause misleading values of relative density from SPT readings (Terzaghi et al., 1996).

SPT data, in the absence of chemical precipitates, also provide a measure of the effects of aging and weathering on liquefaction susceptability. Aging and weathering effects can originate from both mechanical and minor chemical sources (Schmertmann, 1987, 1991; Mesri et al., 1990). In the short term (hundreds to a few thousand years), mechanical effects caused by adjustment of grains are likely to dominate aging (Olson et al., 2001). Fortunately, the total effect of chemical and mechanical aging is relatively minor from a practical viewpoint in some and perhaps many field settings. For example, in glaciofluvial deposits that abound throughout the central United States, the maximum change in SPT blow count resulting from aging is probably on the order of 3 or 4, on the basis of the difference between the loosest sediments of modern ages (with blow counts near 0) and the loosest deposits of early Holocene ages (Pond, 1996). This change in blow count almost certainly decreases

TABLE 1. RELATIVE DENSITY OF SAND AS RELATED TO STANDARD PENETRATION TEST

No. of blows	Relative density or compactness
0–4	Very loose
4–10	Loose
10–30	Medium or moderate
30–50	Dense
>50	Very dense

Note: Data from Terzaghi and Peck, 1967.

substantially with increasing initial relative density because of the diminished opportunity for mechanical adjustment of grains. The possible influence of aging and weathering should be evaluated on a case-by-case basis, depending on the geologic setting, in view of the uncertainty of factors that determine the effects of aging (Olson et al., 2001).

Field conditions favorable for formation of liquefaction features

Dikes and sills cutting a fine-grained host are generally readily visible in vertical section. Dikes and sills form most readily where a thick, sand-rich deposit is capped by a low-permeability deposit and the water table is very near the ground surface. Grain sizes that are generally the most prone to liquefy and fluidize, and to form dikes and sills, range from silty sand to gravelly sand. A thickness of 1 m of liquefied sediment generally suffices to form recognizable clastic dikes, although a much smaller thickness can be adequate, depending on factors such as the severity of liquefaction, the local field setting, and the mechanism that forms the dikes.

For the normal range of Holocene sediments in a river valley, clastic dikes cut readily through a cap about 1 m thick, and can cut through a much greater thickness where shaking has been severe (Ishihara, 1985; Obermeier, 1989). The strength of the cap generally has a minor influence on the development of dikes, and dikes have been observed to have formed in hard, massive silt and clay-rich glacial tills (Obermeier, 1996). Caps having large tensile strengths, however, such as fibrous mats, can greatly inhibit dike formation.

Liquefaction typically occurs at shallow depth, less than a few tens of meters (e.g., Seed, 1979). Paleoliquefaction searches by Obermeier in diverse field settings throughout the United States, where the water table was probably between 1 and 5 m deep at the time of the earthquake, show that dikes are found most often in fine-grained caps that are 1 to 5 m thick. Caps thicker than 10 m rarely host dikes, including small dikes along the base of the cap. Most likely, dikes in thicker caps are scarce because greater depths require exceptionally severe shaking for liquefaction. Liquefaction beneath thicker caps is unlikely because the increasing overburden pressure can increase the shear resistance of a cohesionless sediment beyond the shear stress induced by seismic shaking.

Liquefaction is most pronounced where the water table lies within a few meters or less of the surface. A change in water table depth of 10 m can change the ability of a deposit to liquefy from high to nil (e.g., National Research Council, 1985, Table 4-1).

Ground-failure mechanisms

Dikes in a fine-grained cap are induced chiefly by three ground-failure mechanisms: hydraulic fracturing, lateral spreading, and surface oscillations. All produce tabular dikes in plan view.

Hydraulic fracturing in response to seismic liquefaction was first deduced by Obermeier (1994), following discussion of the process in the failure of earth dams by Lo and Kaniaru (1990). Hydraulic fracturing begins at the base of a fine-grained cap sitting on liquefied sediment. Fracturing of the cap typically occurs in response to the high pore-water pressure entering naturally occurring flaws along the base of the cap, such as small root holes and other openings. The pressure causes vertical, tensile fractures that are tabular and are filled with a fluidized mixture of sand and water driven by the hydraulic gradient. Similarly, vertical tabular defects in the cap that formed by weathering can be opened by the high pore-water pressure, leading to the formation of tabular dikes either parallel to one another or irregular, having a nearly haphazard pattern in both plan and vertical views. Dikes from hydraulic fracturing are typically quite narrow, ranging from a few millimeters to less than 10 cm wide.

Lateral spreads reflect translational movement downslope or toward a stream bank. The movement occurs where there is only minor resistance to lateral translation of the fine-grained cap sitting on liquefied sediment. Dikes originating from lateral spreading, and especially the wider dikes, are the result of fluidized sand and water flowing into breaks through the cap that have been opened by shaking and/or downslope gravity (Bartlett and Youd, 1992). Dikes can be as much as 0.5 to 0.7 m in width even where shaking has been only moderately strong (about 1/4 g). Widths of as much as a few to several meters are not unusual. Lateral spreads are typically defined as occurring on slopes of 3° or less (Youd and Garris, 1995). On steeper slopes, liquefied deposits can flow tens of meters to 1 km (Tinsley et al., 1985).

Surface oscillations can cause tabular clastic dikes to originate in response to the fine-grained cap being strongly shaken back and forth above liquefied sediment. We use the term "surface oscillation" as a generic description of an end effect rather than a driving mechanism. This definition is in the sense commonly used by geotechnical engineers to describe liquefaction-related ground failure that requires, in plan view, large back-and-forth straining of the cap; high accelerations may or may not be involved. Indeed, during strong bedrock shaking at sites of liquefaction, the accelerations in the cap can be deamplified to a lower level even as straining of the cap is greatly augmented and breaks apart the cap (e.g., see analysis in Pease and O'Rourke, 1995). Surface oscillations can originate from either body (S) waves (Pease and O'Rourke, 1995) or surface (Rayleigh or Love) waves (Youd, 1984). The back-and-forth straining in the cap can be in the form of either axial or shear strains (Pease and O'Rourke, 1995, Fig. 2-1). Rayleigh waves are likely involved at sites of severe axial straining, and either S or Love waves are likely involved at sites of severe shear strain. The effects of Rayleigh waves are probably best manifested by dikes that tend strongly to parallel one another with a spacing that can range from tens to hundreds of meters apart; these effects generally are most severe in the *meizoseismal zone* (see Appendix) but can also extend far beyond (T.L. Youd, 1998, personal commun.). Surface oscillations from what are likely to have been Rayleigh waves are often seen by observers as traveling ground waves. Dike widths from surface oscillations may be as much as 15 cm (T.D. O'Rourke, 1998, personal commun.). Sites of severe shear straining may be indicated by lateral offsets along dikes and along fractures at the ground surface.

Factors controlling the ground failure. Different levels of shaking are required to form dikes visible at the ground surface for each of the mechanisms of lateral spreading, surface oscillation, and hydraulic fracturing. For cohesionless deposits that are very loose to moderate in relative density (Table 1) lateral spreads typically occur farthest from the meizoseismal region (if a stream bank is nearby at the time of the earthquake). The factors that determine the most distant occurrence of lateral spreading have not been verified, but such spreading could result when a stream bank offers little or no resistance to lateral movement during shaking, or it could be because the youngest deposits typically border a stream where the water table is shallow. Dikes from surface oscillations (Youd, 1984) can develop considerably beyond the meizoseismal zone, especially where conditions are favorable for developing surface oscillations from surface waves (e.g., broad valleys, alluvium at least tens of meters in thickness, and flat-lying bedrock). Such dikes are likely developed from S waves far beyond the meizoseismal, even at sites of marginal liquefaction and relatively low accelerations, because of the tendency for surface oscillations to develop for S-wave vibrations with longer periods (Pease and O'Rourke, 1995). In most field situations of moderate liquefaction susceptibility, hydraulic fracturing seems to cause only small, scattered dikes to form beyond the meizoseismal zone, even for earthquakes in excess of M_w ~7.

The formation of dikes from lateral spreading predominates near the stream bank, and effects from hydraulic fracturing predominate with increasing distance from the bank (Obermeier, 1996). Farther from the meizoseismal zone, the influence of hydraulic fracturing is often minor.

The thickness of cap penetrated by hydraulic fracturing depends primarily on the thickness of sand that liquefies, apparently because the liquefied thickness controls the magnitude of porewater pressure increase as well as the volume of water expelled; the thickness of the penetrated cap is also strongly dependent on the severity of ground shaking (Ishihara, 1985). Youd and Garris (1995) also found that dikes caused by hydraulic fracturing commonly are much lower in height than those formed by lateral spreading or surface oscillations. They estimate the maximum thickness of cap that has been observed to be ruptured by hydraulic fracturing is about 9 m.

In contrast, the maximum cap thickness that can be ruptured by lateral spreading is commonly much greater than that ruptured by hydraulic fracturing (see Youd and Garris, 1995, Fig. 3). The maximum reported is ~16 m (T.L. Youd, 1997, personal commun.). In many settings, the maximum thickness is controlled by the maximum depth of liquefaction, because of the low tensile strength of the cap in relation to the stresses imposed on it by gravity and seismic shaking. Caps of Holocene and late Pleistocene ages composed of silt and clay sediments typically have very low tensile strengths and thus are easily pulled apart by lateral spreading.

The formation of lateral spreads is not nearly as dependent on the thickness of the liquefied zone as is the formation of hydraulic fracturing. Lateral spreads have not been observed to form on liquefied sand strata only a few centimeters in thickness (J.R. Keaton, 1993, personal commun.). Lateral continuity of the liquefied bed is especially important for lateral spreading, particularly on such a thin stratum.

The thickness of cap ruptured by surface oscillations commonly is greater than that ruptured by hydraulic fracturing (Youd and Garris, 1995), and the effects of oscillations tend to extend much farther from the meizoseismal zone. In general, though, breakage of the cap by surface oscillations is localized away from the meizoseismal zone, even for a very large earthquake.

No data are available concerning the role of thickness of liquefied sediment in development of surface oscillations, although we suspect that 1 m or more suffices for typical fluvial sands, at least near the meizoseismal region of a very large earthquake. This suggestion is based on field observations in the Wabash Valley of Indiana and Illinois (Fig. 1). Preliminary data (Obermeier) also indicate that parallel joints in the cap can develop from seismic shakings, even where no liquefaction has occurred, providing that shaking has been strong enough. Joints from other mechanisms such as weathering and desiccation are commonly much more haphazard and discontinuous than those of seismic origin.

In the previous section we noted that formation of liquefaction features depends on depth to the water table. The influence of water table depth seems to be very dependent on the mechanism primarily responsible for rupturing the cap. Obermeier's data indicate strongly that cap breakage by hydraulic fracturing can be much more sensitive to depth of the water than is cap breakage by lateral spreading.

Field examples of manifestations of liquefaction. Evidence of liquefaction-related ground-failure mechanisms is apparent in aerial photographs of the meizoseismal zone of the great 1811–1812 New Madrid (Missouri) earthquakes (Figs. 1, 2). Within a time span of only three months, numerous strong earthquakes struck along a more than 175-km-long fault zone. One earthquake was probably nearly M_w 8, and two more were nearly as large (Johnston and Schweig, 1996). The earthquakes were centered beneath a huge region of liquefiable deposits and caused tremendous liquefaction. Sand that vented to the surface formed a veneer more than 0.5 to 1 m thick over hundreds of square kilometers. More than 1% of the ground surface was covered by vented sand over thousands of square kilometers (Obermeier, 1989). The meizoseismal zone of the 1811–1812 earthquakes is one of the best in the world to see the effects of liquefaction in both plan and vertical views. The vertical view permits the observer to see dikes that pinch together and do not reach the surface, and also permits viewing of dikes that were later buried by sediments or have been weathered so severely as to not be observable at the surface.

Fissuring and venting during 1811–1812 took place in braid-bar deposits of latest Pleistocene age and in Holocene point-bar

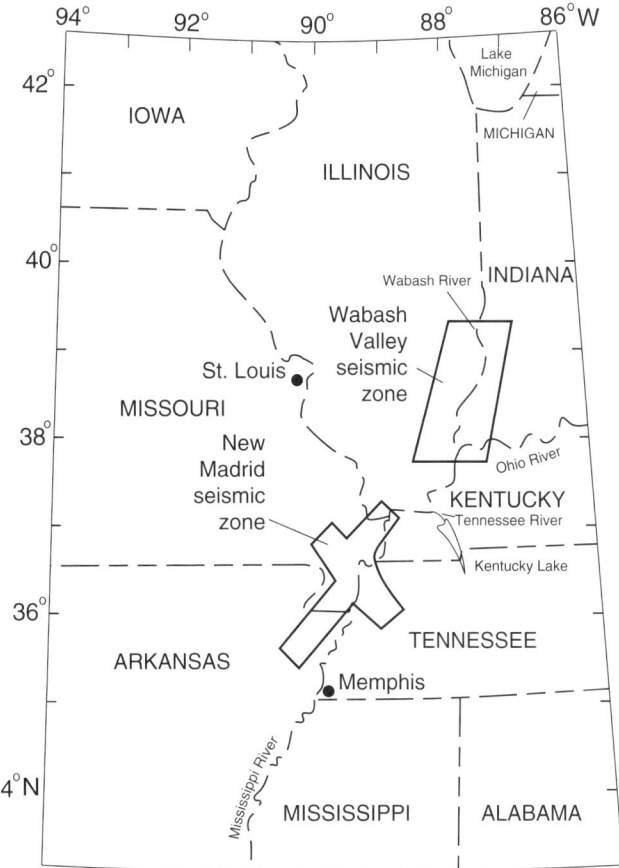

Figure 1. Approximate limits of the New Madrid seismic zone and the Wabash Valley seismic zone. The New Madrid seismic zone is the source area of the great New Madrid, Missouri, 1811–1812 earthquakes; the region continues to have many small and some slightly damaging earthquakes. The Wabash Valley seismic zone is a weakly defined zone of historic seismicity having infrequent small to slightly damaging historic earthquakes.

deposits (Fig. 2). The ground surface there is typically flat, except at stream banks that generally are only several meters high. The light-colored parts of the photos show sand vented to the surface. The dark background is the dark-colored, clay-rich cap onto which the sand vented. The light-colored linear features are long fissures through which sand vented, and the light-colored spots are individual sites of venting. Venting occurred through dikes.

Note the abundance of linear fissures that are more or less parallel to one another in the upper right side of Figure 2. These fissures are of lateral spreading origin and formed near a break in slope. The photo clearly shows that lateral spreading in the area was severe at distances farther than 0.5 km from the stream. Individual sand blows, which are particularly well expressed in the upper left part of the photo, were formed by hydraulic fracturing, clearly indicated by the random "shotgun pattern" of the sand blows.

Hydraulic fracturing can also follow geologic details, as illustrated in the lower part of Figure 2. Sand blows here developed in point bar deposits, as illustrated by the arcuate bands of vented sand. The venting occurred along the crests of scrolls of point bar deposits, where the cap is thinnest (Saucier, 1977). A venting origin for the sand is demonstrated by the irregular, jagged patterns of sand along the arcuate bands, which precludes the possibility of the sand being visible at the surface simply because of the absence of a fine-grained cap along the crest of the scroll.

The development of lateral spreads and individual sand blows is typical of that throughout the meizoseismal region of the 1811–1812 New Madrid earthquakes in that the dikes from lateral spreading commonly extend more than 1 km from any breaks in slope, and the isolated sand blows developed throughout the area, independent of proximity to a stream bank.

There is a widespread perception that wide dikes that form by lateral spreading are restricted to areas very near stream banks. However, in the meizoseismal region of the 1811–1812 New Madrid earthquakes, dikes from lateral spreading as much as 0.5 m wide are plentiful even hundreds of meters from any significant slopes. In another example, in the Wabash Valley of Indiana and Illinois, within the meizoseismal zone of a prehistoric M_w ~7.5 earthquake, dikes up to 0.5 m in width probably formed hundreds of meters from any stream banks when the earthquake struck (Munson and Munson, 1996; Pond, 1996; Obermeier, 1998a).

In both the New Madrid and Wabash regions, liquefaction susceptability is only moderate at most places (Obermeier, 1989; Pond, 1996) and is probably typical of medium-grained, moderately well graded fluvial deposits elsewhere. Data from a worldwide compilation of historical earthquakes by Bartlett and Youd (1992) clearly show that horizontal movements of meters commonly extend hundreds of meters back from stream banks, especially where seismic shaking has been strong.

The probable explanation for the exceptional development of lateral spreading from the 1811–1812 New Madrid earthquakes is that there were very high levels of shaking, caused by large drops in stress in bedrock at depth (Hanks and Johnston, 1992). A major point of relevance, indicated in Figure 2, is that the severity of lateral spreading, including the distance of development from stream banks, can be an indicator of the severity and duration of strong shaking.

WHERE WAS THE SOURCE REGION?

Verification of a seismic origin for suspected liquefaction features typically involves demonstrating that (1) details of individual clastic dikes conform to those of known seismic origin, (2) both the pattern and location of dikes in plan view conform with a seismic origin, on a scale of tens to thousands of meters, (3) the size of dikes on a regional scale identifies a central "core region" of widest dikes, which conforms with severity of effects exposed in the energy source region (the meizoseismal zone), and (4) other possible causes for the dikes, such as artesian conditions and landsliding, are not plausible (Obermeier, 1996, 1998a). As we use the term "core region," we are referring to the region of strongest bedrock shaking. We also

Figure 2. This aerial photograph shows effects of severe liquefaction in the meizoseismal zone of the great 1811–1812 New Madrid (Missouri) earthquakes. White linear features show sand that has vented through breaks in proximity to stream breaks. Isolated white spots are sand that has vented through breaks in the cap by hydraulic fracturing. Black box on map of Arkansas indicates location of photo area.

refer interchangeably to this region throughout the paper as the "source region" or the "energy center."

Two methods have been used to estimate the source region in an excellent study example, in the Wabash Valley (Obermeier et al., 1993; Munson and Munson, 1996; Pond, 1996). Both methods have widespread applicability for paleoseismic studies. One involves direct measurement of dike widths and the other involves back-calculating the strength of shaking. Both require collecting data over a large region in order to see a clear-cut trend in the data. In practical terms, for an earthquake of M_w 6 to 7, the data must be collected over an area of several to many tens of kilometers in radius. Preferably, data are from the region of distal effects of liquefaction, where dikes are small (narrow) and sparse, and also from the area close to the source region, where dikes are much larger (wider) and more plentiful.

Dike width serves as a superior parameter to locate the source region in many field settings (Obermeier, 1996). This width generally reflects the amount of lateral spreading except where dikes are relatively small (say, less than 10 cm wide). Conceptual verification for using dike width to locate the source region is provided from a study of historical earthquakes by Bartlett and Youd (1992). Dike width works well because the development and magnitude of lateral spreading are largely independent of thickness and strength of the cap, at least for sediments that are typical of the Holocene and late Pleistocene. Maximum dike width and the sum of dike widths at a site appear to work equally well to estimate the source zone (Munson et al., 1995). A valid interpretation based on the widths of dikes obviously requires that bank erosion has not been so severe as to have destroyed dikes by lateral spreading. Problems of interpretation due to erosion are generally not serious in the meizoseismal zone of a very large magnitude earthquake because of the tendency for large lateral spreads to develop even relatively far from the stream banks.

Data from historical earthquakes in the Wabash Valley region, in the forms of modified Mercalli intensities and instrumentally located epicenters (Rhea and Wheeler, 1996), suggest that using liquefaction features to locate the source region of prehistoric earthquakes is generally accurate to within a few tens of kilometers, at least for earthquakes of moderate size. The uncertainty in location probably increases with increasing magnitude because of the tendency for the epicenter of larger earthquakes to be farther removed from the area of strongest shaking (e.g., Youd, 1991). Still, it appears that the distribution and severity of liquefaction effects can be used to reasonably estimate the region of strongest bedrock shaking (Pond, 1996).

Other parameters, such as the density of dikes per unit length and density of dikes per unit area, have been used by other researchers in their attempts to locate the region of strongest shaking. There are numerous practical problems in using such an approach to interpret the data, however, because dike density is controlled by different factors for each of the mechanisms of lateral spreading, surface oscillations, and hydraulic fracturing. In many field situations it is impossible to determine which mechanism(s) controlled the density of dikes. Interpretations can be questionable without such a differentiation.

Back-calculation of the strength of shaking at widespread sites can sometimes be used to better locate the source region where dike-width data are sparse. This back-calculation procedure has been verified by comparing this interpretation with that of the dike-width method discussed above; both yielded the same results (Pond, 1996).

A question often asked is whether paleoliquefaction features resulted from a single large earthquake or from a series of small earthquakes that were closely spaced in time. The answer is generally best resolved by analysis of the regional pattern of dike widths. The attenuation pattern of maximum dike widths around a core region should be examined in orthogonal coordinates (preferably along the suspected fault axis and perpendicular to the axis). A monotonic decrease of maximum dike width in orthogonal directions around the suspected core indicates a single large earthquake. In the Wabash Valley, this approach was verified by geotechnical back-calculations of the prehistoric strength of shaking for four prehistoric earthquakes (Pond, 1996). The use of dike widths alone to resolve the issue of the number of events requires that the liquefaction susceptability be reasonably uniform on a regional basis and also that the amplification or attenuation of bedrock motions be similar on a regional basis.

To answer the question of whether a single earthquake or multiple earthquakes caused the observed features, the methods described above usually work best for very large earthquakes, because of the tendency for the regional pattern of liquefaction to become more conspicuous with increasing magnitude. For example, the regional pattern of dike sizes and abundance, in conjunction with radiometric dating, has been used in coastal South Carolina to show that liquefaction effects from prehistoric earthquakes were caused by very large earthquakes rather than multiple small earthquakes closely spaced in time (Obermeier, 1993; Obermeier, 1996). More recently, using basically the same logic, the regional pattern of severity of venting has been used to evaluate whether paleoliquefaction features discovered within the meizoseismal region of the great 1811–1812 New Madrid earthquakes were from a few very large earthquakes rather than a series of much smaller earthquakes (Tuttle, 1999). The New Madrid region is nearly ideal for this type of analysis because the liquefaction susceptibility is remarkably uniform over a huge area, the causative fault system for major earthquakes is likely known, and the regional pattern and extent of liquefaction from the 1811–1812 earthquakes has long been known reasonably well (e.g. Obermeier, 1989).

Using the paleoliquefaction method for determining the timing and strength of shaking of various earthquakes within a relatively small region works best where the large earthquakes are spaced apart sufficiently in time to distinguish different generations of liquefaction features. The techniques for sorting these generations have been developed mainly in a classic study in the Wabash Valley region by Munson and Munson (1996). Their approach is well suited for many field settings and typically uses radiocarbon dating, depth of weathering, pedology, sediment stratigraphy, and archaeological artifacts, in conjunction with the regional pattern of sizes of liquefaction features. They also were the first to note that sand deposits that had been vented to the surface were especially valuable as sites for narrowly bracketing when liquefaction occurred; the vented sand typically formed slightly elevated, dry sections in lowland areas that were otherwise wet and muddy much of the year. The vented sand deposits were frequented by Native Americans, who commonly left behind on the vented sand hearths and artifacts that can be used for dating.

DID STRONG SHAKING OCCUR WITHOUT LEAVING LIQUEFACTION EVIDENCE?

There is a common perception that liquefaction can occur in a region but leave behind no evidence. Discovering effects of liquefaction in the field is usually easy where liquefaction has been severe throughout a region, but it may be difficult where liquefaction has been marginal or highly localized. Below we present some of the major factors that determine the severity of liquefaction.

Effects of strength of shaking and liquefaction susceptibility

Our approach of relating occurrence and severity of liquefaction effects to modified Mercalli intensity (MMI) is used because MMI correlates strongly with both severity of liquefaction and damage to human-made structures (Wood and Neumann, 1931). MMI also correlates reasonably well with peak surface acceleration (Krinitzsky and Chang, 1988). Seed et al. (1985) reported similar correlations developed in China for M_w~7.5 earthquakes. The Chinese correlations emphasize higher earthquake magnitudes than those of Krinitzsky and Chang, whose relations are for a much wider range of magnitudes (Table 2). Relations below by Krinitzsky and Chang (1988, Fig. 7) are for sites in the "far field," which are removed from the region of strongest shaking.

Throughout the meizoseismal region of a very strong earthquake, in which the MMI value is IX or higher, liquefaction features should abound even where the liquefaction susceptibility is only moderate. Any reasonable effort to locate numerous liquefaction features should be successful. Some wide dikes almost certainly exceeding 0.3 m and many small dikes should be discovered.

For moderate liquefaction susceptibility in regions of MMI VII–VIII, small liquefaction features may be sparse but still should be numerous enough that some features would be discovered during the examination of tens of kilometers of stream banks.

Moderate liquefaction susceptibility implies medium relative density (Table 1) as well as a water table within several meters of the surface and a cap thickness less than 8 or 9 m. Moderate liquefaction susceptibility is about the norm for deposits of latest Pleistocene and Holocene age that have been laid down by moderate to large streams in the central and eastern United States. This level of susceptibility applies to streams of both glaciofluvial braid-bar and Holocene point-bar origins. A lower limit of moderate susceptibility requires a bed of silty sand, sand, or gravelly sand (generally less than about 40% gravel) that is at least a few to several meters thick and is capped by at least 0.5 m of lower permeability sediment. Where a cap is underlain by medium-grained sand or coarser sediment, the water table should be at or above the base of the cap at the time of the earthquake; otherwise, unless liquefaction occurs through a large thickness of sediment and has made available a large quantity of water, the high permeability of the material beneath the cap can permit dissipation of pore-water pressure induced by shaking, leaving no evidence of liquefaction.

Effect of grain size

Tsuchida (1970) recognized that liquefaction features predominate in sands containing little or no gravel or fines; uniform fine clean sands are the most susceptible to liquefaction. Since 1970, liquefaction features have been documented in nearly all cohesionless soils, including sandy gravels, silty sands to sandy silts, cohesionless silts, and tailings sands and slimes (e.g., Obermeier, 1996).

Gravelly sand and sandy gravel (as much as 60% gravel content—perhaps even more) can liquefy and form large dikes during earthquakes in conditions of strong shaking, impeded drainage, and a water table near the ground surface (Meier, 1993; Yegian et al., 1994). It appears that the presence of a fine-grained cap controls the formation of liquefaction features in gravel-rich deposits. Both historic (Harder and Seed, 1986; Andrus, 1994; Yegian et al., 1994) and prehistoric liquefaction features have been observed in gravelly soils with caps. Even a thin, fine-grained cap can impede drainage and allow the pore-water pressure to increase during shaking, but it seems likely to us that the areal extent of the cap must also be large in order to prevent dissipation of pore-water pressure during shaking. Earthquake magnitudes of M_w~7 to 7.5 and shaking levels lower than 0.4 to 0.5 g were adequate to trigger liquefaction in many of the cases cited above. However, very gravel-rich deposits without fine-grained caps can withstand strong shaking (on the order of 0.5 to 1.0 g) without forming liquefaction features (Yegian et al., 1994).

Back-analysis of liquefaction cases involving gravelly soils is complicated because of the effect of gravel on the measurement of penetration resistance. Tokimatsu (1988) showed that the penetration resistance for soils with a small percentage of gravel can be artificially increased compared to that of a clean sand at

TABLE 2. CORRELATIONS BETWEEN MODIFIED MERCALLI INTENSITY (MMI) AND PEAK SURFACE ACCELERATION

MMI	Peak surface acceleration (g)	
	Chinese	Krinitzsky and Chang (1988)
VII	~0.1	~0.13
VII	~0.2	~0.2
IX	~0.35	no data

the same relative density and confining pressure, because of the large size of the gravel particles relative to the size of the penetration equipment. Tokimatsu (1988) tentatively suggested a reduction factor to correct the SPT blow count of gravelly soils (based on mean grain size) to that of sandy soil for use in liquefaction analyses. However, the application of such a correction factor raises uncertainty in any back-analysis.

Cohesionless silt will also liquefy and fluidize to form dikes, sometimes extensively (Youd et al., 1989). "Dirty" sands containing as much as 85% fines (silt and clay) have been observed to liquefy (Bennett, 1989), but soils with more than 15% to 20% clay content (<0.005 mm) are unlikely to liquefy (Seed et al., 1983). The effect of fines on liquefaction susceptibility has not been completely resolved, and numerous apparently conflicting data and opinions exist in the literature. The effect of fines on the susceptibility can be separated into two categories: (1) effect on liquefaction resistance of the soil, and (2) effect on penetration resistance.

Recent studies have indicated that the effect of fines on liquefaction susceptibility depends on the nature of the fines (i.e., plasticity and cohesion). Cohesionless silts (e.g., tailings slimes) and some sands with cohesionless silt contents as high as 30% may be more susceptible to liquefaction than clean sands (Ishihara, 1993; T.L. Youd, 1997, personal commun.). In addition, Yamamuro and Lade (1998) noted that at low overburden pressures, uniformly sized sand with a low cohesionless silt content is more likely to collapse and liquefy than the same sand containing no silt; Yamamuro and Lade hypothesized that the silt grains cause the silty sand to form a more honeycombed structure during deposition compared to a clean sand, even at the same global relative density. This causes the silty sand to be more susceptible to collapse and pore-water pressure increase upon shearing.

Field observations vary concerning the influence of silt content on liquefaction effects. In the western United States, M.J. Bennett (2000, personal commun.) has observed that silty sands and sandy silts are more susceptible to liquefaction than clean sands. It has been the experience of Obermeier, however, that dikes and liquefaction-induced features involving silty sands (say, 20% to 30% fines or more) are only rarely observed in paleoliquefaction searches in the central and eastern United States, even where shaking has been very strong; yet nearby, liquefaction features involving clean sand sources are commonly abundant.

Effect of depth to water table

The depth to the water table has a profound effect on the liquefaction susceptibility of a sand deposit and can also have an important bearing on the ground-failure mechanism that develops. Where the water table is more than 4 to 5 m deep, it appears that severe effects of liquefaction, especially due to hydraulic fracturing, become greatly suppressed and can be scarce even where shaking is moderate (~0.2 g).

It is commonly observed that dikes from lateral spreading are the only ones seen in an exposure. Levels of shaking for this situation probably can be as high as 1/4 g in field settings where the source sands are fluvial in origin, medium-grained, moderately well-graded, and moderately compact.

In general, if the water table is ≥10 m below the ground surface, formation of liquefaction features from any failure mechanism is highly unlikely, unless shaking is severe and the field setting is conducive to their formation.

Locating the depth to the water table at the time of the earthquake is very important in estimating the strength of shaking. For clean sands, fine-grained and coarser, this depth can be estimated by observing the highest level of the base of dikes (i.e., at the base of the fine-grained cap at widespread sites). In field situations where the water table is much lower that the base of the cap, for low to even moderate severity of liquefaction, the high permeability of these clean sands would probably allow dissipation of excess pore-water pressure along the base of the cap, thereby precluding the formation of dikes in the cap.

Long bank exposures over a large region, at least kilometers in extent, in which the contact of the fine-grained cap with underlying sand can be observed are especially valuable for the approach discussed above. Where bank exposures are limited in length or in regional extent, confidence in the interpretation of the depth of the water table is increased by measuring the relative liquefaction susceptibilities of sand at various depths. Obermeier et al. (2000) discussed how these factors were incorporated to evaluate the depth of water table in a study area during the 1811–1812 New Madrid earthquakes.

HOW STRONG WAS THE PALEO-EARTHQUAKE?

Much progress has been made in the past few years in the development of techniques to back-calculate strength of shaking and magnitude of paleo-earthquakes. Four methods, each distinct from the rest, and which we believe are especially relevant or promising are: (1) the magnitude bound method, which uses the farthest distance of paleoliquefaction features from the tectonic source to estimate magnitude; (2) the cyclic stress method, which estimates the lower bound peak accelerations at individual sites of liquefaction, and which can be used in conjunction with the regional pattern of acceleration attenuation to estimate the actual magnitude of prehistoric earthquakes; (3) energy-based solutions, which use fundamental parameters of earthquake strength and soil susceptibility to liquefaction, and (4) the Ishihara method, which uses dike height at a site of hydraulic fracturing to estimate the peak acceleration. The first two methods are applicable to many field and tectonic settings, and though existence of these methods is known by many, their strengths and limitations are not widely appreciated. The latter two methods are still in development, but can be useful. Selection of the appropriate method(s) depends on the data available at the field sites, as noted below. Much more detailed discussion for each of the four methods is given in Obermeier et al. (2001).

The first two techniques have been used to determine the prehistoric levels of shaking in the Wabash Valley region of Indiana and Illinois (Munson et al., 1997; Pond, 1996). For comparison, we also used solution still in development, the energy-stress method of Pond (1996). The Wabash Valley region lies in an area of intraplate seismicity in which the largest historical earthquake (during the past 200 yr) has been M_w 5.8. Paleoliquefaction features clearly demonstrate, however, that numerous and much larger Holocene earthquakes have been centered in the region, on the basis of sizes of liquefaction features and regional extent of liquefaction from these earthquakes. This region is typical of many where paleoliquefaction interpretations are especially useful—i.e., there are no surface faults available for study, and the prehistoric earthquakes are spaced widely enough in time to separate their liquefaction effects from one another.

Evaluation of the prehistoric levels of shaking in the Wabash Valley presents challenges, however, because of the absence of seismological records of large earthquakes in the region. The record is available only for small earthquakes (M_w <5). The behavior of the smaller earthquakes has been extrapolated to predict the behavior of much larger events for some of the analyses we discuss in this section, but such an extrapolation may not reflect reality. Similar uncertainty exists in most regions where paleoliquefaction studies have been used as a basis for interpreting the prehistoric record (i.e., central and eastern United States). Unknown seismic factors in the Wabash Valley region include the stress drop, which can have a large effect on the strength of shaking (Hanks and Johnston, 1992) and possibly other factors, such as strength of shaking at various frequencies, in which some frequencies may be too high to induce shear strains large enough to cause liquefaction. A deep focal depth can cause the strength of shaking to be diminished at the ground surface, above the focus. Unlithified sediment of considerable thickness (hundreds to thousands of meters) above bedrock may alter the severity and/or frequency of shaking as it is transmitted from the bedrock.

A preferred orientation of strong shaking (i.e., directionality; see Appendix) can also complicate interpretations of prehistoric strength of shaking. For strike-slip faulting, the effects of directionality can be manifested as higher accelerations along the projection of the fault axis. Other types of faulting have other types of directionality effects. However, a paleoliquefaction search that encompasses a large region should clarify effects of directionality, permitting proper use of back-analysis of strength of shaking and magnitude. These procedures for back-calculations are based on techniques that provide only maximum levels of shaking as a function of earthquake magnitude and distance from the energy center, regardless of orientation from the energy center; this requires, therefore, that back-calculations for paleoseismic interpretations determine the highest level of (bedrock) shaking as a function of distance from the energy center.

The confidence in interpretations of prehistoric levels of shaking is highest where different procedures for back-calculation yield the same results. Even in this case though, there can be some uncertainty because some of the methods may depend similarly on assumed parameters such as stress drop and focal depth. Evaluations of prehistoric magnitudes for four large paleo-earthquakes in the Wabash Valley region are given in Table 3. It is obvious from the table that for each of the paleo-earthquakes, the back-calculated earthquake magnitudes using the different methods are very close to one another. The implication is strong that the magnitude has been reasonably approximated for each of the paleo-earthquakes, at least in terms of destructive potential. This example is the only major extant study for which the various methods have been used for back-analysis, and it is a landmark effort.

Magnitude bound method

The magnitude bound method estimates the magnitude of a paleo-earthquake by using relations between earthquake magnitude and the distance from the tectonic source to the farthest site of liquefaction. The method is based on increasingly stronger earthquakes causing liquefaction at increasing distances from the energy center, in a systematic manner.

Distances from energy centers to the farthest liquefaction features for many worldwide, historical earthquakes have been compiled by Ambraseys (1988). The sites of farthest effects were from locales of minor venting of sand or minor lateral spreading. Sites having only soft-sediment deformation, such as ball-and-pillow structures, load casts, or convoluted bedding, were not included in the data set. The data are from various tectonic conditions and susceptibilities to liquefaction, so it is not surprising that the maximum extent of liquefaction is highly variable for a given earthquake magnitude. This variability makes it essential that the technique be calibrated for the tectonic setting of interest, preferably by using data on the extent of liquefaction from historical earthquakes in the study area, to account for the influence of local factors such as stress drop, focal depth, and liquefaction susceptibility.

Where both the energy source and the outer limits of liquefaction of a paleo-earthquake are well defined and effects of liquefaction from historic earthquakes are available for calibration, a reasonable estimate of prehistoric magnitude can be achieved in many study areas.

TABLE 3. BACK-CALCULATED MAGNITUDES FOR FOUR LARGEST PALEO-EARTHQUAKES CENTERED IN WABASH VALLEY REGION OF INDIANA AND ILLINOIS

Back-analysis method Date*	Paleoearthquake			
	Vincennes 6100 yr	Skelton 12 000 yr	Vallonia 3950 yr	Waverly mid-Holocene
Magnitude bound	7.8	7.2	6.9	6.8
Cyclic stress	7.5–7.7	7.4	6.7	6.9
Energy stress	7.5–7.8	7.3	7.1	6.8–7.1

Note: Data from Pond (1996), Munson and Munson (1996), Munson et al. (1997), and Obermeier (1998a).

*Radiocarbon dates, from Munson and Munson (1996).

Cyclic stress method

Seed et al. (1985) updated a procedure originally proposed by Seed and Idriss (1971) to evaluate the liquefaction susceptibility of sandy soils. The procedure is based on case histories of sites that did or did not develop liquefaction effects during earthquakes worldwide. The occurrence of liquefaction was judged from many types of observations, such as sand blows caused by hydraulic fracturing, lateral spreading, ground cracking or settlement, and damage to structures caused by settling or tilting.

The Seed et al. (1985) method and its predecessors were originally developed as a geotechnical procedure to estimate the strength of cohesionless sediment required to prevent liquefaction during an earthquake, for a given earthquake magnitude and peak acceleration. The method is based on comparing the earthquake-induced (horizontal) cyclic shear stress to the cyclic resistance of the soil (i.e., to the strength of the soil or its resistance to pore-water pressure buildup). The earthquake-induced cyclic shear stress is related to both the strength and duration of shaking. These values, in turn, statistically relate to earthquake magnitude. The influence of the seismically induced horizontal shear stress is incorporated within the parameter of cyclic stress ratio (CSR); CSR is a function of earthquake magnitude, peak surface acceleration, the total and effective overburden stresses, and the depth of the source bed. The strength of the soil is evaluated in terms of the parameter $(N_1)_{60}$, which is the SPT blow count (N) normalized to account for depth of sediment and the water table, as well as for the specific type of SPT test equipment.

The cyclic stress method has been developed from a large database of historic earthquakes having magnitudes $\leq M_w$ 7.5; the database has only limited data for larger earthquakes. Using a technique for analysis developed recently by Pond (1996), we can employ the cyclic stress method to estimate a magnitude for paleo-earthquakes of $M_w \leq 7.5$, which is useful for hazard assessments.

Energy-based approaches

Energy-based approaches to liquefaction analysis are inherently appealing because moment magnitude M_w is a direct measure of seismic energy. Such approaches are all the more attractive because energy is a fundamental physical parameter. Still, energy-based approaches are not at the state of development to be used for routine analysis, although some will doubtless be so in the near future. Two approaches have been used for liquefaction analysis, one based on field case histories and the other based on laboratory testing of sediment from the site of interest.

Field case histories using Gutenberg-Richter relations. Davis and Berrill (1982) first developed an energy-based approach for predicting liquefaction from field data. Similar and extended approaches attempting to relate sediment properties at a site to the Gutenberg-Richter (Gutenberg-Richter, 1956) function for energy at the site were proposed by Berrill and Davis (1985), Law et al. (1990), and Trifunac (1995). Well-defined relations were not observed throughout the distance from the tectonic source. Part of the scatter almost certainly originates from the empirical Gutenberg-Richter (1956) function as a measure of radiated energy, E, defined as

$$E \approx \frac{10^{1.5M}}{R^2}, \qquad (1)$$

where M_w is moment magnitude, and R is either the epicentral distance (Trifunac, 1995) or the distance from the energy center (Davis and Berrill, 1985). As pointed out by Trifunac (1995, 1999), much of the scatter is probably because the model does not account for seismic source mechanisms, directionality of strong motions, or local geologic conditions. It is also likely that part of the scatter is caused by use of epicentral distance rather than distance from the energy center, especially for larger earthquakes. However, we suggest that because body waves are dominant in the development of liquefaction and because surface waves (despite the fact that they can play a large role in the breakup of a cap and determine whether venting takes place at the ground surface) do not extend below shallow depths, and are unlikely to have much influence in the development of liquefaction at many places.

Near the ground surface, breakdown of the sediment grain-to-grain contacts may lead to a considerable loss of energy. This may be the case for some liquefiable deposits; if so, this energy loss could be a source of serious error when the function E is used for analysis.

Laboratory test results. Laboratory testing has clearly demonstrated that a direct relation exists between dissipated energy and buildup of pore-water pressure during undrained cyclic shearing of saturated sands (Nemat-Nasser and Shokooh, 1979; Simcock et al., 1983; Liang et al., 1995; Green et al., 2000). Building on this observation, several researchers have attempted to use laboratory data to correlate normalized energy capacity (i.e., capacity per unit volume, accounting for influence of initial effective confining stress) and relative density for various types of soils (e.g., Al-khatib, 1994; Ostadan et al., 1998). In these studies, normalized energy capacities were computed as the area bounded by the stress-strain hysteresis loops, up to the point of liquefaction. It is now well demonstrated that the normalized energy capacity is a fundamental parameter for evaluating the liquefaction potential of reconstituted samples that are tested in the laboratory (Green, 2001).

Reconstituted laboratory samples such as those used in the studies cited above cannot be used directly for evaluation of the liquefaction potential of naturally occurring samples in the field, because of differences in deposition, overconsolidation, preshearing, or aging (i.e., Terzaghi et al., 1996; Olson et al., 2001). This problem can be circumvented for important projects, including paleoseismic studies, by conducting the laboratory tests on undisturbed frozen samples. However, obtaining undisturbed field samples is very expensive, and therefore this technique has seen limited use.

Arias Intensity method. Kayen and Mitchell (1997) extended preliminary work of Egan and Rosidi (1991) and developed correlations to predict the occurrence of liquefaction as functions of the Arias Intensity of the earthquake motion and penetration resistance of the soil. Arias Intensity, I_h, is defined by the equation

$$I_h = I_{xx} + I_{yy} = \frac{\pi}{2g}\int_0^t a_x^2(t)dt + \frac{\pi}{2g}\int_0^t a_y^2(t)dt, \quad (2)$$

where I_{xx} is the intensity value in the x-direction in response to transient motions in the x-direction, I_{yy} is the intensity in the y-direction, g is the acceleration due to gravity, t is the duration of shaking, and $a_x(t)$ and $a_y(t)$ are the transient accelerations of earthquake motion in the x- and y-directions, respectively. In this approach, the energy applied to the soil (I_{hb}) is:

$$I_{hb} = (I_{xx} + I_{hh})r_b, \quad (3)$$

where r_b is a depth reduction factor that accounts for the variation of Arias intensity with depth.

However, this technique does not explicitly account for the influence of effective confining pressure (e.g., depth of water table), and, as Trifunac (1999) noted, there are basic questions concerning whether the Arias Intensity function represents the actual energy input into an element of soil. Still, the correlations that Kayen and Mitchell (1997) developed appear very good and may be suitable for paleoliquefaction analysis in field situations where the water table was shallow at the time of the earthquake. Acceleration time-history relations that can be used to express Arias Intensity as a function of earthquake magnitude and site-to-source distance are currently available for only a few regions of the world, but this problem possibly may be avoided by using Arias Intensity attenuation relationships as suggested by Kayen and Mitchell (1997) for the western United States and as used tentatively by Schneider (1999) in the central United States.

Ishihara method. For paleoseismic analysis, the Ishihara (1985) method is a technique to estimate peak acceleration at sites of paleoliquefaction. The premise of the method is that the maximum height of dikes (accompanied by venting at the surface) is controlled by two factors: the thickness of liquefied sediment and the peak acceleration. Ishihara (1985) originally developed the bounds using data from only a few earthquakes with magnitudes on the order of ~7.5 and higher, and only limited data have since been added for such large earthquakes. Youd and Garris (1995) showed that the method is not valid for ground failures due to lateral spreading or surface oscillations. The relationships for the Ishihara method probably represent sites where surface effects of liquefaction from hydraulic fracturing were abundant—i.e., liquefaction was severe (T. L. Youd, 1998, personal commun.).

Pease and O'Rourke (1995) also critiqued the Ishihara (1985) method for the M_w 7.1 Loma Prieta earthquake of 1989. They found that the method correctly predicted occurrences of surface effects of liquefaction except at sites of lateral spreading or surface oscillation. Pease and O'Rourke (1995) did not present detailed data concerning the properties of the liquefied source sands, but most appeared to have been loosely compact and some were moderately compact.

The Ishihara (1985) method may be applicable where the cap thickness is reasonably uniform (or at least does not slope much along its base) and for source sands ranging from very loose to moderately compact, at least for M_w~7.5 or larger earthquakes. For lower magnitudes, the method likely applies only for loose deposits. More detailed data regarding site-specific parameters are needed to critique the method more fully.

The Ishihara (1985) method has great potential for paleoseismic analysis at sites where the ground failure can be attributed confidently to hydraulic fracturing. The method is ideally suited for using measurements of dike height and cap thickness, which are observable along stream banks found in many paleoliquefaction searches.

SUMMARY AND COMMENT

The extensive reliance on paleoliquefaction studies in the central and eastern United States is due partly to the abundance of stream valleys in this humid environment containing liquefiable deposits. Similarly, it has been found that such liquefiable deposits occur throughout much of the humid and rainy U.S. Pacific Northwest, revealing the paleoseismic record through at least much of Holocene time (Obermeier and Dickerson, 2000). However, adequate streams are available for paleoliquefaction studies even in many arid conditions. Overall, throughout much of the United States and in many field settings worldwide, there are adequate liquefiable deposits to reveal the record of strong Holocene seismicity.

ACKNOWLEDGMENTS

The section titled "Energy-based approaches" is based in large part on research at Virginia Tech by Russell Green, under the direction of James K. Mitchell, and was funded by the Multi-Disciplinary Center for Earthquake Engineering Research. The funding for Scott Olson at the University of Illinois was provided by National Science Foundation grant 97-01785. We thank Daniel Phelps, who helped greatly in preparation of the manuscript.

APPENDIX: DEFINITION OF TERMS

Directionality—The transmission of seismic energy, as manifest in parameters such as maximum acceleration, in a preferred orientation in plan view. The influence of directionality depends on the orientation and type of fault. For example, strike-slip faults tend to transmit more energy along the plane of the fault; normal faults tend to transmit more to the downthrown block; and reverse faults tend to transmit more to the upthrown block.

Meizoseismal zone—The American Geological Institute *Glossary of Geology* defines "meizoseismal" as "pertaining to the maximum destructive force of an earthquake," from which one could infer that "meizoseismal zone" means the area within, or approximately within, the highest isoseismal (as, for example, the area within the highest Modified Mercalli intensity). Others, however, use the term to refer to the region of higher intensities of the earthquake (note: not highest intensities). We use this term in the latter sense.

REFERENCES CITED

Al-khatib, M., 1994, Liquefaction assessment by strain-energy approach [Ph.D. thesis]: Detroit, Michigan, Wayne State University, 212 p.

Allen, J.R.L., 1982, Sedimentary structures: Their character and physical basis, Amsterdam, Elsevier, v. 2, 663 p.

Ambraseys, N.N., 1988, Engineering seismology: Earthquake Engineering and Structural Dynamics, v. 17, p. 1–105.

Andrus, R.D., 1994 In situ characterization of gravelly soils that liquefied in the 1983 Borah Peak earthquake [Ph.D. thesis]: Austin, University of Texas, 633 p.

Bartlett, S.F., and Youd, T.L., 1992, Empirical analysis of horizontal ground displacement generated by liquefaction-induced lateral spreads: Technical Report NCEER-92-0021, State University of New York at Buffalo, variously paged.

Bennett, M.J., 1989, Liquefaction analysis of the 1971 ground failure at the San Fernando Juvenile Hall: Bulletin of the Association of Engineering Geologists, v. 26, no. 2, p. 209–226.

Berrill, J.B., and Davis, R.O., 1985, Energy dissipation and liquefaction of sands: Revised model: Soils and Foundations, Japanese Society of Soil Mechanics and Foundation Engineering, v. 25, p. 105–118.

Carter, D.P., and Seed, H.B., 1988, Liquefaction potential of sand deposits under low levels of excitation: Berkeley, University of California, College of Engineering, Report No. UCB/EERC-81/11, 119 p.

Castro, G., 1987, On the behavior of soils during earthquakes: Liquefaction, *in* Cakmak, A.S., ed., Soil dynamics and liquefaction: New York, Elsevier, p. 169–204.

Castro, G., 1995, Empirical methods in liquefaction analysis: Proceedings of the First Leonardo Zeevaert Conference, Mexico City, p. 1–41.

Davis, R.O., and Berrill, J.B., 1982, Energy dissipation and seismic liquefaction in sands: Earthquake Engineering and Structural Dynamics, v. 10, p. 59–68.

Dobry, R., 1989, Some basic aspects of soil liquefaction during earthquakes, *in* Jacob, K.H., and Turkstra, C.J., eds., Earthquake hazards and the design of constructed facilities in the eastern United States: Annals of the New York Academy of Sciences, v. 558, p. 172–182.

Egan, J.A., and Rosidi, D., 1991, Assessment of earthquake-induced liquefaction using ground-motion energy characteristics: Pacific Conference on Earthquake Engineering, Wellington, New Zealand, p. 313–324.

Green, R.A., 2001, Energy-based evaluation and remediation of liquefiable soils [Ph.D. thesis]: Blackburg, Virginia, Virginia Tech, 390 p.

Green, R.A., Mitchell, J.K., and Polito, G., 2000, An energy-based pore pressure generation model for cohesionless soils: Proceedings of the John Booker Memorial Symposium, Sydney, Australia, A.A. Balkema, Rotterdam, Netherlands, p. 383–390.

Gutenberg, B., and Richter, C., 1956, Magnitude and energy of earthquakes: Annali de Geofisica, v. 9, p. 1–15.

Hanks, T.C., and Johnston, A.C., 1992, Common features of the excitation and propagation of strong ground motion for North American earthquakes: Bulletin of the Seismological Society of America, v. 82, p. 1–23.

Harder, L.F., and Seed, H.B., 1986, Determination of penetration resistance for coarse-grained soils using the Becker hammer: Berkeley, University of California, Earthquake Engineering Research Center, Report No. EERC-86/06, p. 153.

Ishihara, K., 1985, Stability of natural soils during earthquakes, *in* Proceedings of the Eleventh International Conference on Soil Mechanics and Foundation Engineering, San Francisco, Volume 1: Rotterdam/Boston, A.A. Balkema, p. 321–376.

Ishihara, K., 1993, Liquefaction and flow during earthquakes: Geotechnique, v. 43, no. 3, p. 351–415.

Johnston, A.C., and Schweig, E.S., 1996, The enigma of the New Madrid Earthquakes of 1811–1812: Annual Review of Earth and Planetary Sciences, v. 24, p. 339–384.

Kayen, R.E., and Mitchell, J.K., 1997, Assessment of liquefaction potential during earthquakes by Arias intensity: Journal of Geotechnical and Geoenvironmental Engineering, American Society of Civil Engineers, v. 123, p. 1162–1174.

Krinitzsky, E.L., and Chang, F.K., 1988, Intensity-related earthquake ground motions: Bulletin of the Association of Engineering Geologists, v. 24, p. 425–435.

Law, K.T., Cao, Y.L., and He, G.N., 1990, An energy approach for assessing liquefaction potential: Canadian Geotechnical Journal, v. 27, p. 320–329.

Liang, L., Figueroa, J.L., and Saada, A.S., 1995, Liquefaction under random loading– unit energy approach: Journal of Geotechnical Engineering, American Society of Civil Engineers, v. 121, p. 776–781.

Lo, K.Y., and Kaniaru, K, 1990, Hydraulic fracture in earth and rock-fill dams: Canadian Geotechnical Journal, v. 27, p. 496–506.

Lowe, D.R., 1975, Water escape structures in coarse-grained sediments: Sedimentology, v. 22, p. 157–204.

Martin, J.R., and Clough, G.W., 1994, Seismic parameters from liquefaction evidence: Journal of Geotechnical Engineering, American Society of Civil Engineers, v. 120, p. 1345–1361.

Meier, L.S., 1993, The susceptibility of a gravelly soil site to liquefaction [M.S. thesis]: Blacksburg, Virginia Polytechnic Institute and State University, 74 p.

Mesri, G., Feng, T.W., and Benek, J.M., 1990, Post densification penetration resistance of clean sands: Journal of Geotechnical Engineering, American Society of Civil Engineers, v. 116, no. 7, p. 1345–1361.

Munson, P.J., and Munson, C.A., 1996, Paleoliquefaction evidence for recurrent strong earthquakes since 20,000 yr BP in the Wabash Valley area of Indiana: Final report: U.S. Geological Survey, 137 p.

Munson, P.J., and Munson, C.A., and Pond, E.C., 1995, Paleoliquefaction evidence for a strong Holocene earthquake in south-central Indiana: Geology, v. 23, p. 325–328.

Munson, P.J., Obermeier, S.F., Munson, C.A., and Hajic, E.R., 1997, Liquefaction evidence for Holocene and latest Pleistocene seismicity in the southern halves of Indiana and Illinois: A preliminary interpretation: Seismology Research Letters, v. 68, p. 521–536.

National Research Council, 1985, Liquefaction of soils during earthquakes: Washington, D.C., National Academy Press, 240 p.

Nemat-Nasser, S., and Shokooh, A., 1979, A unified approach to densification and liquefaction of cohesionless sand in cyclic shearing: Canadian Geotechnical Journal, v. 16, p. 659–678.

Obermeier, S.F., 1989, The New Madrid earthquakes: An engineering-geologic interpretation of relict liquefaction features: U.S. Geological Survey Professional Paper 1336-B, 114 p.

Obermeier, S.F., 1993, Paleoliquefaction features as indicators of potential earthquake activity in the southeastern and central United States: Washington, D.C., Highway Research Board Record No. 1411, p. 42–52.

Obermeier, S.F., 1994, Using liquefaction-induced features for paleoseismic analysis, *in* Obermeier, S.F., and Jibson, R.W. eds., Using ground failure features for paleoseismic analysis: U.S. Geological Survey Open-File Report 94-663, Chapter A, 98 p.

Obermeier, S.F., 1996, Use of liquefaction-induced features for paleoseismic analysis: An overview of how seismic liquefaction features can be distinguished from other features and how their regional distribution and properties of source sediment can be used to infer the location and strength of Holocene paleoearthquakes: Engineering Geology, v. 44, p. 1–76.

Obermeier, S.F., 1998a, Overview of liquefaction evidence for strong earthquakes of Holocene and latest Pleistocene ages in the states of Indiana and Illinois, USA: Engineering Geology, v. 50, p. 227–254.

Obermeier, S.F., 1998b, Seismic liquefaction features: Examples from paleoseismic investigations in the continental United States: U.S. Geological Survey Open-File Report 98-488, CD-ROM.

Obermeier, S.F., and Dickenson, S.E., 2000, Liquefaction evidence for the strength of ground motions resulting from late Holocene Cascadia subduction earthquakes, with emphasis on the event of 1700 AD: Bulletin of the Seismological Society of America, v. 90, no. 4, 21 p.

Obermeier, S.F., Martin, J.R., Frankel, A.D., Youd, T.L., Munson, P.J., Munson, C.A., and Pond, E.C., 1993, Liquefaction evidence for one or more strong Holocene earthquakes in the Wabash Valley of southern Indiana and Illinois: U.S. Geological Survey Professional Paper 1536, 27 p.

Obermeier, S., Brack, J., Van Arsdale, R., and Olson, S., 2000, Depth of water table for paleoseismic back-calculations: Geological Society of America Abstracts with Programs, v. 32, no. 7, p. A-367.

Obermeier, S.F., Pond, E.C., and Olson, S.M., with contributions by Green, R.A., Stark, T.D., and Mitchell, J.K., 2001, Paleoliquefaction studies in continental settings: Geologic and geotechnical factors in interpretations and back-analysis: U.S. Geological Survey Open-File Report 01-29, 75 p.; on www at http://pubs.usgs.gov/openfile/of01-029.

Olson, S.M., Obermeier, S.F., and Stark, T.D., 2001, Interpretation of penetration resistance for back-analysis at sites of previous liquefaction: Seismological Research Letters, v. 72, n. 1, p. 46–59.

Ostadan, F., Deng, N., and Arango, I., 1998, Energy-based method for liquefaction potential evaluation, in Proceedings, European Earthquake Engineering Conference, Paris, France: Rotterdam, Netherlands, Balkema, p. 38–54.

Pease, J.W., and O'Rourke, T.D., 1995, Liquefaction hazards in the San Francisco Bay region: Site investigation, modeling, and hazard assessment at areas most seriously affected by the 1989 Loma Prieta earthquake: Final report submitted to the U.S. Geological Survey: Ithaca, New York, Cornell University, School of Civil and Environmental Engineering, 176 p.

Pond, E.C., 1996, Seismic parameters for the central United States based on paleoliquefaction evidence in the Wabash Valley [Ph.D. thesis]: Blacksburg, Virginia Polytechnic Institute, 583 p.

Rhea, S., and Wheeler, R.L., 1996, Map showing seismicity in the vicinity of the lower Wabash Valley, Illinois, Indiana, and Kentucky: U.S. Geological Survey Miscellaneous Investigation Map I-2583-A, scale 1:250 000, 1 sheet.

Robertson, P.K., and Wride, C.E., 1997, Cyclic liquefaction and its evaluation based on the SPT and CPT, in Youd, T.L., and Idriss, I.M., eds., Proceedings of the NCEER Workshop on Evaluation of Liquefaction Resistance of Soils: Buffalo, State University of New York, Technical Report NCEER-97-0022, p. 41–87.

Saucier, R.T., 1977, Effects of the New Madrid earthquake series in the Mississippi alluvial valley: U.S. Army Corps of Engineers Waterways Experiment Station, Miscellaneous Paper S-77-5, 10 p.

Schmertmann, J.H., 1987, Time-dependent strength in freshly deposited or densified sand: Discussion: Journal of the Geotechnical Engineering Division, American Society of Civil Engineers, v. 113, no. 2, p. 173–175.

Schmertmann, J.H., 1991, The mechanical aging of soils: Journal of Geotechnical Engineering, American Society of Civil Engineers, v. 117, no. 8, p. 1288–1330.

Schneider, J., 1999, Liquefaction response of soils in mid-America evaluated by seismic cone tests [M.S. thesis]: Atlanta, Georgia Institute of Technology, School of Civil and Environmental Engineering, 142 p.

Seed, H.B., 1979, Soil liquefaction and cyclic mobility for level ground during earthquakes: Journal of Geotechnical Engineering, American Society of Civil Engineers, v. 105, p. 210–255.

Seed, H.B., and Idriss, I.M., 1971, A simplified procedure for evaluating soil liquefaction potential: Journal of Soil Mechanics and Foundation Engineering, American Society of Civil Engineers, v. 97, p. 1249–1274.

Seed, H.B., Idriss, I.M., and Arango, I., 1983, Evaluation of liquefaction potential using field performance data: Journal of Geotechnical Engineering, American Society of Civil Engineers, v. 109, p. 458–482.

Seed, H.B., Tokimatsu, K., Harder, L.F., and Chung, R.L., 1985, Influence of SPT procedures in soil liquefaction resistance evaluations: Journal of Geotechnical Engineering, American Society of Civil Engineers, v. 111, p. 1425–1445.

Simcock, K.J., Davis, R.O., Berrill, J.B., and Mullenger, G., 1983, Cyclic triaxial tests with continuous measurement of dissipated energy: Geotechnical Testing Journal, American Society for Testing and Materials, v. 6, p. 35–39.

Sims, J.D., 1975, Determining earthquake recurrence intervals from deformational structures in young lacustrine sediments: Tectonophysics, v. 29, p. 141–152.

Stark, T.D., and Olson, S.M., 1995, Liquefaction resistance using CPT and field case histories: Journal of Geotechnical Engineering, American Society of Civil Engineers, v. 121, p. 856–869.

Terzaghi, K., and Peck, R.B., 1967, Soil mechanics in engineering practice (2nd edition): New York, John Wiley, 729 p.

Terzaghi, K., Peck, R.B., and Mesri, G., 1996, Soil mechanics in engineering practice (3rd edition): New York, John Wiley and Sons, Wiley Interscience Publications, p. 193–208.

Tinsley, J.C., Youd, T.L., Perkins, D.M., and Chen, A.T.F., 1985, Evaluating liquefaction potential, in Ziony, J.I., ed., Evaluating earthquake hazards in the Los Angeles region: An earth sciences perspective: U.S. Geological Survey Professional Paper 1360, p. 263–316.

Tokimatsu, K., 1988, Penetration tests for dynamic problems, in Proceedings, 1st International Symposium on Penetration Testing, Orlando, Florida: Rotterdam, Netherlands, Balkema, v. 1, p. 117–136.

Trifunac, M.D., 1995, Empirical criteria for liquefaction in sands via standard penetration tests and seismic wave energy: Soil Dynamics and Earthquake Engineering, v. 14, p. 419–426.

Trifunac, M.D., 1999, Assessment of liquefaction potential during earthquakes by Arias Intensity: Discussion: Journal of Geotechnical and Geoenvironmental Engineering, American Society of Civil Engineers, v. 125, no. 7, p. 627.

Tsuchida, H., 1970, Prediction and countermeans against the liquefaction in sand deposits: Seminar in the Port and Harbor Institute, Yokohama, Japan, Abstracts, p. 3.1–3.33.

Tuttle, M.P., 1999, Late Holocene earthquakes and their implications for earthquake potential of the New Madrid seismic zone, central United States [Ph.D. thesis]: College Park, University of Maryland, Department of Geology, 250 p.

Wood, H.O., and Neumann, F., 1931, Modified Mercalli intensity scale of 1931: Bulletin of the Seismological Society of America, v. 21, p. 277–283.

Yamamuro, J.A., and Lade, P.V., 1998, Steady-state concepts and static liquefaction of silty sands: Journal of Geotechnical and Geoenvironmental Engineering, American Society of Civil Engineers, v. 124, p. 868–877.

Yegian, M.K., Ghagraman, V.G., and Harutinunyan, R.N., 1994, Liquefaction and embankment failure case histories: Journal of Geotechnical Engineering, American Society of Civil Engineers, v. 120, no. 3, p. 581–596.

Youd, T.L., 1973, Liquefaction, flow, and associated ground failure: U.S. Geological Survey Circular 688, 12 p.

MANUSCRIPT ACCEPTED BY THE SOCIETY MAY 11, 2001

Geological Society of America
Special Paper 359
2002

Late Quaternary paleoseismites: Syndepositional features and section restoration used to indicate paleoseismicity and stress-field orientations during faulting along the main Lima Reservoir fault, southwestern Montana

Mervin J. Bartholomew*
Earth Sciences & Resources Institute, School of the Environment, University of South Carolina, Columbia, South Carolina 29208, USA

Michael C. Stickney
Edith M. Wilde
Montana Bureau of Mines & Geology, Montana Tech of the University of Montana, 1300 West Park Street, Butte, Montana 59701, USA

Robert G. Dundas
Department of Geology, California State University, Fresno, California 93740, USA

ABSTRACT

Syndepositional features, interpreted as paleoseismites indicative of six late Quaternary surface ruptures, were exposed in a trench across the Lima Reservoir fault in the Centennial Valley of southwestern Montana. Younger events progressively displaced four loess deposits capped by paleosol, and overstepping unconformities. Older events are marked by paleo–sandblow vents, ejected sand, and a deformed stream-channel deposit. Fossils of horse (*Equus*) and northern pocket gopher (*Thomomys talpoides*), place a 120–10 ka range on the oldest event. Stratigraphy and soils suggest that it occurred during the last Pinedale interstade (ca. 20 ka).

Line-length balancing and progressive retro-deformation of faults show that > 8.8 m of cumulative fault displacement with ~2 m of horizontal extension occurred since then. Orientation of faults, clastic dikes, and slip vectors, along with earthquake focal mechanism solutions, were used to determine stress-field orientations for the six events. Normal faulting associated with northeast-southwest Basin and Range extension produced three events with 4.7 m offset, a displacement rate of ~23.5 cm/k.y., and a recurrence interval of ~6.7 k.y. Oblique-reverse faulting associated with east-northeast–west-southwest compression produced two events with 3.7 m offset, clastic dike injection along joints or joint-reactivated faults in arched areas above the oblique-reverse faults, a displacement rate of ~18.5 cm/k.y., and a recurrence interval of ~10 k.y. Reverse faulting, associated with north-south compression, produced clastic dike injection during one event with 0.4 m offset, a displacement rate of <2 cm/k.y., and a recurrence interval of >20 k.y.

---

*E-mail: jbarth@esri.esri.sc.edu

Bartholomew, M.J., Stickney, M.C., Wilde, E.M., and Dundas, R.G., 2002, Late Quaternary paleoseismites: Syndepositional features and section restoration used to indicate paleoseismicity and stress-field orientations during faulting along the main Lima Reservoir fault, southwestern Montana, *in* Ettensohn, F.R., Rast, N., and Brett, C.E., eds., Ancient seismites: Boulder, Colorado, Geological Society of America Special Paper 359, p. 29–47.

INTRODUCTION

Southwestern Montana and adjacent Idaho is a region of high seismic susceptibility (Hill and Bartholomew, 1999; Bartholomew et al., 1988) where numerous active faults (Slemmons and McKinney, 1977) have been delineated (Stickney and Bartholomew, 1987a, 1987b). Five major (magnitude 6.5 or larger) earthquakes (Fig. 1A) have affected this area during the past 77 years (Pardee, 1926; Steinbrugge and Cloud, 1962; Tocher, 1962; Witkind et al., 1962; Stover, 1984; Barrientos et al., 1987), and it is characterized by a high level of seismic activity (Qamar and Stickney, 1983; Stickney, 1997). This region of latest Quaternary faulting and high seismicity, which defines the Centennial tectonic belt of Stickney and Bartholomew (1987b), is ideal for examination of Quaternary sediments for evidence of paleoseismicity (e.g., Bartholomew, 1989; Bartholomew et al., 1990, 1999). Exposures in a trench across the main Lima Reservoir fault document that strong paleo-earthquakes produced paleoseismites—soft-sediment deformational features related to tectonic events (Seilacher, 1969, 1984).

The Centennial tectonic belt and the Intermountain seismic belt (Sbar et al., 1972; Smith and Sbar, 1974), from Yellowstone southward, define the flanks of the parabolic wake of latest Quaternary faulting, uplift, and seismicity (Anders et al., 1989) associated with the eastward migration of the Yellowstone hotspot (e.g., Morgan, 1972; Iyer, 1984; Rodgers et al., 1990). The stress field associated with the migrating hotspot is superimposed upon the stress field associated with development of the Basin and Range province (Stickney and Bartholomew, 1987b; Rodgers et al., 1990; Parsons et al., 1998). Thus, many of the active faults, such as the Lost River fault (Fig. 1), are reactivated faults (e.g., Reynolds, 1979; Janecke, 1992; Crone and Machette, 1984; Crone et al., 1985; Scott et al., 1985; Sears and Fritz, 1998) that bound Tertiary basins (Hanneman and Wideman, 1991). Others, such as the Red Rock fault (Fig. 1), appear to border Quaternary basins (Stickney and Bartholomew, 1987b; Bartholomew, 1989). Northwest-trending faults such as the Red Rock and Lost River faults (Fig. 1) have orientations consistent with the 1983 Borah Peak earthquake (Smith et al., 1985; Stein and Barrientos, 1985), which is associated with the Basin and Range stress field (Stickney and Bartholomew, 1987b). Others, such as the Centennial fault (Witkind, 1975) (Figs. 1 and 2), have orientations consistent with the 1959 Hebgen Lake earthquake (Witkind, 1964; Doser, 1989), which is associated with the upwardly bulging Yellowstone hotspot (Smith et al., 1989; Savage et al., 1993).

Still other faults, such as the Lima Reservoir fault (Figs. 1, 2, and 3) discussed herein, have intermediate orientations not clearly characteristic of either stress field. As Sibson (1990) pointed out, however, faults can be easily reactivated if they are either favorably or unfavorably oriented relative to the stress field. Hill and Bartholomew (1999) determined that the Lima Reservoir fault zone is favorably oriented for reactivation in Hebgen Lake events, and unfavorably oriented for reactivation in Borah Peak events. Thus, like the minor reactivation of the unfavorably oriented southern part of the Madison fault (Fig. 1B) during the 1959 Hebgen Lake earthquake (Witkind, 1964), different types of movement might have occurred along the Lima Reservoir fault zone at different times.

Geology of the western Centennial Valley

The east-west–trending Centennial Valley is 6–10 km wide and extends 70 km from Red Rock Pass (Fig. 1B), which separates it from Henrys Lake to the east, to the dam of the Lima Reservoir (Fig. 2). The westward-flowing Red Rock River is a low-gradient headwater tributary of the Missouri River system that drains the valley. Downstream from the dam, the river flows westward for 20 km through incised gorges, then at Lima (Fig. 1B) it turns sharply and flows 40 km northwestward to merge with Horse Prairie Creek to become the Beaverhead River. The upper Beaverhead River is roughly perpendicular to the basin-bounding Blacktail fault and is deeply incised for 19 km across Eocene volcanics and older rocks of the Blacktail Range (Fig. 1B). This course was established during the middle Pleistocene and reflects a river-incision rate much greater than the rate of tectonic uplift on the Blacktail fault (Bartholomew et al., 1999). In contrast, over most of its length, the Red Rock River drains elongated valleys that are parallel to the active, basin-bounding Centennial and Red Rock faults (Fig. 1B). This relationship may reflect a higher rate of tectonic uplift, associated with the Yellowstone hotspot (Anders et al., 1989).

Small alpine glaciers were present in the east-west–trending Centennial Mountains along the present Montana-Idaho border south of the Centennial Valley (Witkind, 1975). However, neither glacial till nor outwash are major components of the late Quaternary sedimentary basin fill (e.g., Witkind, 1975; Sonderegger et al., 1982). Thus, the ages of Quaternary sediments in this basin must be inferred from paleosols, fossils, and depositional environments. This Quaternary basin is flanked to the north by the late Pliocene age Huckleberry Ridge Tuff, early Tertiary age gravels, and Cretaceous sedimentary rocks. The Centennial fault bounds the southern side of both the Centennial Valley and Henrys Lake (Witkind, 1972, 1975). South of this fault, the Centennial Mountains also contain Huckleberry Ridge Tuff in addition to Cretaceous to Paleozoic sedimentary rocks. In the vicinity of Red Rock Pass (Fig. 1B), Precambrian metamorphic rocks are exposed.

Lacustrine deposits and large alluvial fans of late Pleistocene age are incised by small stream valleys with Holocene alluvium in the western Centennial Valley (Fig. 2). A late Pleistocene shoreline across a large alluvial fan near the dam is clearly cut by a much smaller Holocene fan (Figs. 2 and 3). A landslide that filled the incised channel of the Red Rock River near the modern dam (Fig. 2) probably created this ephemeral lake of latest Pleistocene age. Numerous landslides are found in this area (Fig. 2). Analogous examples include the landslide, which impounded Quake Lake in the incised gorge along the Madison River (Fig. 1B), associated with the 1959 Hebgen Lake earthquake

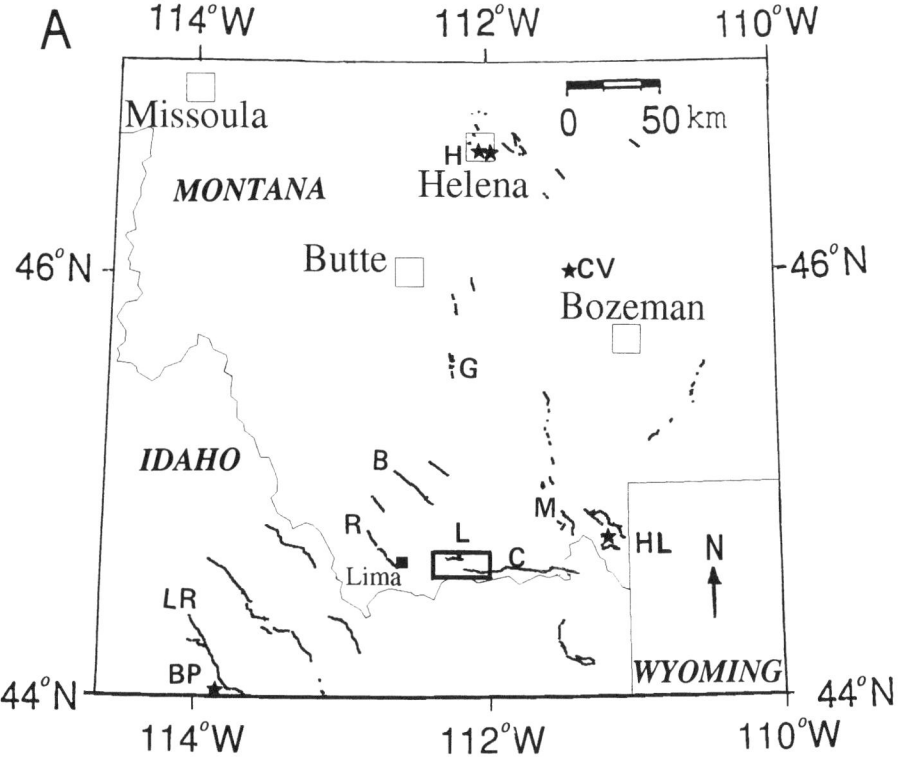

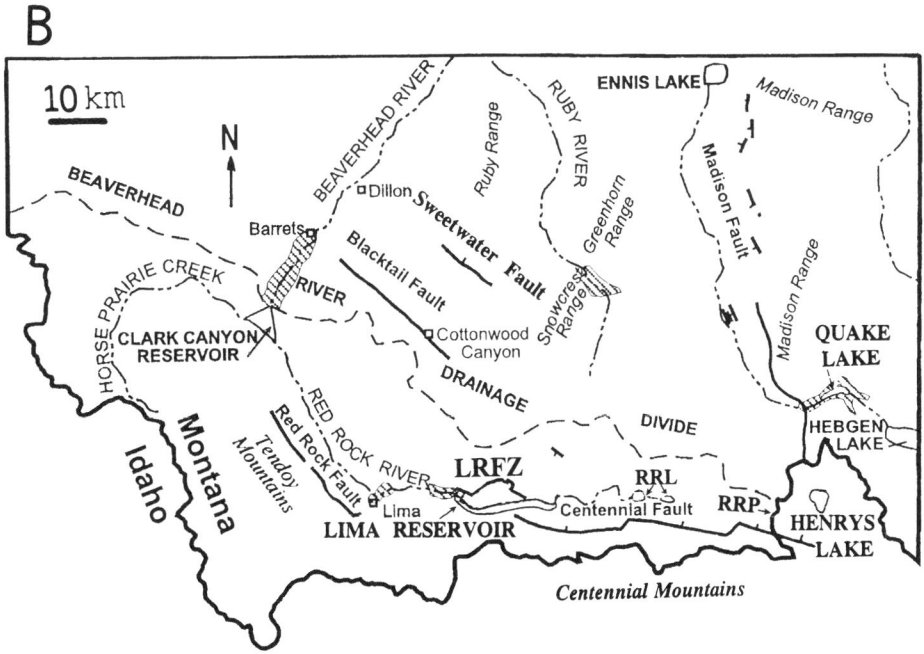

Figure 1. Index map of southwestern Montana showing study area (heavy-line box in A outlines area of Fig. 2) and major features. A: Map showing major earthquake epicenters (stars) and identified active faults (heavy lines) of western Montana and adjacent Idaho (modified after Stickney and Bartholomew, 1987a, 1987b). CV—1925 Clarkston Valley earthquake; H—two 1935 Helena Valley earthquakes; HL—1959 Hebgen Lake earthquake; BP—1983 Borah Peak earthquake; C—Centennial fault; B—Blacktail fault; G—Georgia Gulch fault; L—Lima Reservoir fault zone; LR—Lost River fault; M—Madison fault; R—Red Rock fault. B: Map showing identified active faults, mountain ranges, rivers, and gorges (patterned areas), modified after Bartholomew et al. (1999). LRFZ—Lima Reservoir fault zone; RRL—Red Rock Lakes; RRP—Red Rock Pass.

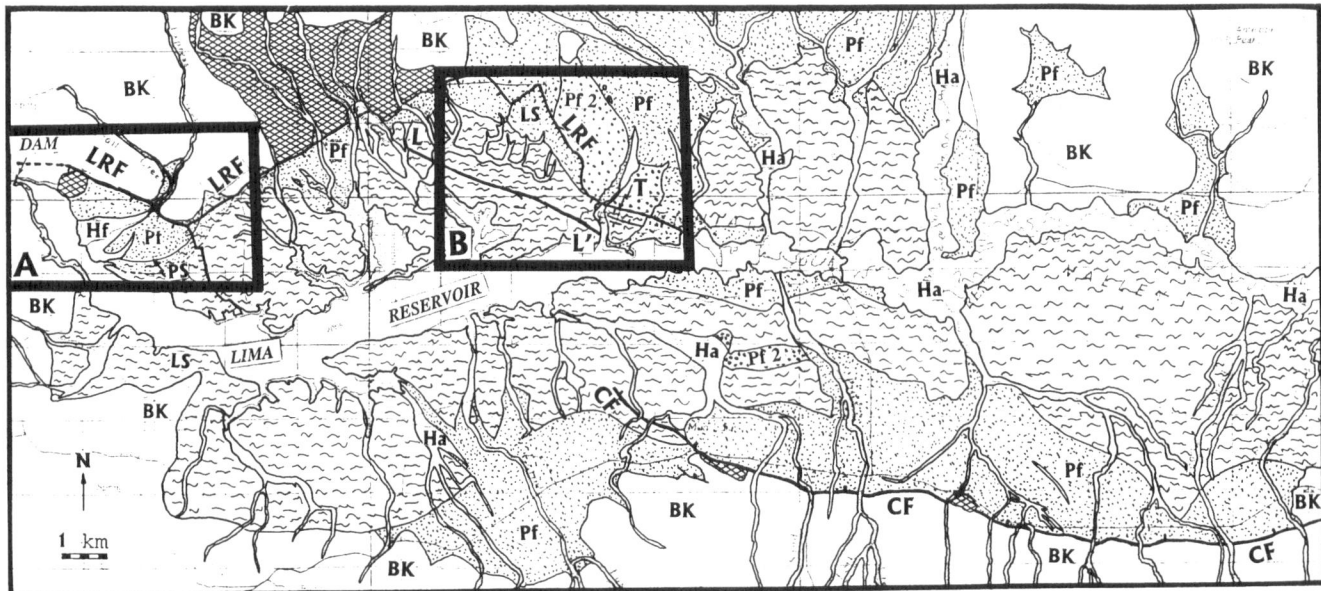

Figure 2. Geologic map of Late Quaternary features of Red Rock River drainage basin in western Centennial Valley of Montana. See Figure 1A for location. Ha—Holocene alluvium; Hf—Holocene alluvial fan; cross-hatch pattern—late Quaternary landslides; LS—Late Quaternary flow landslides or lateral spreads along the Lima Reservoir fault zone (LRF); PS—shoreline of late Pinedale lake; wavy pattern—late Pinedale lacustrine deposits; Pf—late Pinedale alluvial fan; Pf 2—older late Pleistocene alluvial fan; CF—Centennial fault; BK—undifferentiated bedrock; L-L'—linear feature, observed on aerial photograph (Fig. 3B), across late Pinedale lacustrine sediments; DAM—Lima Reservoir dam; T—trench location. Large rectangles show locations of Figures 3A and 3B.

(Witkind, 1964) and landslides that have impinged upon or blocked the Beaverhead River in its incised gorge near Barrets (Bartholomew et al., 1999).

LIMA RESERVOIR FAULT ZONE

The strike of the active, north-dipping Centennial fault changes from approximately east-west along most of its length to about N60°W at its western terminus (Fig. 2). The eastern terminus of the southwest-dipping, main Lima Reservoir fault lies across the valley normal to strike from the terminus of the Centennial fault. Thus, extensional displacement related to the opening of the Centennial Valley along the Centennial fault is probably transferred to the Lima Reservoir fault zone in the western end of the Centennial Valley.

Trench MBMG1986-6 (Figs. 3B and 4) was excavated approximately perpendicular to the N60°W-trending scarp of the main Lima Reservoir fault. The ground surface had 3.4 m of relief along the 32.5-m-long trench, but the northern end of the trench was several meters below the top of the scarp. The trench was excavated with a backhoe to a depth of 2.5 to 3 m. Significant features, such as the main fault and sand blows, were excavated by hand to greater depth where needed. A leveled meter grid was placed on the east wall of the trench, and mapping was done with a metric square marked with 10-cm divisions. Greater accuracy, for significant features such as fossil locations, was obtained with a hand-held 10-cm scale. These features were located with an accuracy of about 1 cm.

The Lima Reservoir fault zone consists of two northwest-trending segments that each merge into landslide features (e.g., flow landslides or lateral spreads) typically associated with liquefaction during earthquakes (e.g., Seed, 1968) (Figs. 2 and 3). A northeast-trending line of springs along a poorly defined, degraded (probably due to mass wasting) scarp, which is roughly along the road, appears to mark the fault between these landslides. Overall, this sinuous trace is approximately 19 km long. The maximum scarp height is about 9 m in the vicinity of the trench. Here, the older Pinedale alluvial fan (Figs. 2 and 3) appears to be partially exhumed along a horst and graben set of scarps. The main fault scarp terminates about 1.5 km east of the trench site at a large stream valley (Figs. 2 and 3). Additionally a northwest trending, 5-km-long, linear feature extends across the lacustrine deposits about 1 km southwest of the trench site (Figs. 2 and 3B). This feature may be another fault scarp of the zone or it may be the frontal fold of the landslide that is northwest of the trench site. In the vicinity of the trench site (Fig. 3B) the main scarp trends about N60°W across the exhumed older Pinedale alluvial fan. At the stream to the northwest, Holocene alluvium is offset approximately 0.5 m (Fig. 3B). Northwest from this stream valley, the main fault scarp trends about N40°W until it reaches the landslide, where its trace is more irregular.

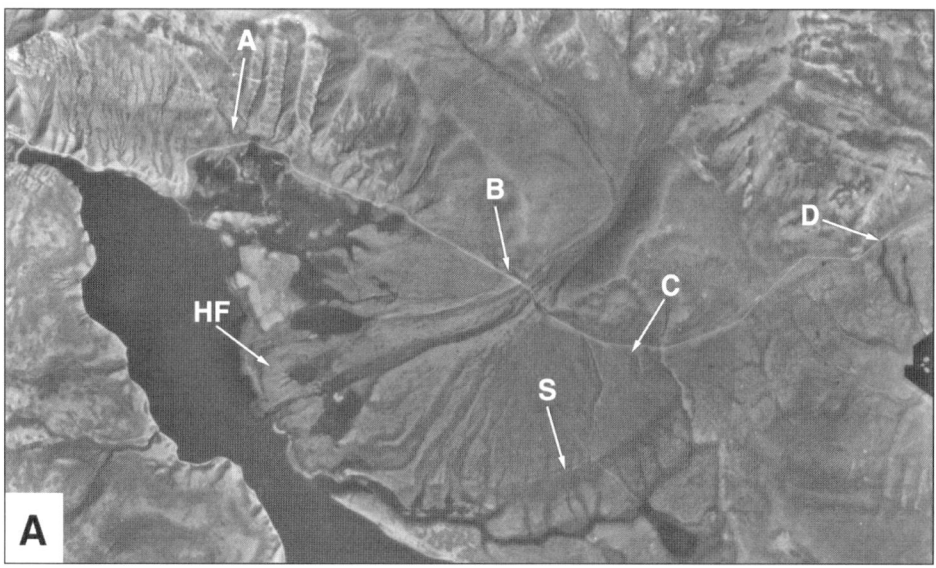
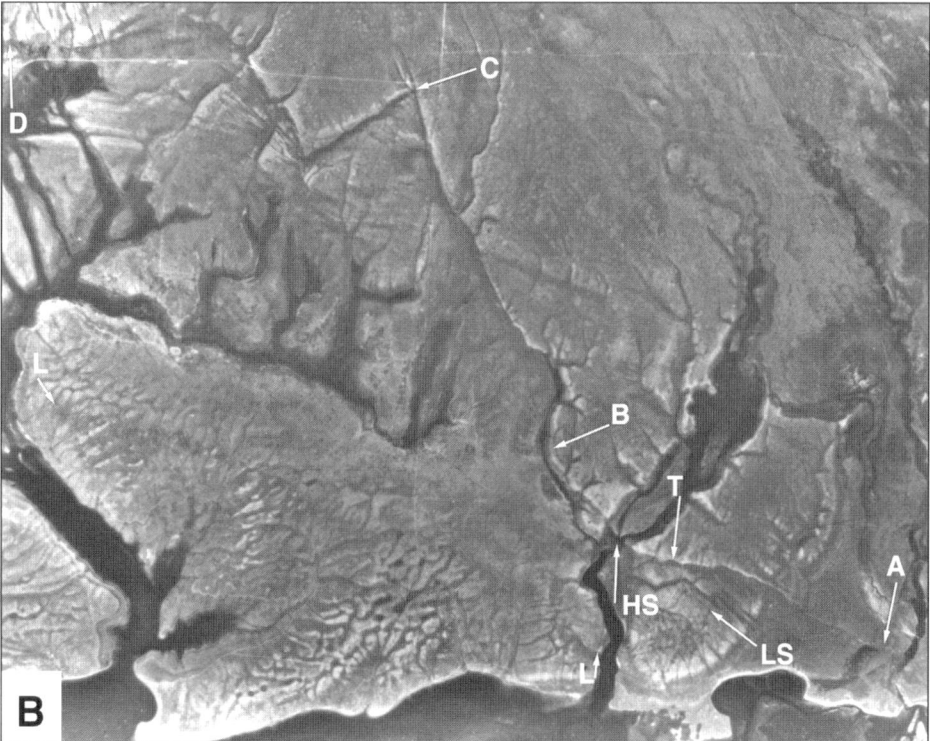

Figure 3. Aerial photographs of northern side of Lima Reservoir, showing Lima Reservoir fault zone (LRF in Fig. 2) trace (A-B-C-D in both photographs); north is to top of each photograph. A: Late Pinedale shoreline (S) along flank of large late Pinedale alluvial fan is truncated by incised Holocene fan (HF); from western terminus (A) near dam, Lima Reservoir fault extends eastward across (B–C) head of large fan and merges into landslide area (south of C) near lake. B: Trench MBMG1986-6 (T) was near eastern terminus (A) of Lima Reservoir fault (Fig. 2); Holocene scarp (HS) offsets Holocene alluvium in stream valley (HA in Fig. 2) just west of trench, then trends more northwestward (to B) and merges northward into landslide area (B-C-D) near road. LS—lacustrine sediments related to late Pinedale shoreline; L-L'—linear feature across late Pinedale lacustrine sediments (see Fig. 2).

TABLE 1. LITHOLOGIC DESCRIPTION OF STRATIGRAPHIC SEQUENCE IN TRENCH MBMG1986-6

	Unit*	Age†	Major Lithology§	Description
Division 1	1	H	BU	Burrows—younger burrows filled with loosely compacted light to medium gray, fine- to coarse-grained sand, silt, and organic material; older burrows filled with compacted dark gray to brown, fine- to coarse-grained sand, silt, and organic material
	2	H	A1	A/O soil horizon—dark gray sand, silt, and organic material
	3	H	B1	B soil horizon—medium gray sand, silt, and clay with organic material
	4	H	L1	Loess—compacted, light brownish gray, massive fine-grained sand and silt
	5	H	A2	Buried A soil horizon—medium brownish gray sand and silt
	6	H	B2	Buried B soil horizon—light brownish gray sand, silt and clay
	7	PLG	L2	Loess—massive, medium gray, fine-grained sand and silt; stage II caliche
	8	PLG	IS1	Injected sand—massive, medium gray, poorly sorted, medium- to coarse-grained quartz sand with scattered larger quartz pebbles up to 1 cm in diameter
	9	PLG	AB3	Buried A/B soil horizons—1–3-cm-thick lenses of dark-gray, fine-grained sand, silt, and clay; stage II caliche
	10	PI	L3	Loess—massive, light-gray, fine-grained sand and silt; stage II caliche
	11	PI	L4	Loess—massive, medium gray, medium- to fine-grained sand and silt; stage III caliche
Division 2	12	PI	C1	Upper channel beds—lenticular beds of medium gray, medium- to coarse-grained sand and medium brown, fine- to medium-grained sand and silt; stage II caliche
	13	PI	LG	Lag gravel—light gray, poorly sorted, fine- to medium-grained conglomeratic sand; matrix-supported, rounded clasts of quartz pebbles are 0.2 to 1 cm in diameter; stage III caliche
	14	PI	ES	Ejected sand—massive, medium brown, medium-grained sand with scattered quartz pebbles up to 0.5 cm in diameter; stage III caliche beneath lag-gravel deposit
	15	PEG	IS2	Injected sand—massive, medium brown, medium-grained sand with scattered quartz pebbles up to 0.5 cm in diameter; stage III caliche beneath lag-gravel deposit
	16	PEG	C2	Lower channel beds—2–10-cm-thick beds of cross-bedded to laminated, medium brown, fine- to medium-grained sand and silt with 1–2-cm-thick lenses of medium gray, medium- to coarse-grained sand; contains large angular rock clasts derived from adjacent overhanging channel wall
	17	PEG	IS3	Injected sand—massive, medium brown, medium-grained sand with scattered grains of coarse sand and quartz pebbles up to 0.5 cm in diameter; contains large angular clasts of sandstone
Division 3	18	PEG	HWS1	Hanging wall sequence 1—moderately well bedded, medium brownish gray mud and silt; well-bedded, light brownish gray, fine-grained sand and silt and medium gray, medium-grained sand; millimeter-laminated marker bed 1; and well-bedded, medium brownish gray, medium-grained sand
	19	PEG	LSM1	Marker bed 1—millimeter-laminated, light grayish brown, fine-grained sand
	20	PEG	FWS2	Footwall sequence 2—interbedded: moderately well bedded, medium brownish gray mud and silt; fining-upward, coarse-to fine-grained, cross-bedded and well-bedded sand; and medium brownish gray, medium-grained sand
	21	PEG	LSM2	Marker bed 2—millimeter-laminated, light grayish brown, fine-grained sand
	22	PEG	FWS3	Footwall sequence 3—interbedded: moderately well bedded, medium brownish gray mud and silt; fining upward, coarse- to fine-grained, cross-bedded and well-bedded sand; and massive to well-bedded, medium brownish gray, medium-grained sand

*See Figure 4.

† H—Holocene; PLG—Pinedale, last glaciation; PI—last Pinedale interstade; PEG—Pinedale, earlier glaciation.

§ See Figure 4 and caption.

Stratigraphic sequence in trench MBMG1986-6

The stratigraphic section in the trench (Table 1) can be divided into three parts: (1) the youngest loess deposits and associated soil horizons, (2) the underlying stream channel and associated lag-gravel deposit, and (3) the older fluvial and lacustrine deposits.

1. The youngest part reflects episodic accumulation of loess, capped by soil horizons, along the down-dropped hanging wall of the main Lima Reservoir fault. Four distinct sequences are distinguished along the main fault, whereas only three are found between the main fault and the principal secondary fault and none north of this secondary fault. The loess-and-soil sequences each postdate successive surface ruptures along the main Lima Reservoir fault and the secondary fault and are interpreted as paleoseismites. The loess sequences disconformably overlie the older stream channel and uncon-

formably overlie the associated lag-gravel deposit as well as the older fluvial and lacustrine deposits.

2. The stream channel developed above a vent of a major sand blow adjacent to the main fault. The sides of the channel cut across the older fluvial and lacustrine deposits. The base of the channel rests directly on the sand filling of the large vent (clastic dike 1, CD1 in Fig. 4). The absence of any soil development over the vent sand indicates a minimal time interval between the sand blow and channel deposition. The lag-gravel deposit appears to be conformably interbedded with the middle of the stream channel, but laterally lies unconformably over small sand-blow vents, ejected sand from one vent, and the older fluvial and lacustrine deposits. The upper part of the channel, above the lag gravel, accumulated in a syndepositional syncline along the main fault and thus reflects a surface rupture and is interpreted as a paleoseismite. The unconformable upper and lower surfaces of the lag gravel (away from the channel) suggest a distinct climatic change associated with more extensive wind erosion than that which produced the underlying, older fluvial and lacustrine deposits. Futhermore, substantial caliche development within the lag gravel appears to indicate prolonged soil development. The caliche, locally, does extend downward into the underlying sediments particularly at sand blows above clastic dikes 2 and 3 (CD2 and CD3 in Fig. 4). The contrast between the overlying thick accumulation of loess and the underlying thick accumulation of fluvial and lacustrine sediment suggests that the channel and lag gravel developed during the last interstade of Pinedale glaciation about 15 to 20 ka.

3. The fluvial and lacustrine deposits in the trench represent distal sedimentation on a large alluvial fan that inter-fingered with lake sediments. Large alluvial fans and lacustrine deposits associated with the last Pinedale glaciation (ca. 13 ka) are younger than these exhumed strata (Figs. 2 and 3). The general lack of thick, clay-rich, reddish soils on the top of these beds in the trench suggests that they are not typical of Bull Lake sediments observed elsewhere, in (1) a nearby trench across the Blacktail fault (Bartholomew et al., 1999), (2) the Tobacco Root Range east of Butte (Hall, 1990; Hall et al., 1990), and (3) the Wind River Range, Wyoming (Chadwick et al., 1997; Phillips et al., 1997; Hall and Jaworowski, 1999). Although an age of youngest Bull Lake (ca. 120 ka) is not precluded from the fossil data discussed below, we feel that assignment of these strata to the peak of Pinedale glaciation (ca. 25 ka) is more consistent with the soil development, major lithologic changes in the trench, and our geologic mapping (Fig. 2). Lateral changes in thickness and lithology of individual beds within these deposits are uniform and do not exhibit effects of syndepositional faulting along the main Lima Reservoir fault.

Age of fossils recovered from sediments in trench MBMG1986-6

Fossil teeth of *Thomomys talpoides* (northern pocket gopher) and *Microtus* (vole) were recovered from stratigraphically lower beds (Table 2 and Fig. 4). A single tooth fragment of *Equus* (horse) was recovered from the upper beds of the footwall (Table 2 and Fig. 4). *Thomomys talpoides* first appeared about 120 ka during the youngest Bull Lake glaciation (e.g., Hall and Jaworowski, 1999; Chadwick et al., 1997; Phillips et al., 1997). *Equus* became extinct about 10 ka, at the end of the Pleistocene. Thus, the fossils only constrain the age of the footwall strata and lower hanging wall strata as Pinedale to youngest Bull Lake. Our mapping (Fig. 2) and interpretation of both the major stratigraphic changes and soil development in the trench (Fig. 4) are consistent with these sediments being part of an incised, partially exhumed, older alluvial fan developed during the older Pinedale glaciation.

Paleoclimatic and depositional conditions indicated by fossils recovered from trench MBMG1986-6

Both Holocene and fossil *Spermophilus* (ground squirrel) bone fragments (Table 2), typically associated with scat, were recovered from burrows in the upper part of footwall strata of the trench (Figs. 4 and 5, A, C, and D). Both *Spermophilus* and especially *Thomomys talpoides* are found only in areas where the ground does not freeze to great depth. Thus, the paleoclimate during the late Pleistocene was probably much like the present-day climate, thick snow accumulation acting as insulation and preventing the ground from freezing to great depth. The large mammal bone fragment (Table 2 and Fig. 4) was heavily weathered on one side, presumably from being partially exposed at the ground surface, before incorporation into the sediment. Strata in both the footwall and lower part of the hanging wall are interpreted as part of an alluvial fan deposit as indicated by (1) the weathered bone, (2) abundant rodents, (3) scat containing rodent fragments, (4) fish fragments, indicating streams or ponds, and (5) numerous channels, cross bedding, and other depositional features (Tables 1 and 2 and Fig. 4).

Fractures filled with massive sand in trench MBMG1986-6

Nontectonic fracture fillings. Three fractures in the northern end of the trench (Fig. 5B) that cut across the fine-grained, laminated sand marker bed 2 (below F2 in Fig. 4) are filled with coarser grained, massive sand. Two of these widen upward and merge upward with overlying coarser grained sand lenses that contain material identical to that of the fracture fillings. Although the fractures themselves may have originated as joints, the clastic fillings are interpreted as having settled by gravity from the top rather than as injected clastic dikes. Thus, these three fractures are not clearly indicative of paleoseismic events and so are not interpreted as paleoseismites.

Injected fracture fillings and ejected-sand deposits. Clastic dikes 1, 2, and 3 (Figs. 4, 6, and 7) are all examples in which overpressured, fluidized sand was ejected onto the land surface from vents above the clastic dikes; thus, they are paleoseismites. Clastic dikes 2 (Fig. 6) and 3 (Fig. 7A) were excavated to the same sand-source bed (Figs. 4 and 6D), which was about 1 m

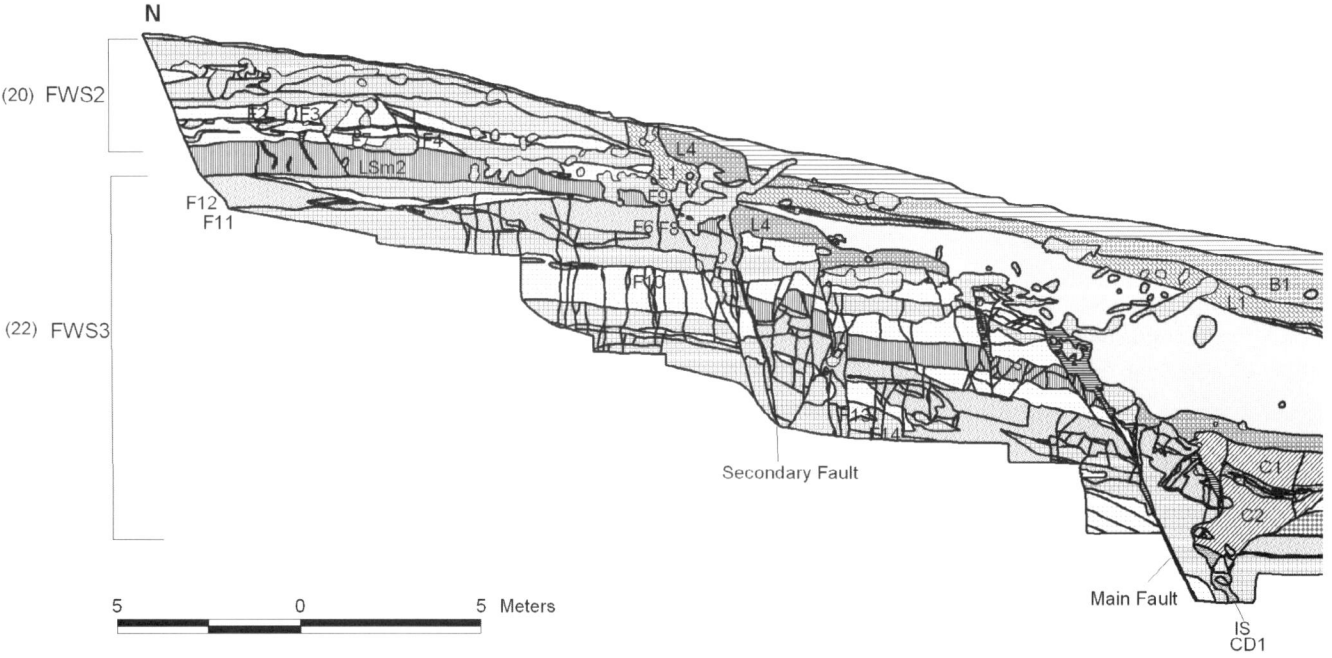

Figure 4. Trench log of east wall (N31°W) of trench MBMG1986-6 across main Lima Reservoir fault. Numbers (1 to 22) correspond to lithologic unit numbers in Table 1; F1 to F14 correspond to fossil-sample numbers in Table 2; L1 to L4 correspond to youngest to oldest loess deposits; C1 and

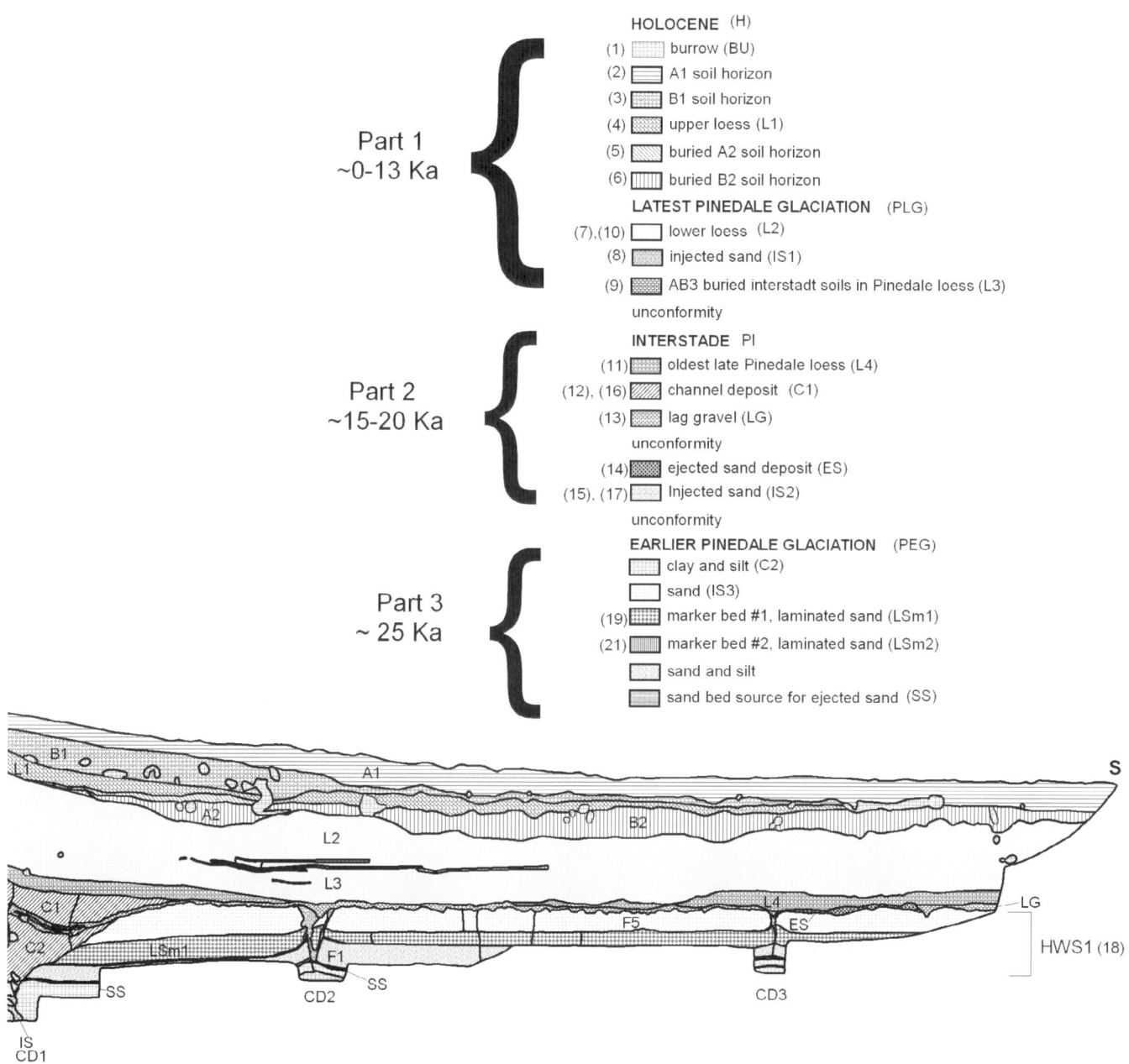

C2 are upper and lower parts, respectively, of channel deposits; MF—main fault; SF—secondary fault; RF1—reverse fault; CD1—clastic dike 1; CD2—clastic dike 2; CD3—clastic dike 3; HWS1—hanging-wall sequence 1; FWS2—footwall sequence 2; FWS3—footwall sequence 3.

TABLE 2. MATERIAL RECOVERED FROM SEDIMENTS IN TRENCH MBMG1986-6

Sample*	Material†	Animal	Description
F1	F	*Thomomys*?	1 tooth fragment; 1 maxillary fragment
		Microtus	7 teeth
		Microtine rodent	2 edentulous left jaws
		Fish	1 vertebra
F2	H	*Spermophilus*	1 skull
F3	H	*Spermophilus*	1 tooth
F4	F	Equus	Tooth fragment
F6	H	Rodent	Vertebra
F7	H	Rodent	Podial element
F8	F	*Spermophilus*	Fragmented skeletal material in coprolites
		Carnivore coprolites	
F9	F	*Spermophilus*	Fragmented skeletal material in coprolites
		Carnivore coprolites	
F10	F	Unknown	Bone fragments
F11	F	*Spermophilus*	Skeletal material
		Fish	Vertebra
F12	F	*Spermophilus*	Skeletal material
		Thomomys talpoides	Teeth—lower right p4, P4, proximal end left ulna
F13	F	Unknown	Bone fragments of large animal
F14	F	Possibly artiodactyl	Distal ulna fragment; heavily weathered on one side

*From Figure 4; F5 is not listed.
† H—Holocene bone fragments; F—fossil bone fragments.

below the ground surface at that time. This 2-cm-thick medium- to coarse-grained sand bed was the basal sand above a more clay-rich unit. The basal part of the channel sequence that is above clastic dike 1, which previously disrupted and exposed this sand source bed, was deposited unconformably over the bed and is believed to have "sealed" it, so that overpressuring was possible during the younger event represented by clastic dikes 2 and 3. Clastic dikes 2 and 3 represent the model for development of clastic dikes in this area. The sand blows are initiated by arching of the sand source bed above a reverse fault (Figs. 4 and 6D). The arching also generates near-vertical, strike-parallel joints, and the overpressured sand is injected upward along a joint, causing beds in the joint walls to deform, and be ejected onto the land surface. Faults (clastic dike 2) can also serve as vents (Fig. 6A).

Clastic dike 1 (Figs. 4 and 7, B and C) was excavated as deep as safety permitted, but its sand-source bed was not exposed. This clastic dike originated at a depth greater than 2 m below the ground surface. By comparison with clastic dikes 2 and 3, this dike was very large, having an upward-flaring surface crater more than 1 m wide and more than 1.5 m deep. Large (10 to 20 cm) clasts (Figs. 4 and 7C) are blocks of the walls that collapsed and mixed with ejected sand during the sand blow. Part of the ejected sand deposit was also preserved beneath the channel sequence on the north side of the crater wall about 1 m above the clastic dike (Fig. 4).

Massive medium grained injected sand is also present along the main fault, along the secondary fault, and along a minor normal fault between these two larger normal faults (Fig. 4). The sand-source beds for these medium- to coarse-grained clastic dikes are unknown; they were probably several meters below the ground surface. Large (20 to 30 cm) clasts within a 0.5-m-wide zone of sand at the top of the main fault suggest that this was collapse material derived from the crater walls, as above clastic dike 1 (Fig. 7C). Moreover, the fissures above both the secondary fault and the largest reverse fault (1 m to the south, Fig. 4) were subsequently filled with the basal loess deposit (L4). The overstepping of the main-fault vent by loess unit 2 and the correlation of the basal loess across both fault vents suggest that these clastic dikes postdated and breached the basal loess unit. Loess unit 3 is interpreted as the associated colluvial wedge that was deposited after the event that produced these clastic dikes.

Fracture sets in trench MBMG1986-6

The central part of the trench (Fig. 4) is dominated by faults of the Lima Reservoir zone. The majority of these faults are located near or between the two large normal faults. Near-vertical joints and minor faults are present along most of the length of the trench, but diminish in frequency away from the main zone of faulting. The orientation of the main fault, measured in the trench, is N63°W, 60°SW (Fig. 8A) and many of the near-vertical joints are strike-parallel. This normal fault and minor, nearby normal faults have offset the basal fluvial-lacustrine sequence by at least 5 m (Table 3). The secondary fault is located about 5 m north of, and parallel to, the main fault (Fig. 4). This normal fault and minor, nearby normal faults offset the basal fluvial and lacustrine sequence by 2.1 m (Table 3). Sand, injected along both major faults, precluded measuring slip vectors directly on either fault surface. However, slickensides were measured on small fault sur-

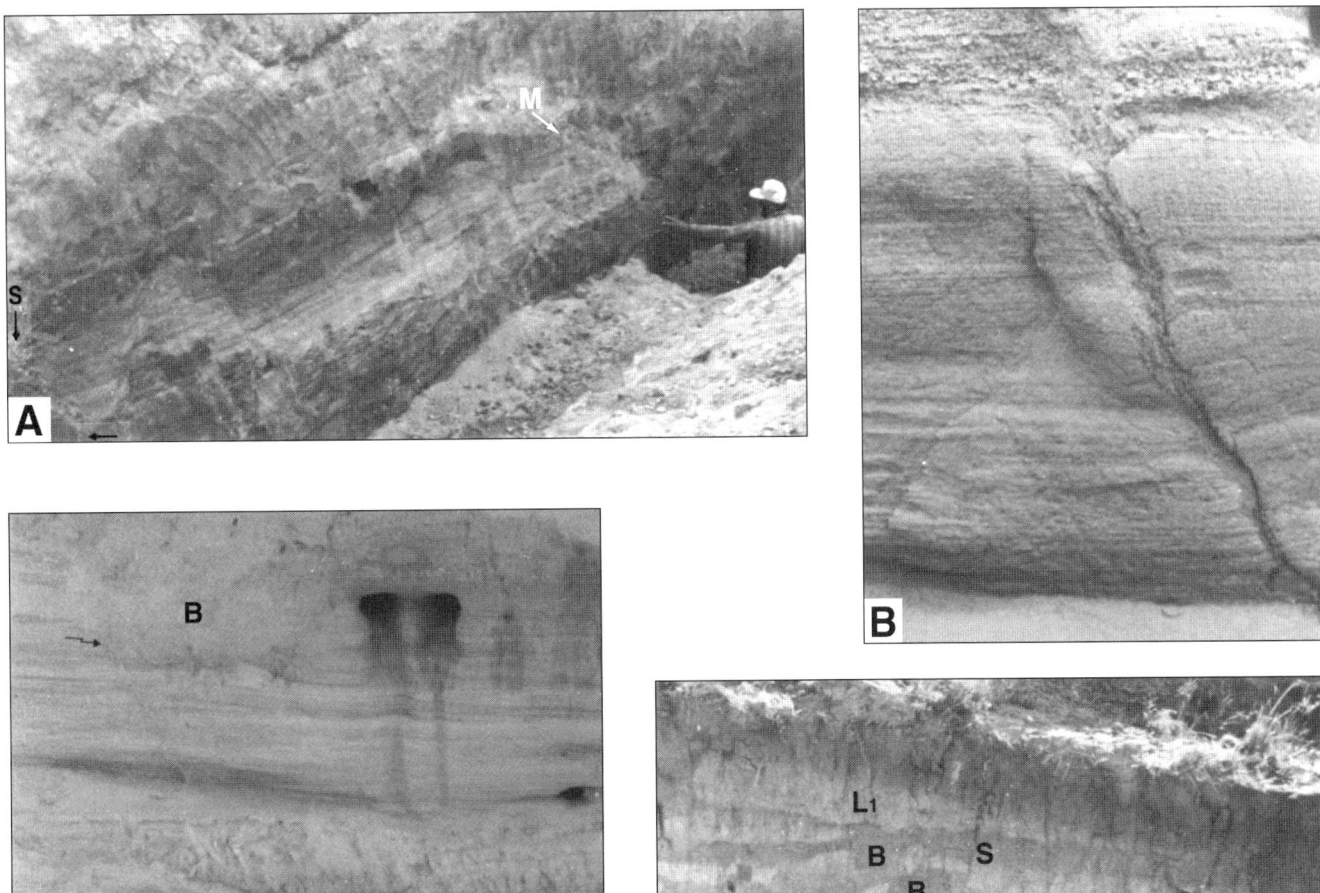

Figure 5. Features observed in trench (Fig. 4) across main Lima Reservoir fault. A: Main fault (M), in east side of trench, juxtaposes fluvial deposits to left of hand against loess to right; arrows indicate principal secondary fault (S); laminated sand marker bed 2 (~1 m thick) is offset by reverse fault horst block and normal fault graben near secondary fault. B: Laminated sand-marker bed 2 cut by 1-cm-wide, downward-tapering fracture (below F2 in Fig. 4) filled by coarser grained sand from overlying bed; fracture width in upper part is ~1 cm. C: Calcite-filled joint (arrows) cutting across laminated sand marker bed 2 (~1 m thick) is truncated by coarse sand-filled burrow (B) containing Holocene ground-squirrel remains (F7 in Table 2 and Fig. 4). D: Buried soil horizon (S) (~10 cm thick) exposed in badger-collapsed area near principal secondary fault after trench mapping was completed; L1—Holocene loess unit; L2—older loess unit; B—burrows.

faces near the main fault. Some of these slip vectors cluster at about S35°W with a plunge of 45° (Fig. 8A). This orientation lies within the range of slip vectors (S45°W +/-15°) reported by Stickney and Bartholomew (1987b) for earthquakes associated with northern Basin and Range extension. It is also consistent with recent focal mechanism solutions (Table 3; Fig. 9) reported by Stickney (1997). Thus, the dominant, episodic, normal displacement (Table 3) on the Lima Reservoir fault zone is typical of northeast-southwest extension of the northern Basin and Range.

The clastic dikes are, however, not related to northern Basin and Range normal faulting. Clastic dike 2 lies above a small, south-dipping reverse fault that arched the sand source bed and overlying strata. Sand was injected along a near-vertical strike-parallel joint in the arched area directly above this reverse fault as well as along a north-dipping reverse fault just south of the main vent (Figs. 4 and 6A). The similar arching of the sand source bed at clastic dike 3 (Fig. 4) suggests that it also developed above a small reverse fault, although this fault was not exposed in the excavation. Clastic dikes 2 and 3 are oriented N83°W, 90° (Fig. 8B). This is consistent with two reverse-fault focal mechanism solutions (Table 3) close to the Lima Reservoir fault zone (Fig. 9).

The injection of sand along both the main fault and the secondary fault are also related to a younger event of reverse faulting when the basal loess unit was deposited (Table 3). Reverse faults related to this event cut across the stream channel and associated lag-gravel deposit (Figs. 4 and 7D). As with clastic dikes 2 and 3,

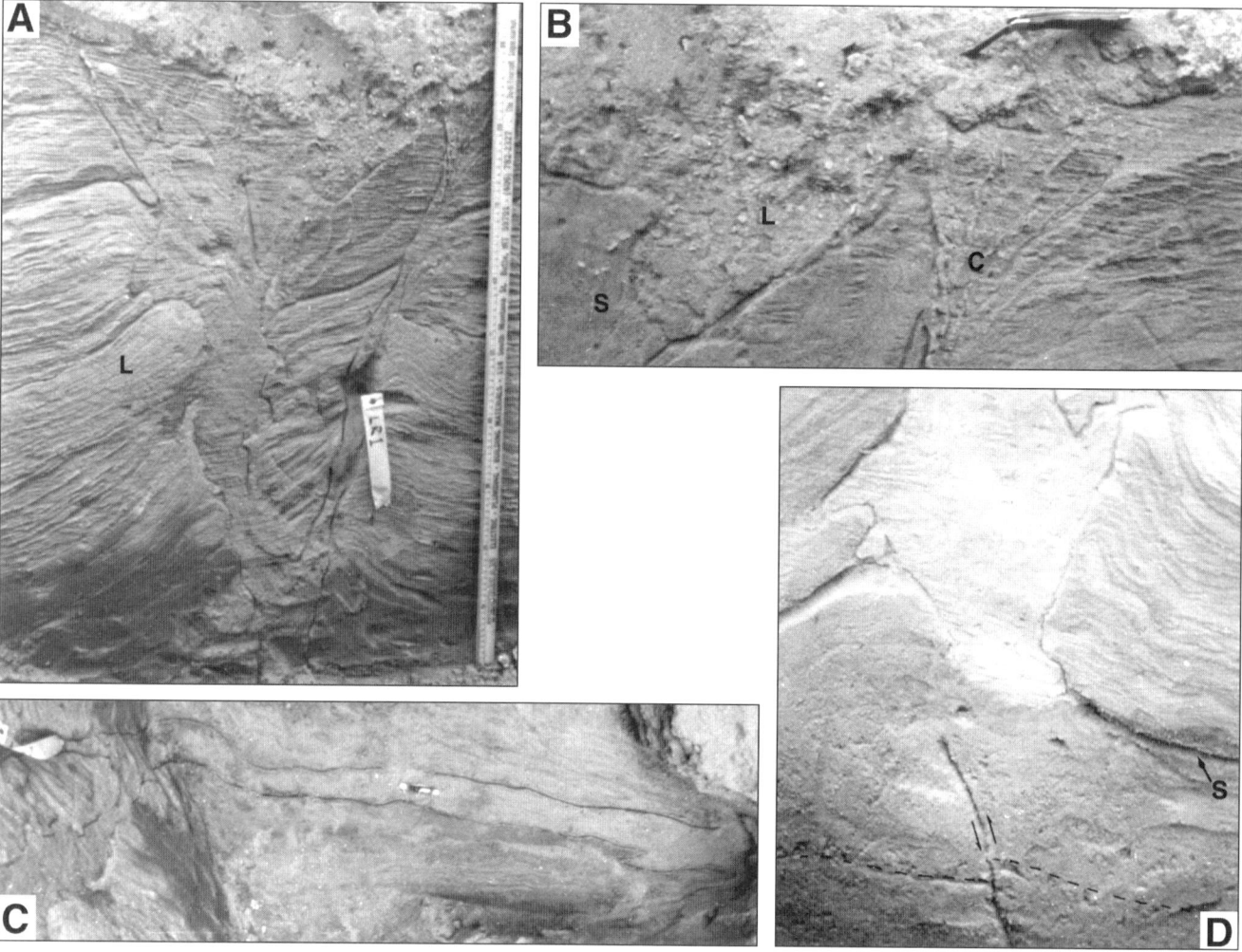

Figure 6. Clastic dike 2 in trench (CD 2, Fig. 4). A: Clastic dike 2 in east side of trench (prior to 0.5 m hand excavation) showing deformed strata adjacent to vent, fossil locality (F1 in Table 2 and Fig. 4), marked by nail (labeled LR1) adjacent to reverse fault with injected sand, and unconformably overlying lag-gravel deposit; meter scale to right. B: Closeup of unconformity between sand (S) of clastic dike 2 and lag-gravel deposit (L) also shows caliche development in lag-gravel deposit (to right of 8 cm long knife) and along steeply inclined surfaces of secondary vent (C) above reverse fault. C: Six- to 12-cm-wide clastic dike 2 (N83°W, 90°SW) in floor of partially excavated trench; fossil locality (F1 in Table 2) and grid nail are same as in A; knife is 8 cm long. D: Reverse fault (N80°W, 74°SW) in east side of hand-excavated trench beneath clastic dike 2; S—sand source bed.

reverse faulting arched the areas around these normal faults, and parts of these faults were reactivated as joints during this event. Because these major normal faults cross many units, the injected sand may have come from one or several source beds. At least part of the sand injected along the main fault came from a depth greater than 5 m below the ground surface (Fig. 4). Clastic dike 1 is the oldest event represented in the trench and is a mirror image of clastic dike 2 in relationship to the nearby faults. It is oriented N53°W, 77°SW, subparallel to sand-injection dikes along the joint-reactivated normal faults; thus, it probably developed during a similar but larger reverse-faulting event (Table 3). A few slickensides are consistent with northwest-southeast reverse faulting with a slip vector of S65°W with a plunge of 45° (Fig. 8C). However, focal mechanism solutions with northwest-southeast–trending reverse faults are lacking (Fig. 9). Instead, northeast-southwest compression is largely accommodated at depth along strike-slip faults (Table 3, Figs. 8 D and E and 9). Thus the episodic, reverse faulting, which caused clastic dike 1 and sand injection along the normal faults, is interpreted as the near-surface, secondary expression of episodic strike-slip faulting at depth.

DISCUSSION

Centennial Valley is seismically active (Fig. 9). Stickney (1997) noted that the linear northeast-trending zone of seismic activity lies north of the seismically quiet Centennial fault that

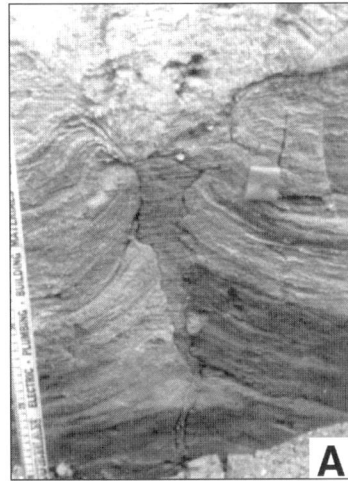
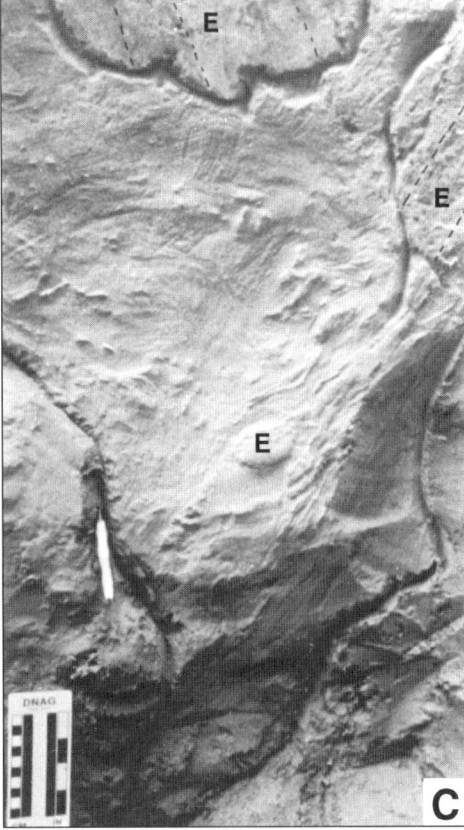

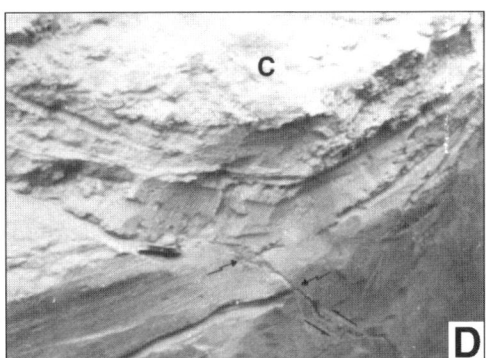

Figure 7. Clastic dikes in trench (Fig. 4). A: Clastic dike 3 (CD3, Fig. 4) in east side of trench (prior to 0.5 m hand excavation) showing deformed strata adjacent to vent; coarse pebbles at top of vent are part of ejected sand beneath L4 loess with stage III caliche; meter scale to left. B: Twelve- to 15-cm-wide clastic dike 1 (N52°W, 77°SW) in floor of trench below channel deposits. C: Upward-flaring vent of clastic dike 1 in west wall of lower, hand-excavated 1 m of trench, showing large exotic clasts (E) of sand (dashed lines indicate bedding) which collapsed into vent. D: Reverse fault offsetting base of channel; 8-cm-long knife is on fault surface at footwall cutoff of channel; arrows mark lower bed.

bounds the southern side of the valley. This zone crosses Centennial Valley in the vicinity of the Lima Reservoir fault zone where the Centennial fault curves northwestward (Fig. 9). Only one recent focal mechanism solution along this zone (October 21, 1982) is consistent with north-south extension related to the stress field of the Yellowstone hotspot. A few others (Table 3; Fig. 9) are consistent with northeast-southwest extension related to the stress field of the northern Basin and Range (Stickney and Bartholomew, 1987b). However, most of the focal-mechanism solutions along this zone of seismicity are consistent with east-northeast–west-southwest compression (Stickney, 1997). Indeed, the majority of these suggest that it is dominantly a zone of right-lateral strike-slip faulting related to east-northeast–west-northwest compression (Fig. 9). Such a

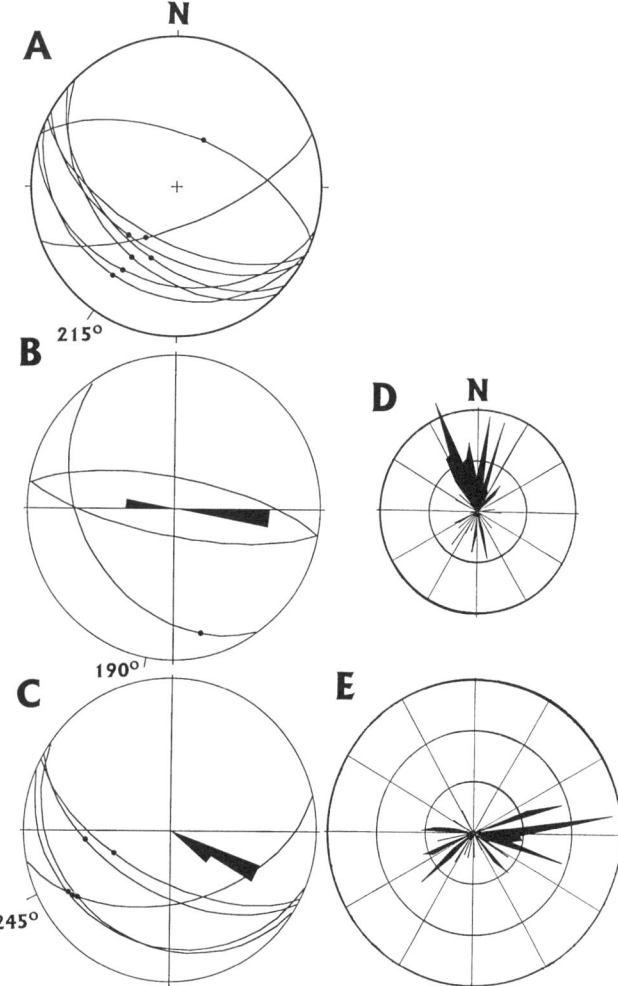

Figure 8. Lower-hemisphere equal-area stereographic projections and rose diagrams. A: Slip vectors on minor faults associated with normal displacement on main fault. B: Faults with slip vectors associated with reverse and strike-slip displacements along Lima Reservoir fault zone and rose diagram of clastic dikes 2 and 3 (CD2 and CD3 in Fig. 4). C: Rose diagram of three clastic dikes (CD1, main fault, and secondary fault in Fig. 4). D: T-axis directions for recorded earthquakes (Fig. 9) in Centennial Valley. E: P-axis directions for recorded earthquakes (Fig. 9) in Centennial Valley.

zone necessarily will also have northwest-trending, left-lateral faulting. Thus, as noted in the introduction, the Lima Reservoir fault zone is indeed located in an area of overlapping stress fields. As such, its displacement history is not a simple, episodic repetition of similar events.

The sequence of six surface-rupturing events (Table 3 and Fig. 10) is constrained by (1) relationships among fault scarps, colluvial wedges and soil horizons, unconformities, and a syn-depositional syncline; (2) reverse- and normal-fault relationships; and (3) clastic dike vents or ejecta and overstepping strata. By utilizing (1) orientations of both faults and clastic dikes in the trench, (2) slip vectors on faults in the trench, and (3) analogous focal mechanism solutions, these six events can be related to the stress fields that produced them (Fig. 11). Episodic normal faulting, related to northeast-southwest extension of the northern Basin and Range accounts for 4.7 m of displacement on the main and secondary faults. Episodic, oblique-reverse faulting, related to east-northeast–west-southwest compression associated with the northeast-trending, right-lateral seismic zone, accounts for 1.3 m and 2.4+ m of displacement, respectively, on the two zones of reverse faults and on the main fault. Finally, one event, which produced clastic dikes 2 and 3, is related to north-south compression. The small amount of displacement, 0.4 m of reverse displacement on the two reverse fault zones, associated with this event suggest that it might be a secondary accommodation event perhaps related to a major north-south extensional event on the western part of the Centennial fault. Minor reactivation of the southern end of the Madison fault during the 1959 Hebgen Lake earthquake (Witkind, 1964) is an analogue.

The importance of distinguishing which events are related to which stress fields is evident when attempting to determine either the rate of slip on the fault zone or seismic hazard susceptibility in this region. Overall, the fault zone has undergone about 8.8 m of apparent displacement and about 2 m of net horizontal extension during the past 20 k.y., for a slip rate of about 44 cm/k.y. This is four times greater than the slip rate of 11.3 cm/k.y. estimated for the Red Rock fault, and eight times greater than the rate of 5.4 cm/k.y. estimated for the Blacktail fault (Bartholomew et al., 1999). The component of displacement due to Basin and Range extension is about half (23.5 cm/k.y.) of the total; the recurrence interval is about 6.7 k.y. The component due to east-west compression is slightly less than half (18.5 cm/k.y.), the recurrence interval being about 10 k.y. The component due to north-south compression is minimal (less than 2.0 cm/k.y.); the recurrence interval is greater than 20 k.y. Determination of seismic hazard susceptibility must separately account for each of these displacement rates and/or recurrence intervals. Another factor that may aid in the determination of seismic hazard susceptibility is that the incised gorges along rivers appear to separate regions where faults exhibit different slip rates (i.e., Basin and Range rates: 5.4 cm/k.y.—Blacktail fault; 11.3 cm/k.y.—Red Rock fault; and 23.5 cm/k.y.—Lima Reservoir fault zone). Thus, the incised gorges of the Beaverhead River and the Red Rock River may help delineate the hinges between regions being uplifted at significantly different rates.

The origin of the east-northeast–west-southwest compressional stress field responsible for the northeast-trending, right-lateral, strike-slip seismic zone is uncertain. Hill and Bartholomew (1999) suggested that the mid-continent stress field significantly overlaps those of the northern Basin and Range and Yellowstone hotspot. They based this suggestion on (1) the occurrence of north-south–trending faults that are favorably oriented in this stress field, and (2) reverse faulting observed in the two trenches across the Georgia Gulch fault (Bartholomew et al., 1990). The north-trending, east-dipping Georgia Gulch fault (Fig. 1) breaks away from the basin-bounding normal fault along the eastern flank of the Jefferson Valley and dies out northward within the

TABLE 3. SEQUENCE OF EVENTS CAUSING SURFACE-RUPTURES AND PALEOSEISMITES IN TRENCH MBMG1986-6

Event	Age*	Displacement (m)[†]				Analogous earthquakes[§]	Comments
		MF	SF	RF1	RF2		
1	H	0.6	0.7			2 Jun 96 8 Jun 96	Buried A and B soil horizons below upper loess unit on main fault and secondary fault; reworked loess colluvial wedge on main fault; rotation of basal loess unit caused fissure (filled with upper loess unit) on secondary fault
2	PLG	0.8				2 Jun 96 8 Jun 96	Buried relict soils in middle loess unit; displaced basal loess unit
3	PLG			0.6	0.7	20 Oct 83 5 Nov 94 27 Dec 96	Displaced basal loess unit; injection of sand along main and secondary faults and other normal faults reactivated as joints
4	PLG	1.2	1.4			2 Jun 96 8 Jun 96	Colluvial wedge on secondary fault loess unit; graben formed along secondary fault; colluvial wedge on main fault included upper channel deposits above folded lag gravel
5	PLI			0.2	0.2	21 Oct 82 4 Nov 82a 3 Jan 85 7 Jun 91 19 Sep 92 11 Dec 92 30 Sep 93	Deposition of lag gravel followed clastic dikes 2 and 3 above minor reverse faults and folding of basal channel deposits; possible arching of secondary fault–reverse fault 1 area
6	PLI	2.4+				20 Oct 83 5 Nov 94 27 Dec 96	Major event accompanied by clastic dike 1; sand source >2 m below ground surface; vent crater cuts across 1+ m of lacustrine and fluvial deposits, later filled by basal channel deposits
T[#]		5.0+	2.1	0.8	0.9		

 * H—Holocene; PLG—Pinedale, last glaciation; PLI—Pinedale, last interstade.
 [†] MF—main fault; SF—secondary fault; RF1 and RF2—reverse faults associated with secondary fault and main fault, respectively.
 [§] Analogous earthquakes from Figure 9.
 [#] T—total displacement on faults.

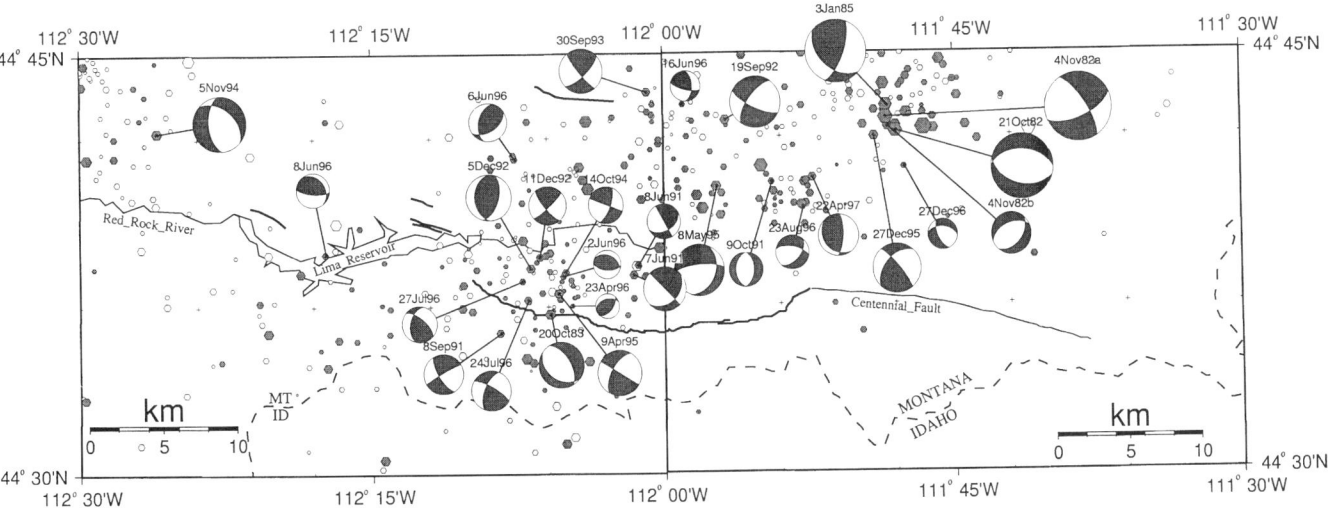

Figure 9. Instrumentally recorded earthquakes in Centennial Valley (modified from Stickney, 1997); lower-hemisphere equal-area stereographic projections showing P wave first arrivals; shaded and unshaded quadrants are areas with compressional and dilatational arrivals, respectively. Sizes of focal spheres are graded by magnitude, from M 2.5 for the April 23, 1996, event to M 4.2 for the November 4, 1982, event. No systematic differences in magnitude, depth, or frequency were noted for normal or strike-slip events; only a few reverse faults were recorded.

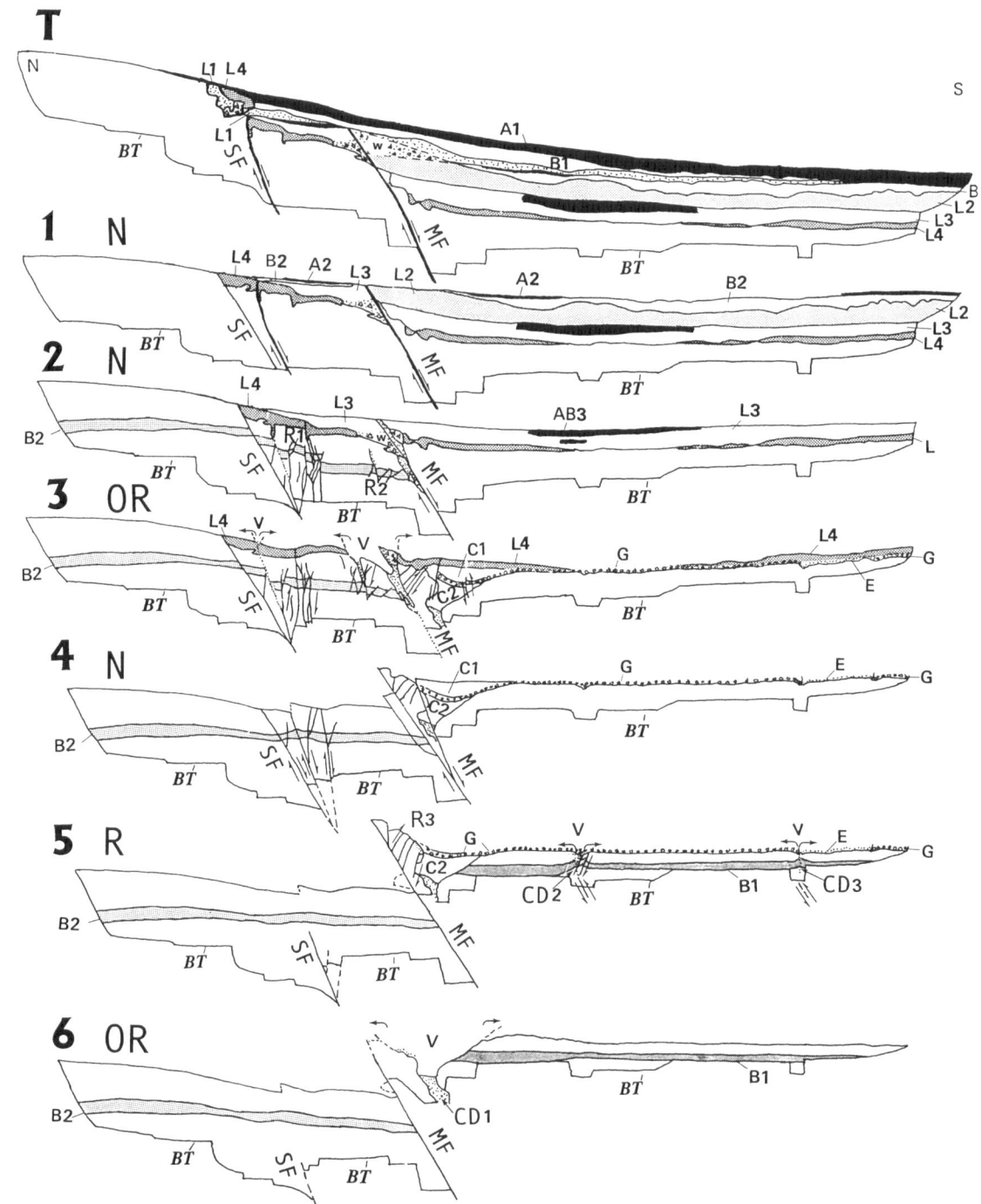

Figure 10. Diagram illustrating trench log with progressive line-length–balanced, retrodeformed displacements (Table 3) along faults of the Lima Reservoir zone (Figs. 2 and 4). T: Trench log showing younger soil and loess units, which are keys for interpreting last four events; main fault (MF) and secondary fault (SF) both moved during event 1; A1—A soil horizon; B1—B soil horizon; L1—youngest loess deposit (stippled); W—possible colluvial wedge loess; B2—older B soil horizon; L2—second youngest loess; L3—next to oldest loess; L4—oldest loess; BT—base of trench excavation. 1: Event 1 normal displacement restored on MF and SF; A1, B1, L1, and W removed; A2—older A soil horizon. 2: Event 2 normal displacement restored on MF; A2, B2, and L2 removed; AB3—oldest soil horizon; W—possible colluvial wedge material in vent; B2—marker bed 2. 3: Event 3 reverse (OR) displacement restored on R1 horst and R2 faults; AB3 and L3 removed; vents (V) for sand blow along MF and SF cut across L4; L4 thickens above graben near SF; lag gravel (G) at surface may indicate slight arching there; E—ejected sand; C1—upper channel deposits; C2—lower channel deposits. 4: Event 4 normal displacement restored along MF and SF graben; L4 removed; syndepositional C1 above folded G. 5: Event 5 reverse displacement restored on RF2; C1 removed; G is unfolded, but still above folded C2; G is unconformable over vents of clastic dikes 2 and 3 and ejected sand (ES); B1—marker bed 1. 6: Event 6 minimum reverse displacement restored on MF; C1 unfolded; G, C1, and E removed; V—vent for clastic dike 1.

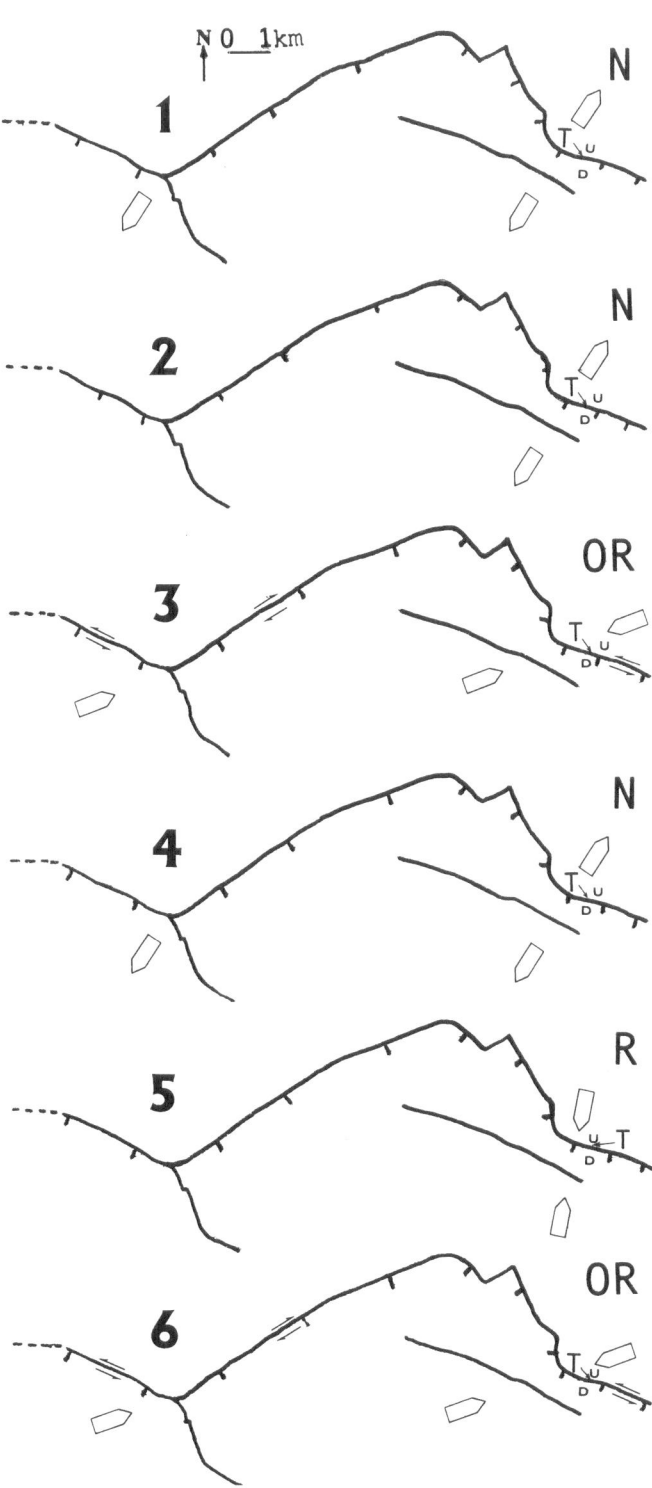

Figure 11. Diagram of Lima Reservoir fault zone showing how events 1 to 6 correspond to events 1 to 6 in Figure 10 and may be related to different segments of the zone. Open arrows indicate directions of compression or extension determined from fault slip-vectors and clastic dike orientations in trench. T—trench location; N—normal-faulting event; OR—oblique reverse-faulting event; R—reverse-faulting event.

basin. The anomalous 1-km-long, 2-m-high scarp near Cottonwood Creek (Fig. 1B) lies 1 km basinward of the main Blacktail fault and moved more recently than the main Blacktail fault (Stickney and Bartholomew, 1987b); thus it might be similar to the Georgia Gulch fault. In trench MBMG1986-6, the reverse faults utilize the preexisting normal faults but then splay off and also migrate basinward as they propagate upward. Recent seismicity outside of the Centennial Valley exhibits little evidence of the mid-continent stress regime, except for the 1925 Clarkston Valley earthquake (Doser, 1989). Nonetheless, the evidence for east-northeast–west-southwest compression along the Georgia Gulch and Lima Reservoir fault and perhaps the Cottonwood Creek scarp may indicate a more diffuse boundary than originally indicated by Stickney and Bartholomew (1987b).

SUMMARY AND CONCLUSIONS

Trench MGMG1886-6 contained the following features, shown by progressive retro-deformation of faults and line-length balancing of the trench strata, indicative of repeated syndepositional paleoseismic events; thus, these features are interpreted as paleoseismites.

Faults progressively offset late Pleistocene to Holocene age sediments, which are exposed in the trench, in six places and are oriented consistent with historic focal mechanism solutions along a nearby zone of seismicity.

Clastic dike 1 lies adjacent to a major active fault and terminated upward at a sand blow vent that was at least 1 m wide and 1.5 m deep. The sand in the vent is mixed with large exotic blocks of bedded sand and overlain by a stream-channel deposit. Ejected sand from this dike is preserved along the wall of the vent more than 1 m above the base of the channel. The orientation of clastic dike 1 is consistent with nearby historic focal mechanism solutions.

Clastic dikes 2 and 3 both taper downward to about 1 m depth and terminate at arches in the same sand source bed, which was sealed by the lower channel (C2, see Fig. 4, Fig. 9) deposits after the event represented by clastic dike 1. The arch below clastic dike 2 was caused by a small reverse fault, indicating near surface faulting at that time. The wall strata of both dikes are deformed consistent with forceful upward injection of the dikes. Ejected sand was deposited on the former ground surface up to 2 m away from clastic dike 3, indicating forceful ejection. The orientations of these dikes are consistent with nearby historic focal mechanism solutions.

The lag-gravel deposit unconformably overlies the vents and ejected sand of clastic dikes 2 and 3. The lag gravel appears conformable within the stream channel deposit where it and the lower part of the channel (C2, see Fig. 4, Fig. 9) are folded adjacent to the main fault. This synclinal fold served as the depositional channel for the upper channel deposits (C1, see Fig. 4, Fig. 9) and the thickness of the channel fill approximated the displacement on the main fault for this event.

The oldest loess deposit (L4, see Fig. 4, Fig. 9) unconformably overstepped ejected sand associated with clastic dike 3, the lag-gravel deposit, the upper stream channel deposits, the main fault, and older fluvial and lacustrine beds of the footwall of the main fault. This loess deposit was disrupted by sand blow vents, which are above clastic dike fillings along both the main fault and the secondary fault.

The next younger loess deposit (L3, see Fig. 4, Fig. 9), which overstepped both L4 and sand ejected along the main fault, accumulated as a colluvial wedge adjacent to the secondary fault.

The next younger loess deposit (L2, see Fig. 4, Fig. 9) accumulated as a colluvial wedge adjacent to the main fault and overstepped L3 and L4 between the main and secondary faults.

The youngest loess deposit (L1, see Fig. 4, Fig. 9) accumulated as colluvial wedges adjacent to both the main and secondary faults.

Additionally, the retro-deformation of the faults and strata in the trench log, combined with analysis of structural data from the trench and historic seismicity, allowed us to determine which stress field caused each of the six events. We then calculated the displacement rates and the recurrence intervals for surface rupture for the stress fields that caused these events on the basis of an estimated age of ca. 20 ka for the oldest event.

Normal faulting associated with northeast-southwest extension of the northern Basin and Range had a displacement rate of ~23.5 cm/k.y. and a recurrence interval of ~6.7 k.y.

Oblique-reverse faulting associated with strike-slip movement along the nearby zone of seismicity had a displacement rate of ~18.5 cm/k.y. and a recurrence interval of ~10 k.y.

Reverse faulting perhaps reflects accommodation associated with major events on the nearby Centennial fault during north-south extension associated with the Yellowstone hotspot. This fault had a displacement rate of less than 2 cm/k.y. and a recurrence interval of more than 20 ka.

ACKNOWLEDGMENTS

Renee A. Greenwell and Elzbieta Covington provided hardware and software support. This study was partially supported by a Montana Tech Research Center grant to Bartholomew and Stickney. We thank Diane I. Doser, James W. Sears, and Edmond G. Deal for reviews that helped us to improve the paper.

REFERENCES CITED

Anders, M.H., Geissman, J.W., Piety, L.A., and Sullivan, J.T., 1989, Parabolic distribution of circum-eastern Snake River Plain seismicity and latest Quaternary faulting: Migratory pattern and association with the Yellowstone hotspot: Journal of Geophysical Research, v. 94, p. 1589–1621.

Barrientos, S.E., Stein, R.S., and Ward, S.N., 1987, Comparison of the 1959 Hebgen Lake, Montana, and 1983 Borah Peak, Idaho, earthquake from geodetic observations: Bulletin of the Seismological Society of America, v. 77, no. 3, p. 783–808.

Bartholomew, M.J., 1989, The Red Rock fault and complexly deformed structures in the Tendoy and Four Eyes Canyon thrust sheets: Examples of late Cenozoic and late Mesozoic deformation in southwestern Montana, in Sears, J.W., ed., Structure, stratigraphy and economic geology of the Dillon area: Tobacco Root Geological Society 14th Annual Field Conference: July 20–22, 1989: Northwest Geology, v. 18, p. 21–35.

Bartholomew, M.J., Stickney, M.C., and Henry, J., 1988, Perspective 28 years after the August 18, 1959 Hebgen Lake earthquake, in A review of earthquake research applications in the National Earthquake Hazard Reduction Program: 1977–1987, Proceedings of Conference 41: U.S. Geological Survey Open-File Report 85-290, p. 236–263.

Bartholomew, M.J., Stickney, M.C., and Wilde, E.M., 1990, Late Quaternary faults and seismicity in the Jefferson Basin, in Hall, R.D., ed., Quaternary geology of the Western Madison Range, Madison Valley, Tobacco Root Range, and Jefferson Valley—Rocky Mountain Friends of the Pleistocene, August 15–19, 1990, Fieldtrip Guidebook: Indianapolis, Indiana University, Department of Geology, p. 238–244.

Bartholomew, M.J., Lewis, S.E., Russell, G.S., Stickney, M.C., Wilde, E.M., and Kish, S.A., 1999, Late Quaternary history of the Beaverhead River canyon, southwestern Montana, in Hughes, S.S., and Thackray, G.D., eds., Guidebook to the geology of Eastern Idaho: Pocatello, Idaho Museum of Natural History, p. 237–250.

Chadwick, O.A., Hall, R.D., and Phillips, F.M., 1997, Chronology of Pleistocene glacial advances in the central Rocky Mountains: Geological Society of America Bulletin, v. 109, p. 1443–1452.

Crone, A.J., and Machette, M.N., 1984, Surface faulting accompanying the Borah Peak earthquake, Idaho: Geology, v. 12, p. 664–667.

Crone, A.J., Machette, M.N., Bonilla, M.G., Lienkaemper, J.J., Pierce, K.L., Scott, W. E., and Bucknam, R.C., 1985, Characteristics of surface faulting accompanying the Borah Peak earthquake, central Idaho, in Workshop 28 on the Borah Peak, Idaho earthquake: U.S. Geological Survey Open-File Report 85-290, p. 236–263.

Doser, D.I., 1989, Source parameters of Montana earthquakes (1925–1964) and tectonic deformation in the northern Intermountain Belt: Bulletin of the Seismological Society of America, v. 79, no. 1, p. 31–50.

Hall, R.D., 1990, Quaternary geology of the Tobacco Root Range, in Hall, R.D., ed., Quaternary geology of the Western Madison Range, Madison Valley, Tobacco Root Range, and Jefferson Valley—Rocky Mountain Friends of the Pleistocene, August 15–19, 1990, Fieldtrip Guidebook: Indianapolis, Indiana University, Department of Geology, p. 183–193.

Hall, R.D., and Jaworowski, C., 1999, Reinterpretation of the Cedar Ridge section, Wind River Range, Wyoming: Implications for the glacial chronology of the Rocky Mountains: Geological Society of America Bulletin, v. 111, p. 1233–1249.

Hall, R.D., Garner, E.A., and Horn, L.L., 1990, A preliminary study of the glacial geology and soils of the South Meadow Creek and North Meadow Creek basins, Tobacco Root Range, in Hall, R.D., ed., Quaternary Geology of the Western Madison Range, Madison Valley, Tobacco Root Range, and Jefferson Valley—Rocky Mountain Friends of the Pleistocene, August 15–19, 1990, Fieldtrip Guidebook: Indianapolis, Indiana University, Department of Geology, p. 194–237.

Hanneman, D.L., and Wideman, C.J., 1991, Sequence stratigraphy of Cenozoic continental rocks, southwestern Montana: Geological Society of America Bulletin, v. 103, p. 1335–1345.

Hill, A.A., and Bartholomew, M.J., 1999, Seismic hazard susceptibility in southwestern Montana: Comparison at Dillon and Bozeman, in Hughes, S.S., and Thackray, G.D., eds., Guidebook to the geology of Eastern Idaho: Idaho Museum of Natural History, p. 131–139.

Iyer, H.M., 1984, A review of crust and upper mantle structure studies of the Snake River Plain–Yellowstone volcanic system: A major lithospheric anomaly in the western USA: Tectonophysics, v. 105, p. 291–308.

Janecke, S.U., 1992, Kinematics and timing of three superimposed extensional systems, east central Idaho: Evidence for an Eocene tectonic transition: Tectonics, v. 11, no. 6, p. 1121–1138.

Morgan, W.J., 1972, Deep mantle convection plume and plate motions: American Association of Petroleum Geologists Bulletin, v. 56, p. 203–312.

Pardee, J.T., 1926, The Montana earthquake of June 27, 1925: U.S. Geological Survey Professional Paper 141-B, p. 43–58.

Parsons, T., Thompson, G.A., and Smith, R.P., 1998, More than one way to stretch: A tectonic model for extension along the plume track of the Yellowstone hotspot and adjacent Basin and Range Province: Tectonics, v. 17, p. 221–234.

Phillips, F.M., Zreda, M.G., Gosse, J.C., Klein, J., Evenson, E.B., Hall, R.D., Chadwick, O.A., and Sharma, P., 1997, Cosmogenic ^{36}Cl and ^{10}Be ages of Quaternary glacial and fluvial deposits of the Wind River Range, Wyoming: Geological Society of America Bulletin, v. 109, p. 1453–1463.

Qamar, A.I., and Stickney, M.C., 1983, Montana earthquakes, 1869–1979: Historical seismicity and earthquake hazard: Montana Bureau of Mines and Geology, Memoir 51, p. 79.

Reynolds, M.W., 1979, Character and extent of basin-range faulting, western Montana and east-central Idaho: Basin and Range Symposium, Rocky Mountain Association of Geology and Utah Geological Association, p. 185–193.

Rodgers, D.W., Hackett, W.R., and Ore, H.T., 1990, Extension of the Yellowstone plateau, eastern Snake River Plain, and Owyhee plateau: Geology, v. 18, p. 1138–1141.

Savage, J.C., Lisowski, M., Prescott, W.H., and Pitt, A.M., 1993, Deformation from 1973 to 1987 in the epicentral area of the 1959 Hebgen Lake, Montana, earthquake (Ms=7.5): Journal of Geophysical Research, v. 98, no. B2, p. 2145–2153.

Sbar, M.L., Barazangi, M., Dorman, J., Scholz, C.H., and Smith, R.B., 1972, Tectonics of the Intermountain Seismic Belt, western United States: Microearthquake seismicity and composite fault plane solutions: Bulletin of the Seismological Society of America, v. 83, p. 13–28.

Scott, W.E., Pierce, K.L., and Hait, M.H., 1985, Quaternary tectonic setting of the 1983 Borah Peak earthquake, central Idaho: Bulletin of the Seismological Society of America, v. 75, p. 1053–1066.

Sears, J.W., and Fritz, W.J., 1998, Cenozoic tilt domains in southwestern Montana: Interference among three generations of extensional fault systems, in Faulds, J.E., and Stewart, J.H., eds., Accommodation zones and transfer zones: The regional segmentation of the Basin and Range Province: Geological Society of America Special Paper 323, p. 241–247.

Seed, H.B., 1968, Landslides during earthquakes due to liquefaction: American Society of Civil Engineers Journal of Soil Mechanics and Foundations Division, v. 95:SM5, p. 1053–1122.

Seilacher, A., 1969, Fault-graded beds interpreted as clastic dikes: Sedimentology, v. 13, p. 155–159.

Seilacher, A., 1984, Sedimentary structures tentatively attributed to seismic events: Marine Geology, v. 55, p. 1–12.

Sibson, R.H., 1990, Rupture nucleation on unfavorably oriented faults: Bulletin of the Seismological Society of America, v. 80, no. 6, p. 1580–1604.

Slemmons, D.B. and McKinney, R., 1977, Definition of "active fault": U.S. Army Engineer Waterways Experiment Station, Soils and Pavements Laboratory, Miscellaneous Paper S-77-8, 22 p.

Smith, R.B., and Sbar, M.L., 1974, Contemporary tectonics and seismicity of the western United States with emphasis on the Intermountain Seismic Belt: Geological Society of America Bulletin, v. 85, no. 2, p. 1205–1218.

Smith, R.B., Richins, W.D., and Doser, D.I., 1985, The 1983 Borah Peak, Idaho earthquake: Regional seismicity, kinematics of faulting, and tectonic mechanism Workshop 28 on the Borah Peak, Idaho earthquake: U.S. Geological Survey Open-File Report 85-290, p. 236–263.

Smith, R.B., Reilinger, R.E., Meertens, C.M., Hollis, J.R., Holdahl, S.R., Dzurisin, D., Gross, W.K., and Klingele, E.E., 1989, What's moving at Yellowstone? The 1987 crustal deformation survey from GPS, leveling, precision gravity, and trilateration: EOS (Transactions, American Geophysical Union), v. 70, no. 8, p. 113–128.

Sonderegger, J.L., Schofield, J.D., Berg, R.B., and Mannick, M.L., 1982, The upper Centennial Valley, Beaverhead and Madison counties, Montana: Montana Bureau of Mines and Geology Memoir 50, 53 p.

Stein, R.S., and Barrientos, S.E., 1985, Planar high-angle faulting in the Basin and Range: Geodetic analysis of the 1983 Borah Peak, Idaho, earthquake: Journal of Geophysical Research, v. 90, no.3, p.11355–11366.

Steinbrugge, K.V., and Cloud, W.K., 1962, Epicentral intensities and damage in the Hebgen Lake, Montana earthquake of August 17, 1959: Bulletin of the Seismological Society of America, v. 52, no. 2, p. 181–234.

Stickney, M.C., 1997, Seismic source zones in southwestern Montana: Montana Bureau of Mines and Geology Open-File Report 366, 52 p.

Stickney, M.C., and Bartholomew, M.J., 1987a, Preliminary map of late Quaternary faults in western Montana: Montana Bureau of Mines and Geology Open-File Report 186, scale 1:500 000, 1 sheet.

Stickney, M.C., and Bartholomew, M.J., 1987b, Seismicity and Quaternary faulting of the northern Basin and Range province, Montana and Idaho: Bulletin of the Seismological Society of America, v. 77, no. 5, p. 1602–1625.

Stover, C.W., 1984, The Borah Peak, Idaho earthquake of October 28, 1983 : Isoseismal map and intensity distribution: Earthquake Spectra, v. 2, p. 11–16.

Tocher, D., 1962, The Hebgen Lake, Montana, earthquake of August 17, 1959, MST: Bulletin of the Seismological Society of America, v. 52, no. 2, p. 153–162.

Witkind, I.J., 1964, Reactivated faults north of Hebgen Lake, in The Hebgen Lake, Montana earthquake of August 17, 1959: U.S. Geological Survey Professional Paper 435, p. 37–50.

Witkind, I.J., 1972, Geologic map of the Henrys Lake quadrangle, Idaho and Montana: U.S. Geological Survey Map I-781-A, scale 1:62 500, 1 sheet.

Witkind, I.J., 1975, Geology of a strip along the Centennial fault, southwestern Montana and adjacent Idaho: U.S. Geological Survey Map I-890, scale 1:62 500, 1 sheet.

Witkind, I.J., Myers, W.B., Hadley, J.B., Hamilton, W., and Fraser, G.D., 1962, Geological features of the earthquake at Hebgen Lake, Montana, August 17, 1959: Bulletin of the Seismological Society of America, v. 52, no. 2, p. 163–180.

MANUSCRIPT ACCEPTED BY THE SOCIETY MAY 11, 2001

Stratigraphic evidence of coseismic faulting and aseismic fault creep from exploratory trenches at Mt. Etna volcano (Sicily, Italy)

Luca Ferreli
*Agenzia Nazionale per la Protezione dell'Ambiente (Italian Environmental Protection Agency),
via Vitaliano Brancati 48, 00144 Rome, Italy*

Alessandro Maria Michetti
Dipartimento di Scienze Chimiche, Fisiche e Matematiche, Università dell'Insubria, via Lucini 3, 22100 Como, Italy

Leonello Serva
Eutizio Vittori
*Agenzia Nazionale per la Protezione dell'Ambiente (Italian Environmental Protection Agency),
via Vitaliano Brancati 48, 00144 Rome, Italy*

ABSTRACT

Recognition of coseismic and aseismic slip in trench exposures is a major goal in paleoseismology. To define stratigraphic criteria for discriminating between (1) fast earthquake-related slip and (2) slow quasi-continuous creep, we carried out several exploratory trenches along the eastern flank of Mt. Etna, where capable faults (active faults producing displacement at or near the surface) show both modes of movements with high slip rates and short recurrence intervals. Our sites have experienced predominant coseismic (Fondo Macchia) and aseismic (Mandra del Re) fault slip during historical times. At the Fondo Macchia site we trenched a normal fault scarp where ~20 cm of vertical offset occurred in 1971 and three other similar earthquakes repeated in the past 150 yr. Several erosional surfaces close to the fault zone in the footwall indicate that (1) a distinct and recognizable fault scarp free face retreated repeatedly, shaped by erosion and fault activity, and (2) the observed vertical displacement is a result of repeated scarp-forming earthquakes. At the Mandra del Re site a left-lateral, strike-slip fault with a large vertical component dams the drainage of a small valley. A vertical fault slip rate of ~2 cm/yr and consequent high deposition rates of ponded, mainly well-layered, fine-grained sediments allow to reconstruct with excellent stratigraphic resolution the fault growth in the past few centuries. More than 3 m of vertical displacement has accumulated in the fault zone, leaving no indication of scarp-related erosion in the footwall deposits or of colluvial wedges in the hanging wall. This unequivocal stratigraphic evidence of "aseismites" (i.e., sedimentary features and relations generated by continuous fault creep) shows that earthquake surface faulting and aseismic creep generate completely different sedimentary responses.

INTRODUCTION

Until now, very little attention has been paid to the stratigraphic signature produced by creeping faults (Sieh and Williams, 1990; Pavlides, 1996; Bell and Helm, 1998; Stenner and Ueta, 2000). However, the coseismic versus aseismic nature of slip events observed in fault trench exposures is a major issue in the paleoseismic analysis of surface faulting. To define stratigraphic criteria for discriminating the two end-member modes of fault movement, rapid earthquake-related slip and slow quasi-continuous creep, we conducted trench investigations on the eastern flank of Mt. Etna volcano in Sicily (Fig. 1). There, a dense set of capable faults (active faults producing displacement at or near the surface; e.g., Azzaro et al., 1998; Azzaro, 1999; Machette, 2000) shows both modes of deformations and sometimes a mixture of two modes either in different sections of the same fault or alternating in time, probably due to the interaction of volcano-tectonic and purely tectonic processes. Typically, these faults are characterized by high slip rates (up to several centimeters per year), kinematics ranging from entirely dip slip to entirely strike slip, and recurrence time of stick-slip events on the order of a few decades. Earthquake surface faulting in this area is associated with very shallow (less than 5 km deep) and low-magnitude (M_w ~3 to 5) events, and may produce end-to-end surface rupture lengths of up to 5-7 km and vertical displacement of up to 90 cm (Azzaro, 1999, and references therein). Therefore, due to the variety of fault behavior, the rapid evolution of morphogenetic processes, the large amount of sediment constantly supplied by volcanic activity, and the rich availability of datable material, the Mt. Etna volcano represents an ideal natural laboratory for refining paleoseismological techniques.

The Italian Environmental Protection Agency has carried out, with the scientific and financial involvement of the Italian Volcanology International Institute, Italian Volcanology Group, and the Italian Group for Protection against Earthquakes, a paleoseismological research project along the eastern side of the volcano (Fig. 1), based on field survey, geomorphic analysis, interpretation of air photos, and excavation of fault trenches. Paleoseismology in volcanic environments can provide invaluable information on, for example, the association between earthquakes and eruptions, and the related assessment of natural hazards (e.g., Smith et al., 1996), the rates of recent fault activity within the volcanic zone (e.g., Azzaro et al., 2000; Villamor, 2000), and the nature of surface faulting processes and the characterization of fault creep (e.g., Rasà et al., 1996, and references therein).

We focus here on the last issue, which is essential for the correct interpretation of deformed features observed in trenches and consequently for improving the reliability of geological seismic hazard assessment in general.

The following discussion is based on trench sites at two of the most active faults affecting the Etna volcano, (1) the Moscarello fault, characterized by dip slip, high morphological relief, frequent coseismic reactivation (four events during the past two centuries; e.g., Azzaro et al., 2000; Ferreli et al., 2000), and (2) the Pernicana fault, a mainly sinistral-slip fault with a high slip rate, along which segments producing stick-slip low-magnitude events alternate with other segments showing stable-sliding motion (Azzaro et al., 1998, and references therein). The results obtained from the stratigraphic study of trench exposures along these two faults provide evidence that specific sedimentary structures (or the lack of them) characterize the two end modes of slip. This conclusion strengthens our confidence in paleoseismological inference, particularly in those Italian regions of highest seismic hazard, such as the central and southern Apennines (e.g., Vittori et al., 1991; Michetti et al., 1995, 1996, 1997, 2000; Pantosti et al., 1993, 1996).

TECTONIC FRAMEWORK

The Etna volcano (Fig. 1) has developed since the middle-late Pleistocene at the inland intersection between the Hyblean-Malta escarpment (north-northwest–south-southeast–trending), a crustal discontinuity separating the Hyblean continental plateau from the oceanic floor of the Ionian Sea, and the north-northeast–south-southwest–trending faults bordering the northeastern coast of Sicily (Scandone et al., 1981; Lo Giudice et al., 1982; Monaco et al., 1995; Azzaro and Barbano, 2000).

The northern inland extension of the Hyblean-Malta fault escarpment is the Timpe fault system, a belt of north-northwest–south-southeast–trending normal faults. The Timpe system is commonly interpreted as the result of volcano-tectonic processes, which induce the gravitational instability and eastward sliding of the sea-side flank of the volcano (Lo Giudice and Rasà, 1986; Borgia et al., 1992; Lanzafame et al., 1996; McGuire et al., 1997). In historical times activity on the fault has caused tectonic creep dislocations affecting roads and buildings (Rasà et al., 1996) and frequent seismic events with low to moderate magnitudes (m_s <5.0). Despite their modest size, these earthquakes can cause significant damage (macroseismic Intensity I_{max} = X Medvedev–Sponheuer–Karnik (MSK) scale; note that use of the Modified Mercalli (MM) scale would yield approximately the same values) and surface fault displacement due to the superficial hypocenters (H <5 km) (Azzaro et al., 1989; Lo Giudice and Rasà, 1992).

One of the most active faults of the Timpe system, in terms of slip rates and seismicity, is the 340° trending, east-dipping Moscarello normal fault, 10 km long (Figs. 1 and 2) (Azzaro et al., 2000). During the past two centuries, this fault has generated four seismic events (in 1855, 1865, 1911, and 1971) with intensities of VII–VIII to X MSK. All of these events were accompanied by surface ruptures up to 6 km long and with maximum dip-slip displacements between 25 and 90 cm. In contrast to most of the other faults of the Timpe system, there is no evidence of ongoing creep on the Moscarello fault (Rasà et al., 1996; Gresta et al., 1997). Holocene slip rates along the Moscarello fault have been reassessed by Azzaro et al. (2000) and Ferreli et al. (2000). The vertical component of the slip rate along the most active sector of the fault at Fondo Macchia basin (Fig. 2) is on the order of 2 to 3 mm/yr during the past ca. 6 ka, in good agreement with the short-term value inferred from the four historical ruptures.

The Timpe system interrupts to the north against the arch-shaped Provenzana-Pernicana-Fiumefreddo fault system, trending east-west to west-northwest–east-southeast for ~18 km; this fault system crosses the volcanic edifice from the central crater to the sea (Figs. 1 and 3). This very active system plays an important role in the geodynamic and morphologic evolution of the area, and it is interpreted by some authors (Borgia et al., 1992; Lo Giudice and Rasà, 1992; McGuire et al., 1996; Groppelli and Tibaldi, 1999) as the northern boundary of the unstable eastern flank of the volcano, sliding toward the sea. Its kinematics are essentially left-lateral, with local extensional components lowering the southern and eastern sides. The slip rate, >2 cm/yr, is the highest known in the Etna volcanic area. Slip occurs as a result of both coseismic rupture and continuous or episodic creep (Rasà et al., 1996). During the past few decades of local instrumental monitoring, slip along the Provenzana fault has been accompanied by frequent seismic events with very shallow sources (H <1 km), where H represents hypocentral depth, and magnitudes not exceeding m_s 4.2 (Azzaro et al., 1989; Azzaro, 1997). The central segment, corresponding to the Pernicana fault, moves essentially by creep without appreciable seismicity, continuously deforming roads and houses (Rasà et al., 1996; Azzaro et al., 1998). The Pernicana's average slip rate exceeds 2 cm/yr, as indicated by offsets of 80 cm on 50-yr-old walls and 350 cm on 150-yr-old farming roads. Fault behavior is also dominated by creep deformations along the contiguous eastern segment, the Fiumefreddo fault. There, the high slip rate produces significant damage to several roads, including the Messina-Catania toll way, and a gas pipeline (Azzaro et al., 1998).

MOSCARELLO FAULT: COSEISMIC MOVEMENT

The best morphological evidence accompanied by the largest stratigraphic offsets of the Moscarello fault are observed in its northern part, where the fault hanging wall hosts the Fondo Macchia basin (Fig. 2). There, fault movement has generated a prominent and steep (30° to 40°) escarpment up to 125 m high. At the foot of this escarpment, we carried out exploratory trenching and geological and geophysical prospecting in the years 1996 and 1997.

In this chapter, we focus on trench A (Fig. 2; for detailed analyses of both trenches A and B, see Azzaro et al., 2000), located where eyewitnesses of the April 1971 earthquake had observed a vertical rejuvenation of ~20 cm. The surface fault rupture, which had evidently followed a preexisting fault scarp up to 270 cm high, was carefully mapped in the field for a total length of ~3 km (Riuscetti and Distefano, 1971).

Trench A (Figs. 4 and 5) exposed a rupture zone several meters wide, bounded and crossed by several fault branches; the main branch strikes 350° to 355° and dips eastward 75° to 90°. No striations or other lineations were found on the fault planes, but along the fault zone the long axes of oblate clasts parallel the fault dip, suggesting a dip-slip motion.

The fault separates two different stratigraphic sequences, whose ages have been constrained by means of ^{14}C dating of charcoal and organic matter samples, and by archaeological analyses of pottery fragments. The footwall consists of unconsolidated sandy gravel including a few large rounded boulders, interbedded with sandy and silty-clayey horizons (unit A in Fig. 5). The deposits are interpreted to be part of an alluvial fan within the latest Pleistocene to Holocene Chiancone formation (Calvari and Groppelli, 1996). Calibrated ages range between 6250 and 3250 yr B.P. Near the rupture zone, unit A is locally backtilted (level 3 in northern wall and level 2a in southern wall; see Fig. 5) and shows several erosional surfaces that we relate to episodes of

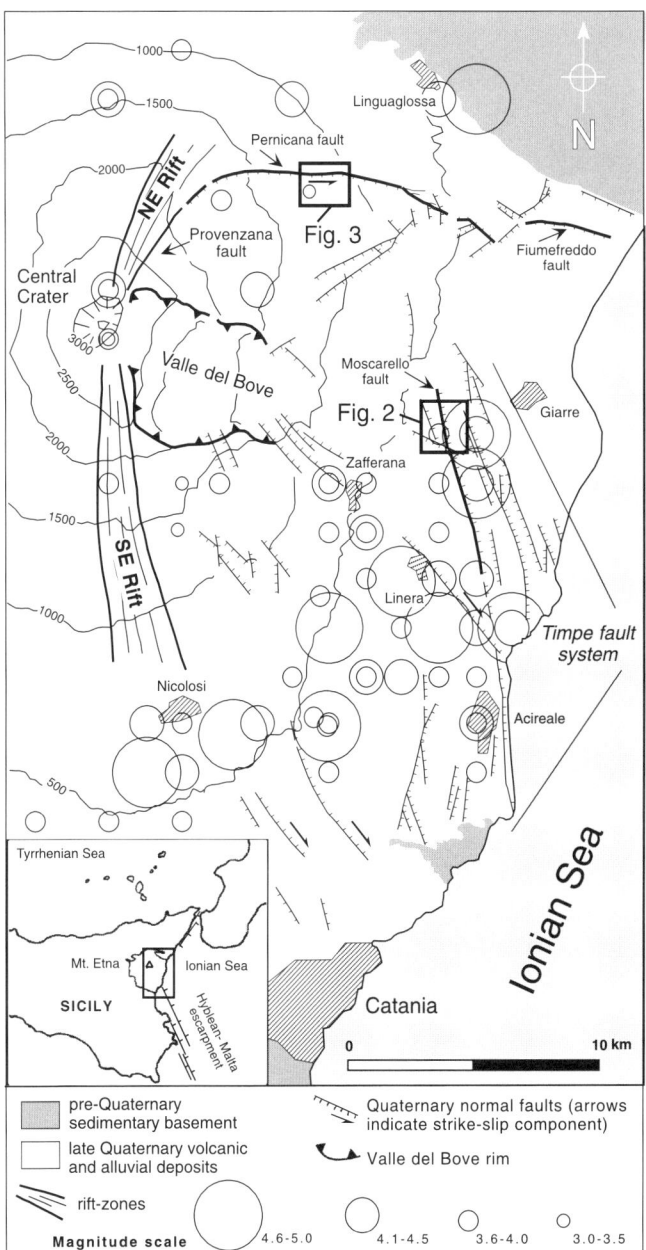

Figure 1. Tectonic sketch map of eastern flank of Mt. Etna and distribution of M >3.0 earthquakes (from Camassi and Stucchi, 1997). The volcano edifice corresponds to the intersection zone of the two main fault trends in eastern Sicily: north-northeast–south-southwest in the northern sector of the coast and north-northwest–south-southeast in the southern sector. The trench sites along the Moscarello and Pernicana faults are located in the areas labeled Fig. 2 and Fig. 3, respectively.

scarp retreat triggered by the formation of a fault-scarp free face. The erosional contacts exist only at the fault, and they are unlikely to be the results of stream action.

In the hanging wall, we observed a sequence, ~4 m thick, of massive, fine-grained colluvial deposits (unit C in Fig. 5) containing pottery sherds of pots made on a wheel, and tiles. On the

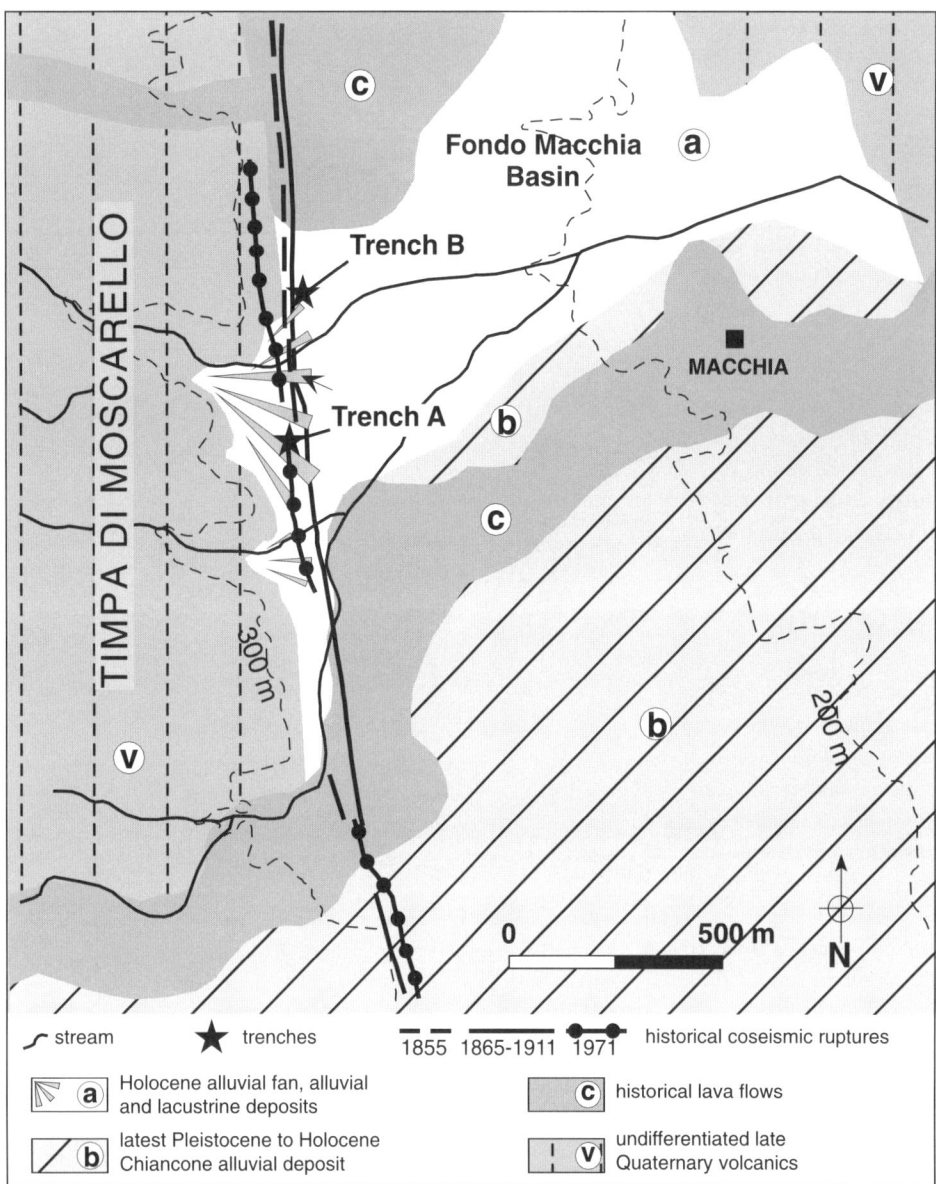

Figure 2. Geologic and topographic setting of the central part of the Moscarello fault, showing patterns of coseismic surface ruptures along this part of the fault since 1855 and locations of trenches A and B at the Fondo Macchia site. See Figure 1 for location.

basis of their typological and workmanship features, the fragments may be dated between 1100 B.C. and the modern (post–Middle Ages) period. Calibrated ages range here between 4570 and 1330 yr B.P. This hanging-wall sequence lacks any sedimentary structures related to single events of scarp formation and erosion, including the 20 cm of coseismic offset due to the 1971 earthquake. It is likely that 1971-like scarps formed during many previous earthquakes, but the subsequent colluvial wedges were obliterated due to the continuous agricultural activities of the past millennia (for discussion of this point, see Azzaro et al., 2000). Hand drilling to 1.5 m depth in the hanging-wall side of the trench bottom proved that the footwall unit A deposits are here deeper than 5.6 m. A standard penetration test (SPT) performed near this trench and a second trench dug ~500 m to the north (see Ferreli et al., 2000) have shown that the facies change from unit A to unit C should not be deeper than 7 m in the hanging-wall section of trench A (for discussion, see Azzaro et al., 2000). On the basis of these data, the dip-slip component of the Moscarello fault slip rate at this site ranges between 1.4 mm/yr (minimum offset of 5.6 m and maximum age of 4 ka for horizon 2b, which is the youngest alluvial layer) and 2.7 mm/yr (minimum offset of 8 m and maximum age of 3 ka for horizon 2b).

Historical seismicity and data collected at Fondo Macchia prove that morphological scarps produced by surface faulting formed during the deposition of unit A. Such scarps were smoothed away by erosional processes during the nondeposition periods and drowned by subsequent alluvial events. The absence of exposed correlatable layers between footwall and hanging wall and of colluvial wedges in the subsiding side does not allow us to quantify the vertical displacements during each event and, conse-

Stratigraphic evidence of coseismic faulting and aseismic fault creep from exploratory trenches 53

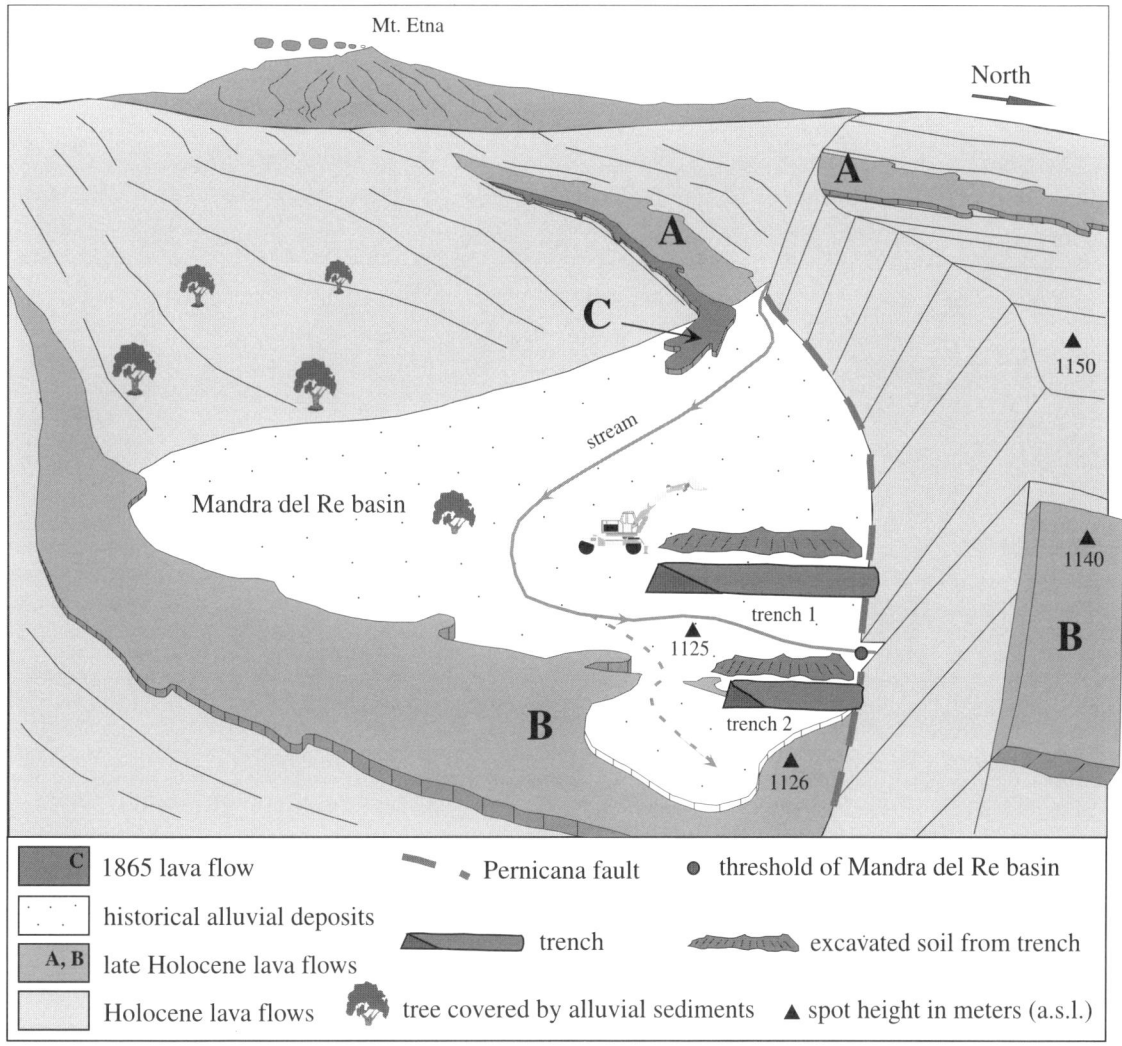

Figure 3. Sketch of the Mandra del Re trench site along the Pernicana fault; see Figure 1 for location.

quently, the number of such slip events. Nevertheless, the sedimentary structures, fault geometry, erosional surfaces near the main fault trace, and several upward terminations clearly demonstrate the occurrence of repeated scarp formation related to instantaneous surface faulting during the deposition of unit A and provide sufficient information to reconstruct the evolution of Holocene displacement along the fault scarp. Figure 6 summarizes our interpretation of fault scarp growth in the last 6 kyr at the site of trench A.

The heights of the scarps sketched in Figure 6, A to F, which formed during the growth of alluvial unit A, are constrained by the geometry of the erosional surfaces. Only in Figure 6C could we refer to the dimension of the sedimentary wedge preserved on top of fault F2. Consequently, in Figure 6 we cannot discriminate whether the inferred heights are the result of a single event or cumulative surface rupture events. It is certainly possible, however, to rule out fault creep as a major contributor to scarp growth during the deposition of unit A. Surface faulting events and scarp growth in this period occurred mostly because of recurrent earthquakes. Slow and continuous movements cannot generate the observed sequence of erosional surfaces in the uplifted side of trench A.

The 360 cm offset shown in Figure 6G is also the result of several discrete events of surface rupture. The lack of sedimentary structures and clasts from unit A in the colluvium of unit C near the fault plane makes unattainable the recognition of the most recent phases of scarp growth. The absence of clastic wedges from unit A in the downthrown side of the trench is a probable effect of disturbance due to human activities. Farming the area, plowing and clearing the soil of stones, followed the shift from the alluvial to colluvial sedimentary environment ca. 3 to 4 kyr ago. Moreover, as suggested by the 20 cm offset observed during the last earthquake in 1971, it is also possible that coseismic slip events during the past 2 to 3 kyr were always smaller (<30 to 50 cm, which is about the present-day thickness of the organic soil along the trench) than the thickness of the col-

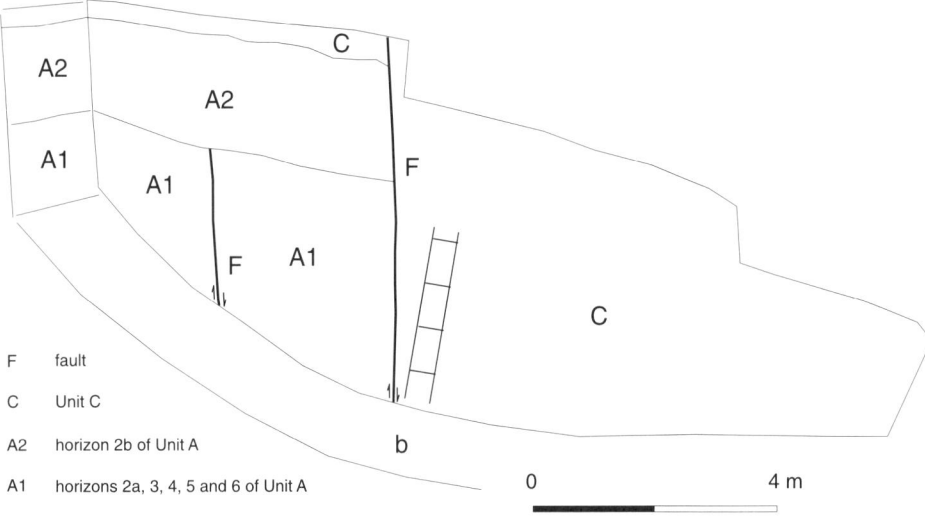

Figure 4. Fondo Macchia site; photo shows the northern wall of trench A at the foot of the Moscarello fault escarpment; see Figure 2 for trench location.

luvial and soil deposits resting on unit A in the footwall, never exposing and therefore never allowing the erosion of the underlying alluvial sediment.

PERNICANA FAULT: ASEISMIC MOVEMENT

The Mandra del Re site is located where the Pernicana fault generates a small, approximately east-west–striking basin (Fig. 3). Geomorphological and structural observations have revealed a significant vertical component of fault motion, as illustrated by a steep scarp 20 to 30 m high. At Mandra del Re, this scarp has recently disrupted the regular pattern of Holocene lava flows and the related drainage network, which follow the natural slope of the volcano (in the study area it descends from south-southwest to north-northeast). Two apparently simultaneous, latest Holocene, north-northeast–trending lava flows are displaced by the fault (A and B in Fig. 3). Therefore, the Mandra del Re scarp formed a few thousand years ago. In contrast to the Fondo Macchia site, there is no evidence of significant human disturbance of the morphogenic and sedimentary processes.

Using as a pre-faulting datum features of lava flows A and B that are clearly recognizable on both sides of the fault, the vertical offset is on the order of 20 to 30 m, and the left-lateral offset is less than 30 to 40 m. It is impossible to evaluate in better detail the lateral component of slip, because the width of the offset lava channels is on the order of 10 m and the fault zone is 20 to 30 m wide where the scarp was trenched.

The growth of the Mandra del Re scarp diverted the drainage path along strike of the fault toward the east, as occurred, for instance, to the 1865 A.D. lava flow (C in Fig. 3). Figure 3 also shows that at the eastern end of the Mandra del Re basin the downthrown section of lava flow B dams the drainage again. At this location, the water flowing into the Mandra del Re depression finds its way to the north through a threshold incised across the south-verging fault scarp, where the erosional downcutting of a narrow gully strives to keep pace with the continuous uplift of

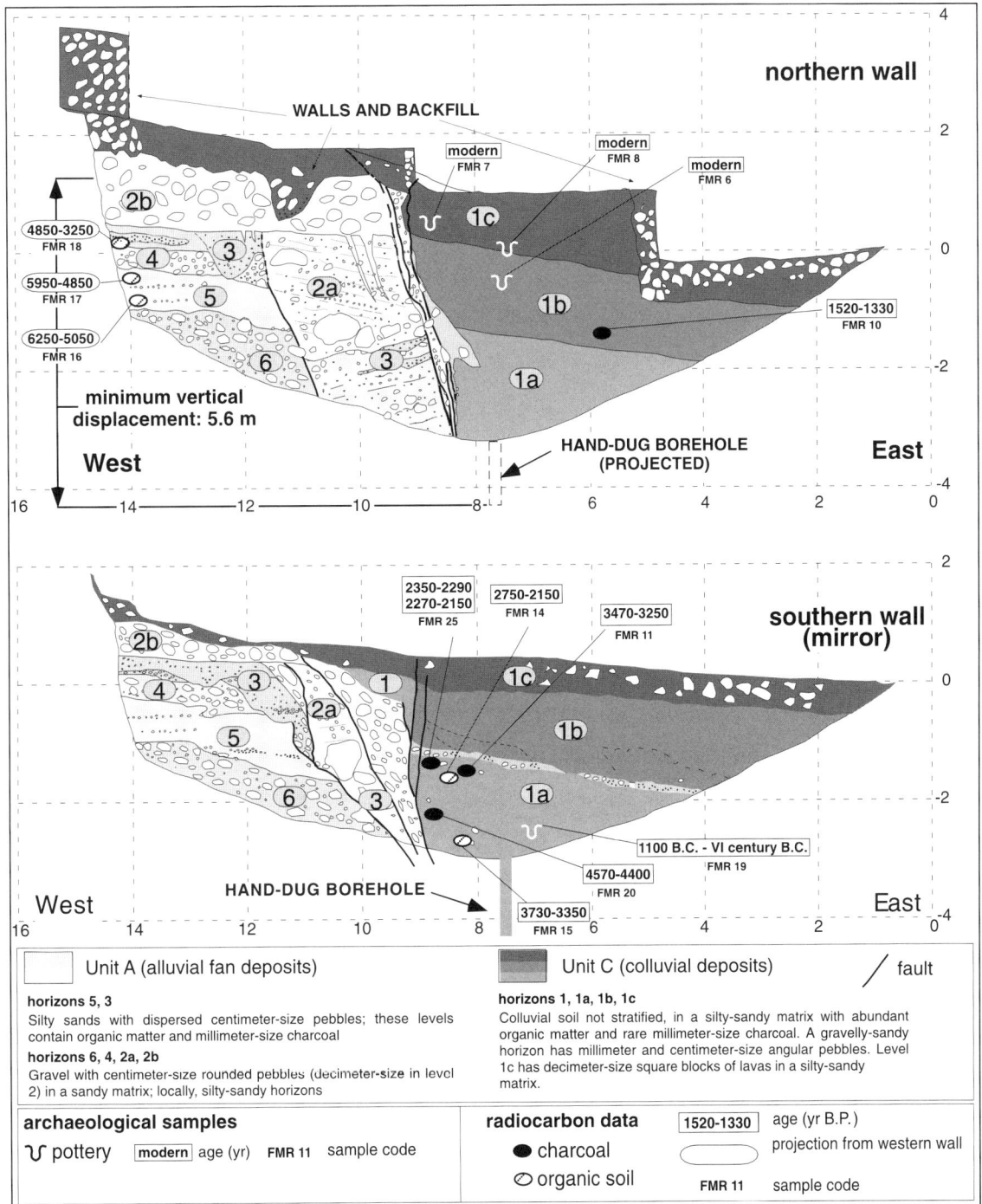

Figure 5. Logs of trench A north and south walls (mirror image of the original log) at the Fondo Macchia site; see Figure 2 for trench location. Reference grid is in meters. Circled numbers 1a–6 refer to stratigraphic horizons described in the text.

the northern side of the fault. The small basin is therefore a sedimentary trap controlled by the uplift of the threshold. At present, the top of lava flow B is only about 65 cm higher than the threshold, and the lower trunks of ~50-yr-old oak trees in the valley floor are buried by alluvial sediments (Fig. 3). This demonstrates that the vertical component of the fault slip rate is greater than the erosional downcutting rate. Indeed, the Mandra del Re basin is now almost completely filled. During flooding events, a second drainage threshold above the lava flow B toward the east was already active (dashed-line arrow in Fig. 3), as proved by a thin, discontinuous layer of alluvial sand and silt deposited over the lava flow.

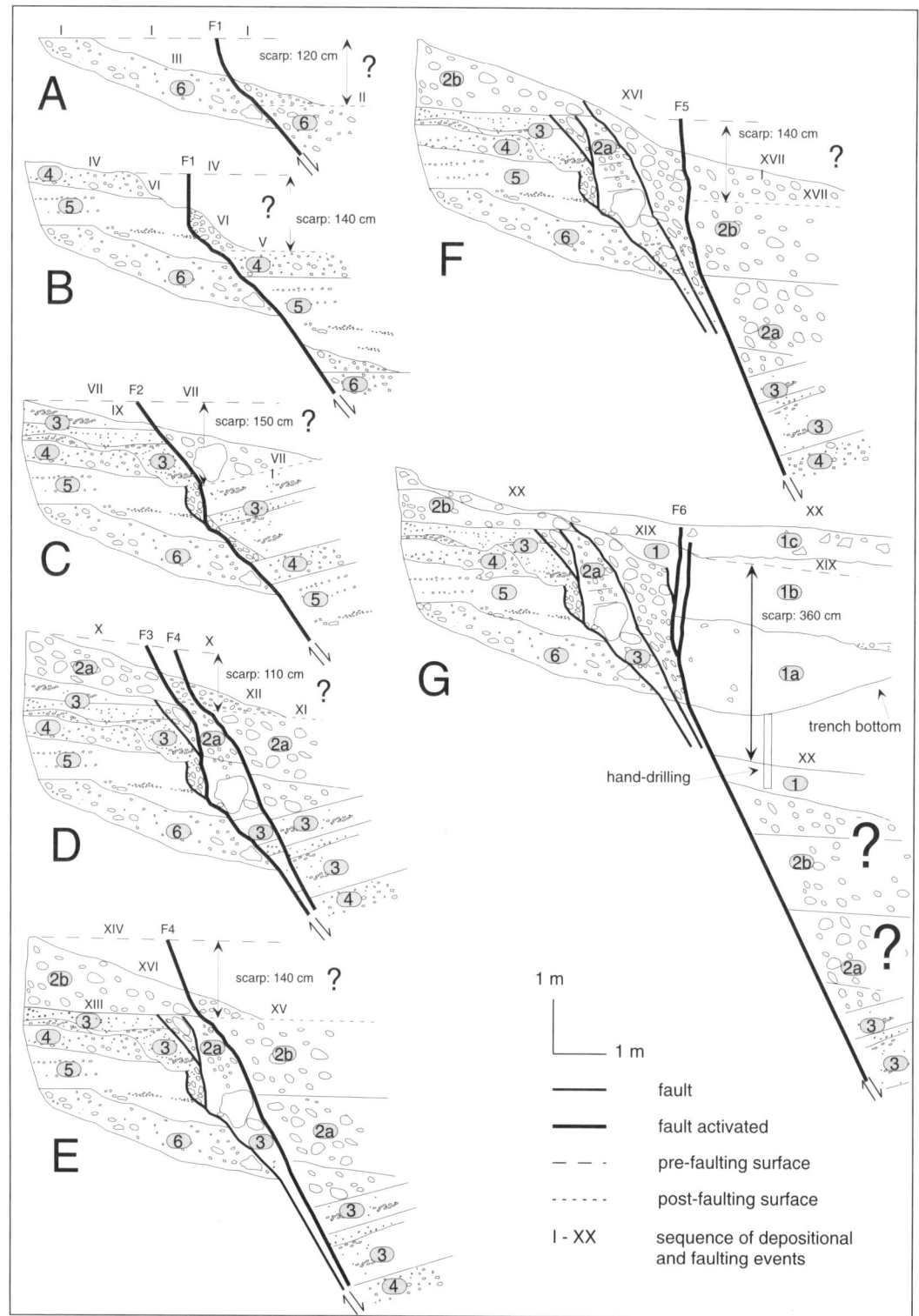

Figure 6. Trench A, south wall exposing the Moscarello fault zone. Schematic diagrams show the inferred sequence of deposition and faulting events (indicated by Roman numerals from I, oldest, to XX, the present-day geometry) recorded by the stratigraphic horizons (circled numbers) as defined in Figure 8: A: Deposition of horizon 6 (I), faulting event along F1 (II), and formation of a thin colluvial wedge (III). B: Deposition of horizons 5 and 4 (IV), faulting event along F1 (V), erosion of horizons 5 and 4, and related formation of a small colluvial wedge (VI). C: Deposition of horizon 3 (VII), faulting and backtilting along F2 (VIII), likely due to a gravity-graben-like feature and growth of the associated colluvial wedge with fault affecting coarser material (IX). D: Deposition of horizon 2a (X), faulting and backtilting along F3 and F4 (XI), erosion of horizon 2a, and related formation of a colluvial wedge (XII). E: Further erosion of the upper part of horizons 2a and 3 (XIII), deposition of horizon 2b (XIV), faulting along F4 (XV), erosion of horizon 2b, and related formation of a colluvial wedge (XVI). F: Faulting and dragging of coarser material along F5 (XVII), further erosion of horizon 2b, and related formation of a colluvial wedge (XVIII). G: Deposition of horizon 1 (XIX), repeated faulting events and related deposition in the fault hanging wall of horizons 1a, 1b, and 1c, reworked by agricultural activities, to reach the present-day setting (XX).

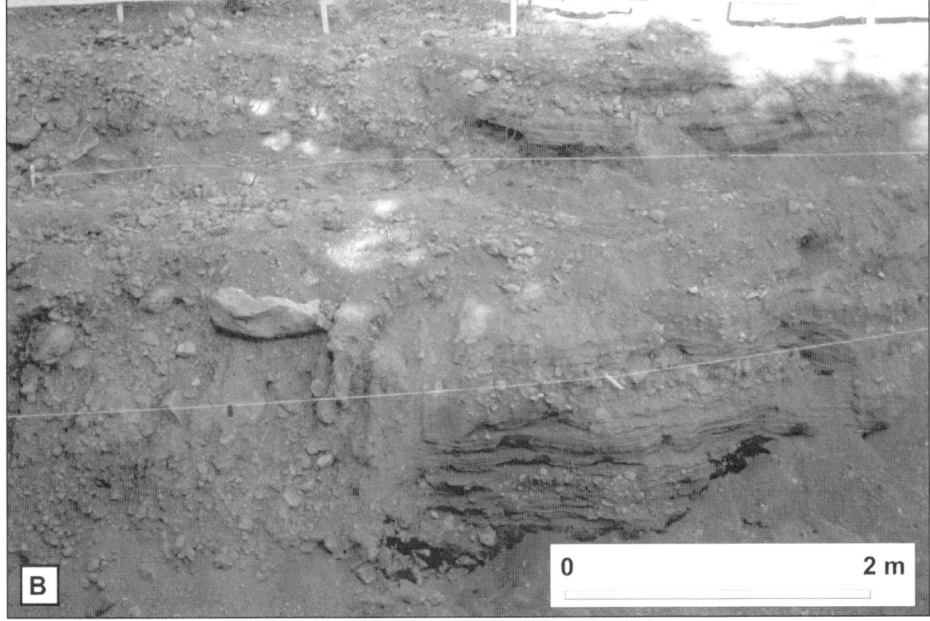

Figure 7. Trench 1 west (A) and east (B) walls at the Mandra del Re site along the Pernicana fault. See Figure 3 for trench location.

We excavated a wide exploratory trench, trending approximately north-south, in the valley floor a few meters west of the drainage outlet (Figs. 7, 8, and 9; trench 1 in Fig. 3). A smaller trench was excavated at the eastern border of the valley floor (Fig. 10; trench 2 in Fig. 3) where alluvial deposits overlie the lava flow B. As expected from the different location relative to the fault scarp, trench 1 (sited where the valley floor is right above the fault trace) cut almost the complete fault zone, while trench 2 (sited to one side of the valley floor at the foot of the 15-m-high scarp) intersected only some of the fault traces.

Trench 1 revealed a fault zone ~5 m wide with three main fault traces: F1, F2, and F3 (Figs. 7, 8, and 9). These parallel faults strike N80°; their dip is subvertical with a clear reverse component for faults F2 and F3, and downthrow is progressively toward the south. Near fault planes F2 and F3, drag zones up to 30 cm wide are indicated by well-oriented clast fabrics.

The trench walls are ~5 m apart (Fig. 8). The fault zone juxtaposes highly fractured lavas alternating with volcanic breccias (Lv), a chaotic debris flow deposit ~3 m thick, composed of lava blocks sometimes >1 m (Df), a reworked unstratified colluvial soil (Rw), and finally a subhorizontal or very gently dipping alluvial stratigraphic sequence (unit A), with a visible thickness of 5 m, composed mainly of gravel and sand channels. This unit can be further subdivided into several horizons, as shown in Figure 8.

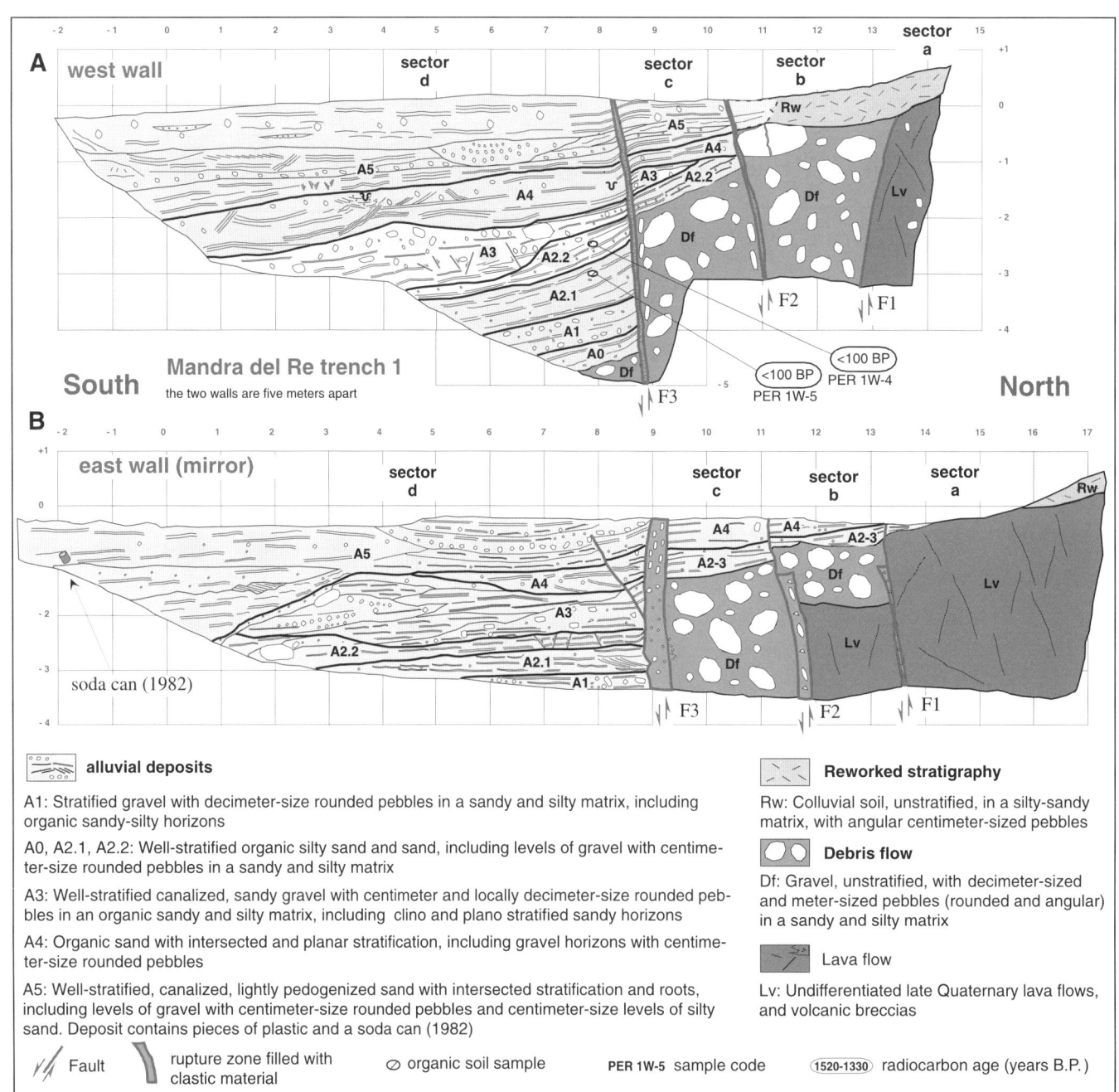

Figure 8. Logs of the Mandra del Re trench 1 west (A) and east (B; mirror image of the original log) walls. See Figure 3 for trench location on the Pernicana Fault. Reference grid is in meters.

In sectors a, b, and c, a distinction between horizons A2 and A3 was not clearly recognized; consequently, they were considered as a unique horizon, A2-3.

Two radiocarbon age determinations on samples taken in horizons A2.1 (3.2 m depth) and A2.2 (2.7 m depth) have provided ages younger than 100 yr (Fig. 8). A soda can dated 1982, found at ~70 cm depth within the lower part of horizon A5, and the above cited burial of oak trunks both confirm the high rate of present-day sedimentation. On the basis of the dated can, we can assume that the base of horizon A5 is about 20 to 30 yr old.

In the east wall, the vertical displacement generated by F3 is 150 cm for unit A2.1 (thus giving a vertical component of slip rate higher than 1.5 cm/yr) and more than 50 cm for unit A5 (vertical component slip rate higher than 2.0 cm/yr). In the same wall, F2 and F1 have produced, at the bottom of unit A2-3, displacements of 25 cm and 20 cm, respectively (vertical component of

Figure 9. Details of the Pernicana fault zone after deepening of trench 1 at Mandra del Re.

slip rate >0.45 cm/yr). These data provide a total vertical component of slip rate of >2.45 cm/yr along the east wall of trench 1.

On the west wall, the distribution of the displacement over the three fault traces is different from that of the east wall, but the general number for the cumulative vertical slip rate is confirmed (Fig. 8). Because the Pernicana fault is chiefly a strike-slip structure, the transfer of displacement from one fault strand to another seems reasonable.

Assuming an offset of ~20 to 30 m, a vertical slip rate >2.45 cm/yr implies an age younger than ca. 1.0 ka for the lava flows A and B, which seems reasonable on the basis of the fresh morphological features shown by these lava flows. Additionally, the radiocarbon dating of alluvial deposits overlying the top of lava flow B in trench 2 is in agreement with this age (Fig. 10).

It should be noted that trenches 1 and 2 both are perpendicular to the fault strike (Fig. 3). Thus, we could not evaluate the left-lateral component of slip at this site. We do know that at the Rocca Campana site, a few kilometers east of Mandra del Re, offset roads of different ages consistently show that over the past 150 yr the left-lateral slip rate has been ~2 cm/yr (e.g., Azzaro et al., 1998). Also, at Piano Pernicana, a few kilometers west of Mandra del Re, a volcanic cone dated at 13.7 ± 2.4 ka is left-laterally offset by ~370 m, yielding a Holocene slip rate of ~2.7 cm/yr (Groppelli and Tibaldi, 1999). This should be taken into account when looking at the logs in Figures 8 and 10. However, because the stratigraphy is very young, the amount of lateral slip is relatively small. For example, using the strike-slip rate seen at Rocca Campana, in trench 1 the cumulative lateral slip should be on the order of 40 to 60 cm for unit A5 and 160 to 180 cm for unit A2.

The stratigraphy exposed in trench 1 confirms that the vertical component of the fault slip rate is much greater than the erosional downcutting rate at the threshold of the Mandra del Re basin. The stratigraphic section in the downthrown side of F3, therefore, preserves a very detailed record of the sedimentary response to the faulting processes along the Mandra del Re scarp during the past few hundred years. If coseismic surface faulting had occurred, wedges of redeposited material should have been clearly recognizable in this section, especially because F3 has a reverse component, and therefore any coseismic scarp would be overhanging the valley floor. However, there is no evidence of these wedges on the downthrown side of the fault, even where the grain size of most alluvial units is sandy to silty, as seen in trench 2. Trench 2 is located to the side of the valley floor, so that the stratigraphy is finer grained than that in trench 1, allowing for finer stratigraphic resolution (Fig. 10). Within trench 1, coarse gravel alluvial deposits are included in units A3 and A1, and so sedimentary structures thinner than ~10 cm would probably not be noticeable. This implies that in trench 1 we may have missed the fossil record of some coseismic scarps up to ~20 cm high, but this is impossible in trench 2, where no sign of scarp-rim erosion and scarp-related deposits are seen.

It should also be noted that over a short distance within the same alluvial unit, there are frequent horizontal variations in grain size. For example, unit A3 is composed of coarse gravel near F3 in the trench 1 west wall, but it is composed of coarse sand near F3 in the east wall, ~5 m away. We believe, therefore, that is very unlikely that any coseismic scarp higher than a few centimeters went undetected in all the excavated trench walls, because of the fine detail in the sedimentation record.

CONCLUSIONS

The first paleoseismological investigations of active faults on the eastern flank of Etna volcano have permitted refinement of the interpretative tools for studying such a complex setting, where coseismic slip and creep can alternate in space and possibly also in time along a single fault.

We have analyzed two faults with different kinematics and slip rates – Moscarello, a normal fault with a slip rate of 1.5 to 2.0 mm/yr, and Pernicana, an oblique strike-slip fault with a vertical slip rate >2.45 cm/yr at locations characterized by different geomorphic and sedimentary environments. In addition, compar-

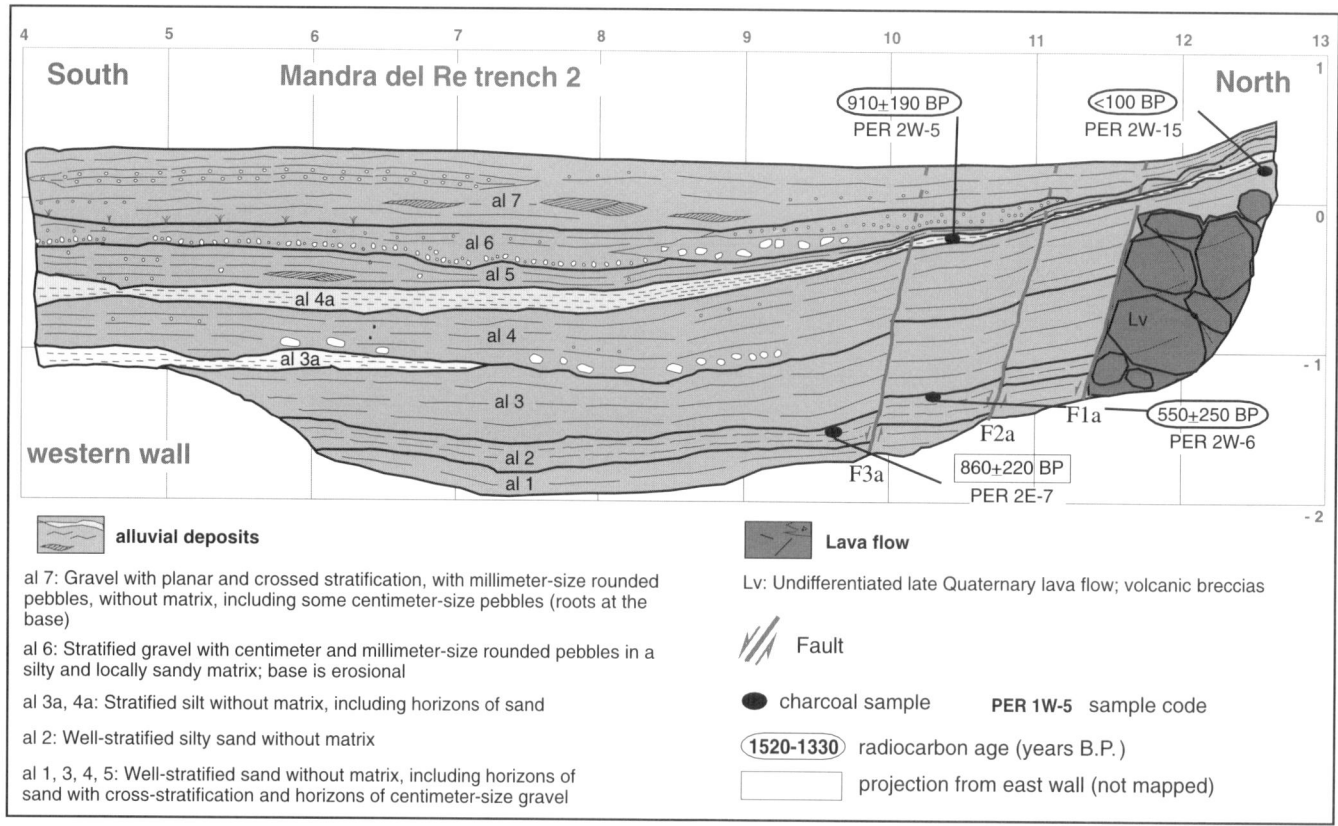

Figure 10. Log of trench 2 west wall at the Mandra del Re site. Note how the whole alluvial sequence, and in particular silt horizon 4a, is progressively dragged upward toward the north (i.e., against the fault scarp); this suggests that the main fault trace is located upslope and is not exposed in the trench.

ing fault trench stratigraphy from the Fondo Macchia (Moscarello fault) and Mandra del Re (Pernicana fault) paleoseismic sites allows us to propose some specific criteria for discriminating between coseismic and aseismic slip along capable faults.

Observations of footwall erosional surfaces at the Fondo Macchia normal fault scarp indicate near-instantaneous scarp formation during the late Holocene. This is in agreement with historical data, which record repeated events of coseismic scarp formation in the past 150 yr along the Moscarello fault, and no modern record of tectonic fault creep.

In contrast, evidence drawn from the stratigraphic logs at the Mandra del Re site shows (1) a lack of erosional surfaces that are unconformable relative to the depositional surfaces in the upthrown block, excluding those features clearly related to channel margins, and (2) the absence of colluvial wedges in the downthrown side. Sudden formation of a scarplet would have induced erosion of its rim and the development of a colluvial wedge. Instead, there is no colluvium at the foot of the fault and no upper rim erosion, even within fine-grained alluvial units. Another remarkable feature is the continuous and uniform increase in deformation and dislocation along the fault trace as progressively older sedimentary units are encountered down from the surface (Fig. 8 and Fig. 9).

This setting is only compatible with slow, aseismic slip along the fault planes. Such a hypothesis is supported by the deformation style of the sedimentary layers near the faults. The deformation pattern of the deposits (see, for example, fault F3 in the western wall, Fig. 8) suggests ductile behavior, which, considering their composition (loose sand and gravel with a poorly sorted sandy-silty matrix), is not compatible with coseismic slips. Stratigraphic features and relations like those illustrated above can be defined as "aseismites," in contrast to "seismites," which are earthquake-induced sedimentary structures (Vittori et al., 1991; McCalpin, 1996).

Aseismites such as those described for the Mandra del Re site have previously never been found in trench exposures along capable normal faults in the Apennines (e.g., Giraudi and Frezzotti, 1995; Pantosti et al., 1993, 1996; Michetti et al., 1995, 1996, 1997; Galadini et al., 1997; Galadini and Galli, 1999). This seems to us an important observation in terms of the reliability of seismic hazard assessments based on surface-faulting trench studies in Italy.

This chapter provides some criteria for recognizing the geologic signature of the aseismic formation of fault scarp. Our case history is somewhat special, because the Pernicana fault is a tectonic structure in a volcanic environment, having a very high slip rate and displacing at the surface finely stratified alluvial sediments (silt, fine sand and gravel, loose and locally without matrix), typically characterized by very brittle geotechnical behavior.

We believe, however, that these criteria, combined with the wise use of expert judgment, can also be applied in more general terms (i.e., for faults with lower slip rates and relatively less brittle sediments). This combination should permit paleoseismic analysis in assessing the nature of fault behavior on structures that we know from geological and geomorphological data are affected by recent movements, without the need for historical evidence or modern observations to establish to what extent the fault motion is due to earthquake ruptures rather than to tectonic (aseismic, preseismic or postseismic) creep.

ACKNOWLEDGMENTS

We are indebted to R. Azzaro, D. Bella, B. Antichi, F. Barbano, S. Rigano, M. Cosentino, F. Santagati, and M. Coltelli, who provided assistance in the field and discussion during interpretation of the results. We are grateful to M. Guerra and H. Stenner, who helped in the revision of the text. The Ente Parco dell'Etna kindly authorized the excavation of the trenches at Mandra del Re.

REFERENCES CITED

Azzaro, R., 1997, Seismicity and active tectonics along the Pernicana fault, Mt. Etna (Italy): Acta Vulcanologica, v. 9, no. 1, p. 7–14.

Azzaro, R., 1999, Earthquake surface faulting at Mount Etna volcano (Sicily) and implications for active tectonics: Journal of Geodynamics, v. 28, p. 193–213.

Azzaro, R., and Barbano, F., 2000, Analysis of the seismicity of southeastern Sicily: A proposed tectonic interpretation: Annali di Geofisica, v. 43, no. 1, p. 171–188.

Azzaro, R., Lo Giudice, E., and Rasà, R., 1989, Catalogo degli eventi macrosismici e delle fenomenologie da creep nell'area etnea dall'agosto 1980 al Dicembre 1989: Bollettino Gruppo Nazionale Vulcanologica, v. 1, p. 13–46.

Azzaro, R., Ferreli, L., Michetti, A.M., Serva, L., and Vittori, E., 1998, Environmental hazard of capable faults: The case of the Pernicana fault (Mt. Etna, Sicily): Natural Hazards, v. 17, p. 147–162.

Azzaro, R., Bella, D., Ferreli, L., Michetti, A.M., Santagati, F., Serva, L., and Vittori, E., 2000, First study of fault trench stratigraphy at Mt. Etna volcano, Southern Italy: Understanding Holocene surface faulting along the Moscarello fault: Journal of Geodynamics, v. 29, p. 187–210.

Bell, J.W., and Helm, D.C., 1998, Ground cracks on the Quaternary faults in Nevada: Hydraulic and tectonic, in Borchers, J.W., ed., Proceedings of the Dr. Joseph F. Poland Symposium, Land subsidence case studies and current research: Association of Engineering Geologists Special Publication 8, p. 165–173.

Borgia, A., Ferrari, L., and Pasquarè, G., 1992, Importance of gravitational spreading in the tectonic and volcanic evolution of Mount Etna: Nature, v. 357, p. 231–235.

Calvari, S., and Groppelli, G., 1996, Relevance of the Chiancone volcanoclastic deposit in the recent history of Etna Volcano (Italy): Journal of Volcanology and Geothermal Research, v. 72, p. 239–258.

Camassi, R., and Stucchi, M., eds., 1997, NT4.1, A parametric catalogue of damaging earthquakes in the Italian area (Release NT4.1.1): Milano, Gruppo Nazionale Difesa Terremoti Consiglio Nazionale delle Ricerche (GNDT–CNR), Open-File Report, 93 p., URL: http://emidius.itim.mi.cnr.it/NT/home.html.

Ferreli, L., Azzaro, R., Bella, D., Filetti, G., Michetti, A.M., Santagati, F., Serva, L., and Vittori, E., 2000, Analisi paleosismologiche ed evoluzione olocenica della fagliazione superficiale lungo la Timpa di Moscarello, Mt. Etna (Sicilia): Bollettino Società Geologica Italiana, v. 119, p. 251–265.

Galadini, F., and Galli, P., 1999, The Holocene paleoearthquakes on the 1915 Avezzano earthquake faults (central Italy): Implications for active tectonics in the central Apennines: Tectonophysics, v. 308, p. 143–170.

Galadini, F., Galli, P., and Giraudi, C., 1997, Geological investigations of Italian earthquakes: New paleoseismological data from the Fucino plain (Central Italy): Journal of Geodynamics, v. 24, no. 1-4, p. 87–103.

Giraudi, C., and Frezzotti, M., 1995, Palaeoseismicity in the Gran Sasso Massif (Abruzzo, Central Italy): Quaternary International, v. 25, p. 81–93.

Gresta, S., Bella, D., Musumeci, C., and Carveni, P., 1997, Some efforts on active faulting processes (earthquakes and aseismic creep) acting on the eastern flank of Mt. Etna: Acta Vulcanologica, v. 9, no. 1/2, p. 101–108.

Groppelli, G., and Tibaldi, A., 1999, Control of rock rheology on deformation style and slip-rate along the active Pernicana Fault, Mt. Etna, Italy: Tectonophysics, v. 305, p. 521–537.

Lanzafame, G., Neri, M., and Rust, D., 1996, Active tectonics affecting the eastern flank of Mount Etna: Structural interactions at a regional and local scale, in Gravestock, P.J., and McGuire, W.J., eds., Proceedings of the International conference: Etna: Fifteen years on: Cheltenham, U.K., p. 25–33.

Lo Giudice, E., and Rasà, R., 1986, The role of the NNW structural trend in the recent geodynamic evolution of north-eastern Sicily and its volcanic implications in the Etnean area: Journal of Geodynamics, v. 25, p. 309–330.

Lo Giudice, E., and Rasà, R., 1992, Very shallow earthquakes and brittle deformation in active volcanic areas: The Etnean region as example: Tectonophysics, v. 202, p. 257–268.

Lo Giudice, E., Patanè, G., Rasà, R., and Romano, R., 1982, The structural framework of Mount Etna: Memorie Società Geologica Italiana, v. 23, p. 125–158.

Machette, M.N., 2000, Active, capable, and potentially active faults: A paleoseismic perspective: Journal of Geodynamics, v. 29, p. 387–392.

McCalpin, J.P., ed., 1996, Paleoseismology: London, Academic Press, 583 p.

McGuire, W.J., Moss, J.L., Saunders, S.J., and Stewart, I.S., 1996, Dyke-induced rifting and edifice instability at Mount Etna, in Gravestock, P.J., and McGuire, W.J., eds., Proceedings of the International Conference, Etna: Fifteen years on: Cheltenham, U.K., p. 20–24.

McGuire, W.J., Stewart, I.S., and Saunders, S.J., 1997, Intra-volcanic rifting at Mount Etna in the context of regional tectonics: Acta Vulcanologica, v. 9, no. 1/2, p. 147–156.

Michetti, A.M., Brunamonte, F., Serva, L., and Whitney, R.A., 1995, Seismic hazard assessment from paleoseismological evidence in the Rieti region, central Italy, in Serva, L., and Slemmons, D.B., eds., Perspectives in paleoseismology: Association of Engineering Geologists Special Publication 6, p. 63–82.

Michetti, A.M., Brunamonte, F., Serva, L., and Vittori, E., 1996, Trench investigations of the 1915 Fucino earthquake fault scarps (Abruzzo, Central Italy): Geological evidence of large historical events: Journal of Geophysical Research, v. 101, no. B3, p. 5921–5936.

Michetti, A.M., Ferreli, L., Serva, L., and Vittori, E., 1997, Geological evidence for strong historical earthquakes in an "aseismic" region: The Pollino case (Southern Italy): Journal of Geodynamics, v. 24, no. 1-4, p. 67–86.

Michetti, A.M., Ferreli, L., Esposito, E., Porfido, S., Blumetti, A.M., Vittori, E., Serva, L., and Roberts, G.P., 2000, Ground effects during the September 9, 1998, Mw = 5.6, Lauria earthquake and the seismic potential of the aseismic Pollino region in Southern Italy: Seismological Research Letters, v. 71, no. 1, p. 31–46.

Monaco, C., Petronio, L., and Romanelli, M., 1995, Tettonica estensionale nel settore orientale del Monte Etna (Sicilia): Dati morfotettonici e sismici: Studi Geologici Camerti, Volume Speciale 1995/2, p. 363–374.

Pantosti D., Schwartz, D., and Valensise, G., 1993, Paleoseismology along the 1980 surface rupture of the Irpinia Fault: Implications for earthquake recurrence in the Southern Apennines, Italy: Journal of Geophysical Research, v. 98, no. B4, p. 6561–6577.

Pantosti D., D'Addezio, G., and Cinti, F.R., 1996, Paleoseismicity of the Ovindoli-Pezza fault, Central Apennines, Italy: A history including a large previously unrecorded earthquake in Middle Ages (890–1300): Journal of Geophysical Research, v. 101, no. B3, p. 5937–5959.

Pavlides, S., 1996, First palaeoseismological results from Greece: Annali di Geofisica, v. 39, no. 3, p. 545–555.

Rasà, R., Azzaro, R., and Leonardi, O., 1996, Aseismic creep on faults and flank instability at Mt. Etna volcano, Sicily, *in* McGuire, W.C., Jones, A.P., and Neuberg, J., eds., Volcano instability on the earth and other planets: Geological Society [London] Special Publication 110, p. 179–192.

Riuscetti, M., and Distefano, R., 1971, Il terremoto di Macchia (Catania): Bollettino Geofisica Teorica Applicata, v. 13, no. 51, p. 150–164.

Scandone, P., Patacca, E., Radoicic, R., Ryan, W.B.F., Cita, M.B., Rawson, M., Chezar, H., Miller, E., Mckenzie, J., and Rossi, S., 1981, Mesozoic and Cenozoic rocks from Malta Escarpment (Central Mediterranean): American Assoication of Petroleum Geologists, AAPG Bulletin, v. 65, p. 1299–1319.

Sieh, K.E., and Williams, P.L., 1990, Behavior of the southernmost San Andreas Fault during the past 300 years: Journal of Geophysical Research, v. 95, no. B5, p. 6629–6645.

Smith, R.P., Jackson, S.M., and Hackett, W.R., 1996, Paleoseismology and seismic hazards evaluations in extensional volcanic terrains: Journal of Geophysical Research, v. 101, no. B3, p. 6277–6292.

Stenner, H.D., and Ueta, K., 2000, Looking for evidence of large surface rupturing events on the rapidly creeping Southern Calaveras fault, California, *in* Active Fault Research for the New Millennium, Proceedings of the Hokudan International Symposium and School on Active Faulting: Hokudan-cho, Awaji Island, Japan, p. 479–486.

Villamor, P., 2000, Characterisation of normal faults in a volcanic area, the Taupo volcanic zone, New Zealand, *in* Active Fault Research for the New Millennium, Proceedings of the Hokudan International Symposium and School on Active Faulting: Hokudan-cho, Awaji Island, Japan, p. 537–541.

Vittori, E., Serva, L., and Sylos Labini, S., 1991, Palaeoseismology: Review of the state-of-the-art: Tectonophysics, v. 193, p. 9–32.

MANUSCRIPT ACCEPTED BY THE SOCIETY MAY 11, 2001

Geological Society of America
Special Paper 359
2002

Mid-Tertiary paleoseismites: Syndepositional features and section restoration used to indicate paleoseismicity, Atlantic Coastal Plain, South Carolina and Georgia

Mervin J. Bartholomew*
Earth Sciences & Resources Institute, School of the Environment, University of South Carolina, Columbia, South Carolina 29208, USA

Brendan M. Brodie
EnviroSouth, Suite 1, 104 Mauldin Road, Greenville, South Carolina 29605, USA

Ralph H. Willoughby
South Carolina Geological Survey, 5 Geology Road, Columbia, South Carolina 29212, USA

Sharon E. Lewis
Earth Sciences & Resources Institute, School of the Environment, University of South Carolina, Columbia, South Carolina 29208, USA

Frank H. Syms
Bechtel Savannah River, Inc., Savannah River Site, Aiken, South Carolina 29808, USA

ABSTRACT

Synclinal troughs that served as depositional channels adjacent to small reverse faults are examples of paleoseismites. Section restoration on one syncline demonstrates progressive fault growth, during three seismic events with 1.5 m of relative uplift. Normal faults, which are bent or offset, preceded reverse faults. Faults are truncated by unconformities progressively upward through the exposed stratigraphic section. One *Ophiomorpha nodosa* burrow cuts a reverse fault, indicating that faulting occurred within a few meters of the ground surface in a marine environment. Near-surface, small-scale reverse faults on the flanks of the second syncline accompanied folding, and record two events with ~3 m of relative uplift. At a third locality, 21 near-vertical clastic dikes are subparallel to nearby joints and small normal faults. They are 2 to 10 cm wide, 1 to 2 m long, 1 to 2 m high, relatively planar with parallel walls and sharp contacts, and filled with medium- to coarse-grained quartz sand. The dikes cut perpendicular to bedding across sandy clay beds containing *Ophiomorpha nodosa* burrows indicating their formation within a few meters of the ground suface in a marine environment. Collectively, extensional structures here are interpreted to have developed across a broad fold above the Pen Branch fault.

The stratigraphic and geographic association of these features suggests that all are secondary structural features, interpreted herein as paleoseismites, that accompanied major paleo-earthquakes on nearby large reactivated faults like the Pen Branch fault during late Eocene to late Miocene tectonism.

*E-mail: jbarth@esri.esri.sc.edu.

Bartholomew, M.J., Brodie, B.M., Willoughby, R.H., Lewis, S.E., and Syms, F.H., 2002, Mid-Tertiary paleoseismites: Syndepositional features and section restoration used to indicate paleoseismicity, Atlantic Coastal Plain, South Carolina and Georgia, *in* Ettensohn, F.R., Rast, N., and Brett, C.E., eds., Ancient seismites: Boulder, Colorado, Geological Society of America Special Paper 359, p. 63–74.

INTRODUCTION

Passive margins, such as the Atlantic Coastal Plain (Sheridan, 1989), are not generally seismically active along their entire lengths. However, the 1886 Charleston, South Carolina, earthquake (Fig. 1A) was estimated to have had a modified Mercalli intensity of X (Dutton, 1889; Bollinger, 1977), a magnitude of about 7 (Bollinger, 1977), and a moment magnitude of 7.3 (Johnston, 1996) and signifies a nearby major seismogenic zone beneath this passive margin. Numerous sand blows were associated with this earthquake (Dutton, 1889), and subsequent work on paleo–sand blows (e.g., Amick and Gelinas, 1991; Obermeier et al., 1990) demonstrated that the seismic zone near Charleston (Tarr, 1977; Talwani, 1982; Tarr and Rhea, 1983; Madabhushi and Talwani, 1993) has repeatedly undergone strong earthquakes during late Quaternary time. However, because no fault scarp was observed and the sand blows occurred over a large area (Dutton, 1889), the fault that generated the 1886 earthquake is unknown. Possible causative faults (e.g., Ashley River and Woodstock faults) have been inferred from later seismicity (e.g., Tarr, 1977; Talwani, 1982; Tarr and Rhea, 1983; Madabhushi and Talwani, 1993).

Obviously, the lower Coastal Plain near Charleston was affected by large earthquakes within the modern stress field. Both historic earthquakes, other than those associated with Charleston (e.g., Talwani and Rajendran, 1991; Domoracki et al., 1999), and late Pleistocene–Holocene fracture sets (Bartholomew et al., 2000) in the coastal area southwestward from Charleston into Georgia also reflect the modern stress field. Bartholomew et al. (2000) documented that Holocene to late Pleistocene orthogonal joint sets have formed in strata related to the Pamlico, Princess Anne, and Silver Bluff shorelines of the lower Coastal Plain (Fig. 1A); they range in age from about 130 to about 33 k.y. (Willoughby et al., 1999). These orthogonal joints reflect extension in near-surface strata. Because the joints cut through beds, they are potential conduits for injection of clastic dikes during earthquakes as well as for fluids during burial and diagenesis. Similarly, fracture sets in older strata of the middle and upper Coastal Plain (Fig. 1A) might also have served as conduits for injection of clastic dikes during paleoseismic events.

Is there, however, evidence of earlier, major paleoseismic events during the Tertiary that are not related to the Charleston seismic zone and the modern stress field? In particular, the Atlantic Coastal Plain, northwest of the Orangeburg scarp (Fig. 1A), which is the boundary between the middle and upper Coastal Plain in South Carolina, was uplifted sufficiently during late Eocene to late Miocene time that it has since remained above sea level regardless of sea-level fluctuations. On the basis of thickness changes in the middle to upper Eocene section near the Garner-Edisto fault in South Carolina (e.g., section E–E′ of Colquhoun et al., 1983), this major uplift began before the Oligocene Epoch (D.J. Colquhoun, 1997, personal commun.). Thus, the late Eocene appears to have been a time when stronger tectonism affected this passive margin; therefore, strata of this age

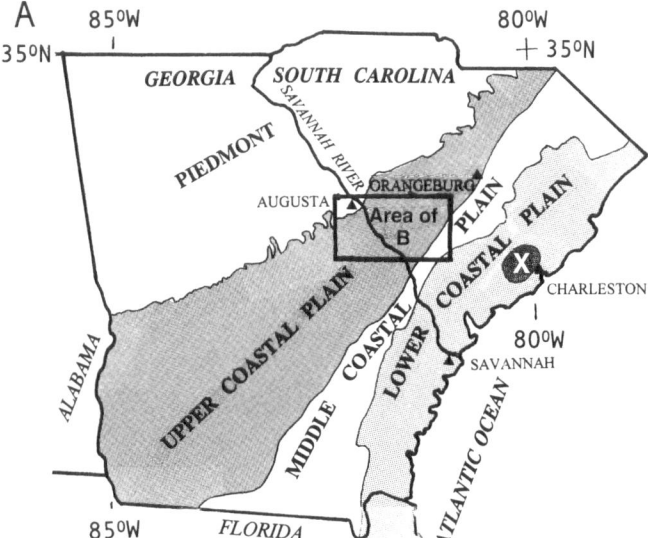

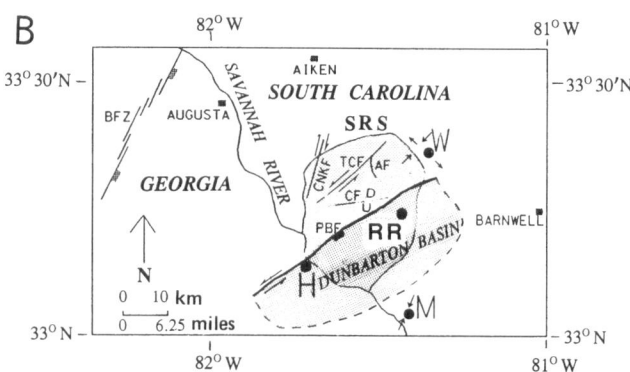

Figure 1. (A) Location map of the study area (rectangle is area of B) relative to the subdivisions of the Atlantic Coastal Plain, the Orangeburg scarp, and the epicenter of the 1886 Charleston, South Carolina, earthquake (dark shaded area with X indicates the area of modified Mercalli intensity of X; Dutton, 1889; Bollinger, 1977). (B) H—Hancock Landing, W—Williston, M—Martin, and RR—SRS (Savannah River Site) railroad relative to the Belair fault zone (BFZ) of Prowell and O'Connor (1978) and Bramlett et al. (1982) and selected faults (PBF—Pen Branch fault; CNKF—Crackerneck fault; TCF—Tinker Creek fault; AF—Atta fault; CF—Crouch Branch fault) identified beneath the U.S. Department of Energy Savannah River Site (SRS) (Cumbest et al., 1992; Domoracki, 1995; Snipes et al., 1993a, 1993b; Fallaw and Price, 1995; Stieve and Stevenson, 1995). Arrows indicate the main extension and compression directions associated with the principal features at the Williston and Martin localities. Strike-slip arrows on SRS faults show probable slip directions associated with the Williston and Martin phases of late Eocene to late Miocene faulting.

are good candidates for recording evidence of paleoseismicity related to near-surface faulting at that time.

Major faults are not commonly exposed in the Atlantic Coastal Plain. However, Howell and Zupan (1975), Inden and Zupan (1975), Zupan and Abbott (1975), Prowell and O'Conner (1978), Prowell (1980, 1983, 1988, 1989a, 1989b), Colquhoun et al. (1982), and McDowell and Houser (1983) all discussed exposures with faulted Coastal Plain sediments, and Secor et al. (1998) demonstrated offset strata of the upper Coastal Plain of South Carolina. More commonly, the tectonic history of the Coastal Plain has been inferred from drill-hole data or seismic lines (e.g., Colquhoun et al., 1983; Snipes et al., 1993a, 1993b, 1995; Aadland et al., 1995; Domeracki, 1995; Domeracki et al., 1999). Cretaceous to Eocene syndepositional faulting was previously inferred from subsurface data along the Belair fault in Georgia (Fig. 1B) (e.g., Prowell and O'Conner, 1978; Bramlette et al., 1982), along some faults in South Carolina (e.g., Colquhoun et al., 1983), and, in particular, along major faults at the U.S. Department of Energy Savannah River Site (SRS in Fig. 1B). At the SRS (Fig. 1B), the Pen Branch faults bounds the northern margin of the buried Triassic Dunbarton basin. The Pen Branch fault was reactivated and underwent syndepositional faulting during the late Eocene (e.g., Aadland et al., 1995). The buried Martin fault, which is along or near the southern margin of the Dunbarton basin, was reactivated after deposition of the Altamaha Formation between late middle to early late Miocene time (Snipes et al., 1993a, 1993b; Nystrom and Willoughby, 1992; Nystrom, 1998). These interpretations of displacement on the Pen Branch and Martin faults are based on changes in both thickness and elevation of strata across these faults. Generally, this comparison of subsurface thickness changes in cross sections is based on an assumption of dip-slip motion on the faults in question. If a significant component of strike-slip motion characterized a fault at any time during its movement history, it is essential to know both the vector of displacement along the fault and the timing of that episode of movement characterized by a particular stress field (e.g., Bartholomew et al., 1997a; Brodie and Bartholomew, 1997). This structural information is needed before either isopach-thickness or structural-contour changes across that fault are used to infer syndepositional faulting. Thus, documentation of syndepositional paleoseismic indicators (e.g., paleoseismites) in roadside exposures has important implications for proper interpretation of subsurface syndepositional faulting near the Savannah River Site and elsewhere in the Atlantic Coastal Plain.

Four exposures (Williston, Martin, Railroad, and Hancock Landing; Fig. 1) are used here to document that strong paleoearthquakes, probably associated with late Eocene to late Miocene oblique-slip reactivation of the Triassic, basin-bounding Pen Branch fault or the nearby Crackerneck fault (Fig. 1B), produced paleoseismites—that is, soft-sediment deformational features related to tectonic causes (Seilacher, 1969, 1984). These mid-Tertiary paleoseismites include both (1) small syndepositional folds with local unconformities, orthogonal sets of normal faults, and a colluvial wedge or slump along small near-surface, reverse faults, indicating both late Eocene and middle to late Miocene paleoseismicity and (2) syndepositional near-surface, marine clastic dikes associated with a subparallel joint set and strike-parallel small normal faults that are also consistent with late Eocene paleoseismicity. The synclinal troughs, which formed adjacent to these reverse fault scarps, then served as depositional channels. Faulting continued during filling of these channels.

Exposures, which were sketched, were marked with a leveled 5 ft (1.5 m) grid. A 5 ft (1.5 m) square, marked with 1 ft (30.5 cm) intervals, was then placed over the grid. Within the 1 ft by 1 ft (30.5 cm by 30.5 cm) squares, small features were located using 1 ft (30.5 cm) scales with 1 in. (2.5 cm) divisions for greater accuracy. Thus, the accuracy of the sketches is about 1 in. (2.5 cm). Photographs were taken of cleaned and scraped exposures after the grid-marks were placed on the exposure for reference. Measurements of strike, dip, and pitch follow the right-hand convention: strike is measured from 1° to 360°, with the plane dipping to the right, dip is perpendicular to strike, and pitch is measured from 1° to 180° on the plane from the strike direction.

UPPER EOCENE SYNDEPOSITIONAL STRUCTURES

The exposure near Williston, South Carolina (Figs. 1 and 2), is in a roadcut through the upper Eocene Tobacco Road Sand. The principal feature at this exposure is a broad syncline (Colquhoun et al., 1982), which is faulted along the northwest flank (McDowell and Houser, 1983; Willoughby and Nystrom, 1994). Poles to bedding (Fig. 3A) in both the lower clay-rich part and the upper sand part of the section define a fold axis of ~145°, ~12°. This fold trend is parallel to regional folds such as the northwest-trending, late Eocene South Georgia embayment (Galloway et al., 1991) and thus is an integral part of the regional tectonic picture. The main reverse fault at this exposure has a maximum offset of ~0.3 m of the lower beds, and offset decreases upward to ~0.2 m in the upper beds. Measured strikes, dips, and pitches (of slickensides) of the main fault are 138°, 085°, 088° and 149°, 083°, 094° (Fig. 3B). The normal faults primarily are part of two sets (Fig. 3C). The northwest-southeast–striking set is strike parallel to the main reverse fault, and the northeast-southwest–striking faults are orthogonal to it. Both sets of normal faults are primarily dip slip. At least some of the strike-parallel normal fault set preceded the orthogonal set, as indicated by oblique movement on the earlier set consistent with dip slip on the later set. At point Z (Fig. 2) normal faults are bent or offset by reverse faults (Fig. 7C). At point X (Figs. 2 and 4) a clay-lined *Ophiomorpha nodosa* burrow cuts across the trace of a reverse fault. The more clay-rich lower beds (units E, F, G, and H; Fig. 2) are laterally of uniform thickness and lithology and contain numerous *Ophiomorpha nodosa* burrows. These burrows are inferred, from comparisons with modern and Quaternary ghost shrimp, to have been dug underwater by ghost shrimp at depths of 3–5 m below the ground surface (Bishop and Brannen, 1993). By contrast, the conglomeratic cross-bedded upper sands (units A, B, C, and D; Fig. 2) exhibit lateral variations in both thickness and lithology between the trough of the syncline and the faulted northwest flank of the fold.

The outcrop sketch (Fig. 2) was retrodeformed by line-length balancing in order to better illustrate the local interaction of sedimentation and tectonics at this locality (Fig. 5). The retrodeformed section demonstrates that at least three faulting or folding events are associated with the main reverse fault zone (Fig. 5). The youngest reverse event is reverse faulting that post-

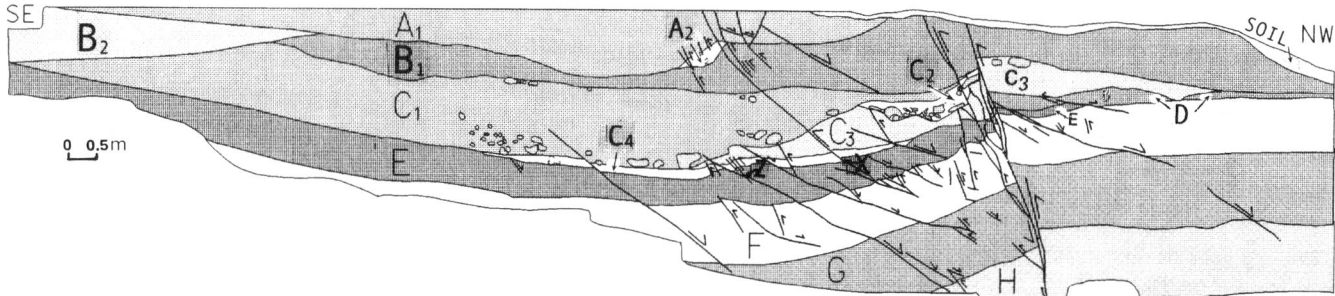

Figure 2. Sketch of exposure near Williston, South Carolina, showing syncline associated with syndepositional faults in upper Eocene shallow marine strata of the Tobacco Road Sand. Subscript numbers indicate subdivisions of units A, B, and C. The basal boundaries of units A, B, C, and D are erosional unconformities. X marks the location of an *Ophiomorpha nodosa* burrow cutting a reverse fault. Z marks the location of a normal fault cut by a reverse fault. Unit A_1—extensively cross-bedded, medium reddish brown, conglomeratic (clay and pebbles) sandstone interbedded with well-bedded coarse-grained sandstone. Unit A_2—well-laminated light grayish brown sandstone. Unit B1—well-bedded to cross-bedded, reddish brown, coarse-grained sandstone with sandy conglomerate at the base northwest of the main fault and at the southeastern end. Unit B_2—well-bedded, reddish brown, coarse-grained sandstone. Unit C_1—extensively cross-bedded, medium reddish brown, conglomeratic (clay and pebbles) sandstone graded upward from coarse to fine; basal bed is sandy conglomerate with large clasts of claystone and sandstone. Unit C_2—massive, unsorted medium- to coarse-grained reddish brown sandstone with abundant subangular clay clasts; upper 10 cm on southeastern side is laminated coarse-grained sandstone. Unit C_3—well-bedded, medium-grained sandstone with clay and pebble conglomerate at base and with interbedded lenses of coarse-grained sandstone, conglomeratic near main fault. Unit C_4—laminated, grayish red, fine-grained sandstone with thin beds of medium red claystone. Unit D—well-bedded, grayish brown, medium-grained sandstone. Unit E—burrowed, light reddish gray, fine-grained clayey sandstone. Unit F—well-bedded, medium gray, sandy claystone with thin beds of medium red claystone. Unit G—extensively burrowed, massive-bedded, light reddish gray, medium-grained, clayey sandstone. Unit H—burrowed interbedded, light pinkish gray, fine-grained sandstone and medium gray, sandy claystone with thin beds of medium grayish red claystone.

dated deposition of unit B (and probably unit A) and is shown by removal of displacement on faults cuttings units A and B (outcrop to stage 1 in Fig. 5). The younger reverse faults (Z in Fig. 2) are subparallel to the main fault here, but they are parallel to reverse faults at the Martin, South Carolina, locality (Fig. 1), discussed below, and the displacement vector (Fig. 7C) is more consistent with the Martin faults. The main fault zone has 0.2 m of displacement of the basal contact of unit B. The channel for unit A lies in the syncline (stage 1 in Fig. 5). The basal contact of unit B cuts unconformably down section from unit C_1 in the synclinal trough northwestward across the main fault to unit D on the uplift (stage 1 to stage 2). However, if the base of unit B is restored across the fault, the bases of units E, F, and G still are offset approximately 0.1 m and unit C_3 is arched slightly above the main fault (stage 1 in Fig. 5).

The middle event is also reverse faulting that postdated deposition of units C and D and caused the 0.1-m offset of units E, F, and G (stage 1 to stage 2). Reverse faults, which are strike-parallel to the main reverse fault here, are also present at both the Hancock Landing, Georgia, and the SRS, South Carolina, railroad localities (Fig. 3, D and E), discussed below. The basal contacts of units C and D also cut unconformably down section from unit E in the synclinal trough northwestward across the main fault to unit F on the uplift (stage 3). In particular, unit C thins northwestward and pinches out from the axial part of the syncline across the main fault to the uplift (stage 2 in Fig. 5). Within this sand, one lens of sandy breccia (unit C_2 in Fig. 2) is present only on the downthrown side of the main fault. This lens consists of unsorted angular clay and sandstone clasts with an unsorted sand matrix. Toward the synclinal trough, the upper 10 cm of this lens is thinly laminated. The composition and location of this lens suggest that it is either a small slump or a colluvial wedge formed adjacent to the scarp on the downthrown side of the main fault. The upper part of C_2 was then slightly modified by fluvial processes.

The earliest event is folding accompanied by normal faulting. Displacement along the normal faults was removed separately (stage 2 to stage 3 in Fig. 5) from unfolding the fold (stage 3 to stage 4); however, the normal faults are related to extension across the fold arch and thus formed during or slightly after the fold. If the base of unit E is restored to horizontal by retrodeforming this fold (stage 3 to stage 4), then the bases of units F and G are also horizontal, and their horizontality, combined with a lack of lateral changes in either thickness or lithology, suggests that units E, F, G, and H were not affected by a tectonic event during their deposition. Thus, the earliest event postdated deposition of unit E, which pinches out northwestward onto the uplift due to erosion (stage 3 to stage 4).

Willoughby and Clendenin (1995) suggested that the large channel at this locality was controlled by syndepositional, listric normal faults, whereas McDowell and Houser (1983) believed that this syncline was associated with the high-angle reverse fault. The small-scale normal faults here are a consequence of the folding, not the cause of it (stages 2, 3, and 4 in Fig. 5). The parallelism of the synclinal axis, as defined by bedding (Fig. 3A), and the strike of the reverse fault (Fig. 3B) confirm the interpretation of McDowell and Houser (1983). Slickensides with pitches near 90° on the main fault indicate that the fault is a dip-slip fault, not a strike-slip fault as interpreted by Willoughby and Clendenin (1995).

Collectively the features at this locality are interpreted as seismites for the following reasons.

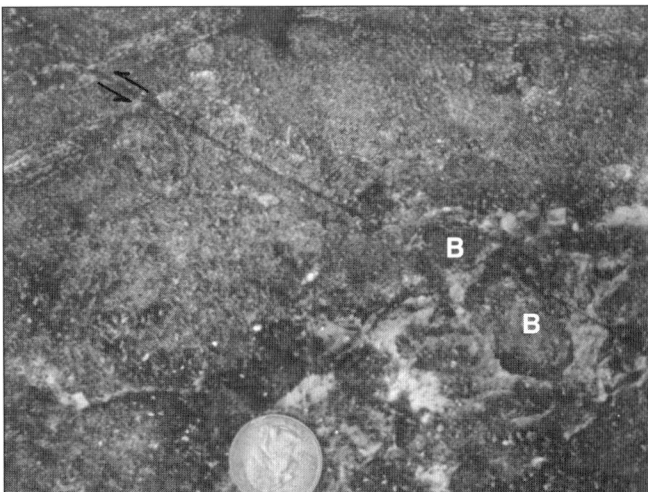

Figure 4. *Ophiomorpha nodosa* burrow (B) cutting reverse fault (point X in Fig. 2) at Williston, South Carolina (W in Fig. 1B); arrows indicate reverse 2-cm offset of 1-cm-thick bed. Coin diameter is 2.25 mm.

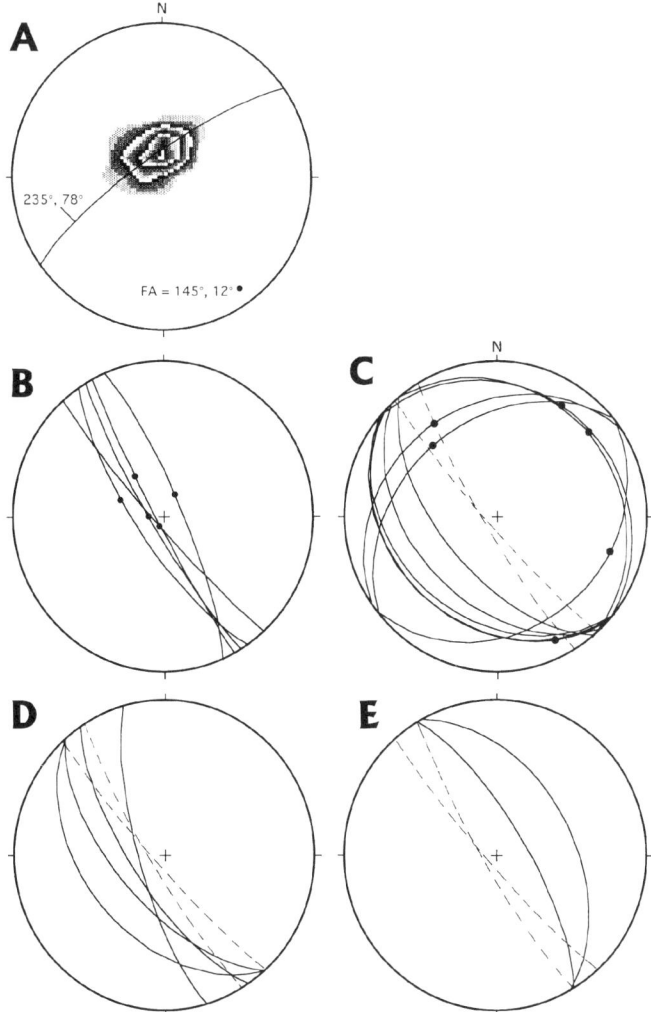

Figure 3. Structural data from bedding, northwest-southeast–striking reverse faults, and orthogonal sets of normal faults in exposure near Williston, South Carolina (W in Fig. 1B), and northwest-southeast–striking reverse faults at other locations. A: 79 poles to bedding (1% area contour) define fold axis (FA) as pole to best fit curve. B: Five northwest-southeast–striking, steeply dipping, dip-slip, reverse faults with slip vectors; two nearest the center are on the main fault. C: Ten gently dipping, dip-slip normal faults defining two sets of faults that are strike parallel and orthogonal to the main reverse fault (see 3B). D: Four southeast-striking reverse faults at the SRS (Savannah River Site) railroad locality that are subparallel to the main reverse fault set at Williston. E: Two northwest-striking, reverse faults at Hancock Landing, Georgia, that are subparallel to the main reverse fault set at Williston, South Carolina.

1. The presence of numerous *Ophiomorpha nodosa* burrows in units E, F, G, and H indicates a shallow-marine, intertidal depositional environment at the time that the earliest event occurred.

2. The retrodeformed sections (stage 1 to 4) show that the depositional-erosional surface remained approximately horizontal during the three events, indicating that the folding and faulting affected the ground surface, deposition occurring in the synclinal trough and erosion occurring on the uplifted northwest flank.

3. Some faults are truncated by unconformities at the bases of C_1, C_2, and C_3 (Fig. 2), indicating exposure at the ground surface shortly after their formation.

4. An *Ophiomorpha nodosa* burrow cuts a reverse fault (X in Fig. 2). Thus, this fault developed at a very shallow depth (~5 m or less) in an intertidal environment.

5. Some reverse faults cut or bend normal faults (as at Z in Fig. 2) or reactivated them as reverse faults, indicating that normal faulting preceded reverse faulting in both the lower (units C to H) and upper (units A and B) parts of the section.

6. Displacement on the main reverse fault is different in units A and B than in the older units (C through H), indicating that an event occurred after deposition of unit C but before deposition of unit B.

7. The presence of the slump or colluvial wedge (C_2) adjacent to the main fault is an indication that faulting was very near the ground surface and either produced a scarp or tilted the strata sufficiently to precipitate a slump. In any event, the tilting produced the more steeply dipping fold limb at the ground surface that would have appeared as a linear surface feature or scarp at that time.

MIDDLE TO UPPER MIOCENE SYNDEPOSITIONAL FEATURES

The exposure (Figs. 1 and 6) near Martin, South Carolina, is in a railroad cut through the upper Eocene Tobacco Road Sand (units C, D, and E), which locally contains lenses of silica cemented, leached, formerly calcareous sandstone and which is overlain by a nonmarine unit (A and B) interpreted as Altamaha Formation (Huddlestun, 1988) of late-middle to early-late Miocene age (Nystrom and Willoughby, 1992; Nystrom, 1998; Willoughby, 2000). The principal feature here is a small syncline that is faulted along the flanks and filled with a channel sand (unit B) of the nonmarine

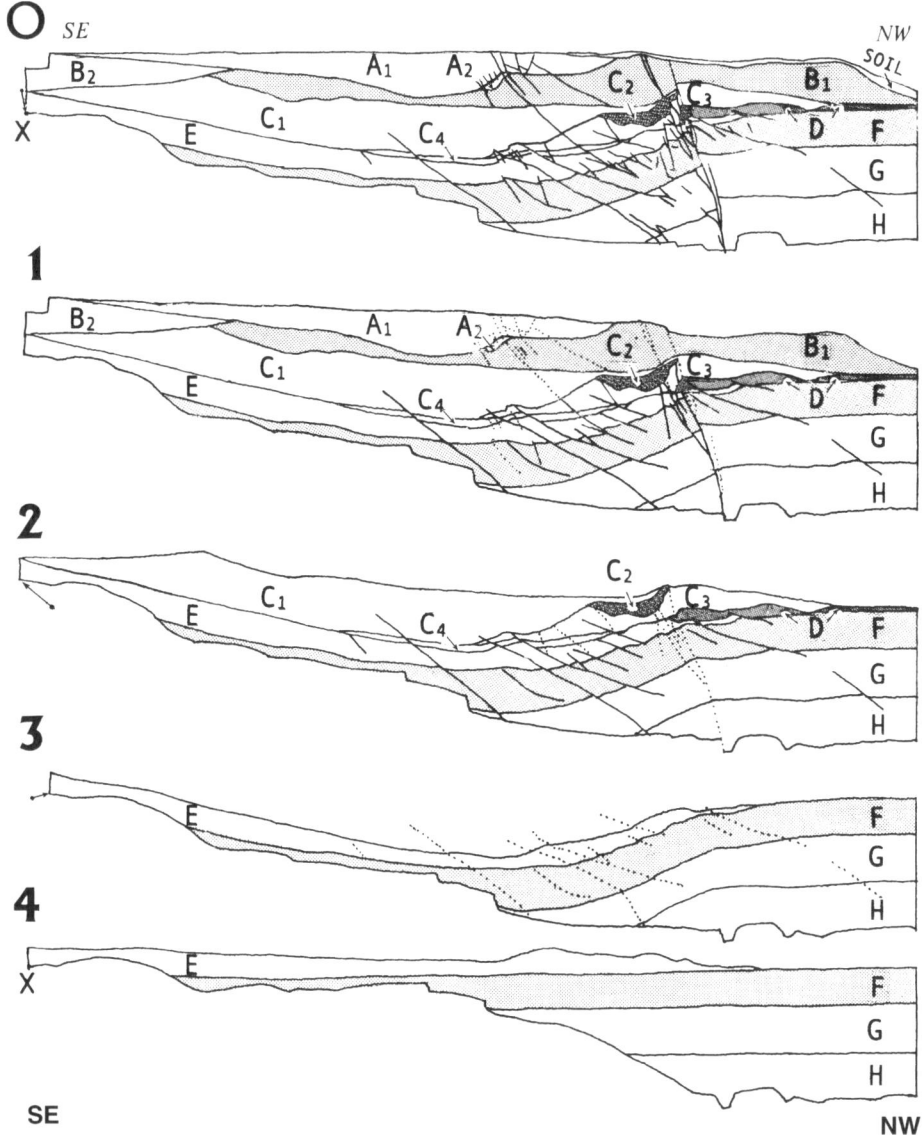

Figure 5. Line-length balancing and progressive restoration of sketch of exposure at Williston, South Carolina (Fig. 2), with right-hand edge fixed as a pin line. Units are designated with the same letters and subscript numbers as in Figure 2. O—outcrop from Figure 2. Stage 1—removal of displacement along faults offsetting units A and B. Stage 2—removal of remainder of displacement along reverse faults offsetting units C to H. Stage 3—removal of displacement along normal faults offsetting units C to H. Stage 4—removal of displacement due to folding and restoration of units E to H to horizontal. Arrows show progressive changes in the location of the lower left corner of the exposure relative to the pin line (right-hand edge) from the restored original location (X) to the present-day location in the outcrop (X′ in O); 2.3% contraction due to folding (stage 4 to stage 3); 3.6% extension due to normal faulting (stage 3 to stage 2); 0.5% contraction due to reverse faulting (stage 2 to stage 1); 0.3% contraction due to reverse faulting (stage 1 to outcrop).

unit. Reverse faults, each having a few centimeters of apparent offset, are concentrated along both flanks (at Z and X) of the small syncline; fewer faults are present beneath the trough. Slickensides were not observed in the loose sand that makes up unit D. Several small reverse faults also occur south of this exposure across a much larger, broad anticline, but none were observed in the horizontal strata north of this exposure. All of the small reverse faults on the southern limb of the syncline either terminate upward at the thin (5–15-cm-thick) clay bed (C_2) or die out in the sand below this clay bed. On the northern limb of the syncline, most of the small reverse faults also terminate at or below the clay bed; however, one fault (Z in Fig. 6) does offset this clay bed. This fault had ~20 cm of apparent reverse offset, the largest amount of offset observed on any of these faults. A plot of poles to bedding (Fig. 7A) measured below this clay bed shows that fold-axis orientation is very consistent with the intersection of the average reverse fault surfaces (Fig. 7B). Offset of the clay bed by the largest fault indicates that the faulting postdated unit C_2 and occurred either during or after deposition of unit C_1. Termination of smaller faults at the clay is probably related to the greater strength of the clay compared to the uncemented sand. The thickness of units C, D, and E appears to reflect only uniform depositional changes across an area much larger than the syncline. By contrast, the synclinal sag, formed by the reverse faults, is the site of localized deposition of a lens of conglomeratic sand (unit B). The thickness of this sand lens approximates the cumulative offset of the reverse faults along either flank of the syncline. Thus, the timing of the faulting here coincides with the rapid transition from marine to nonmarine deposition during middle to late Miocene time. Reverse faults that are subparallel to the Martin set are present at Williston, Hancock Landing, and the SRS railroad locality (Fig. 7, C, D, and E).

UPPER EOCENE CLASTIC DIKES

Near Hancock Landing, Georgia (Fig. 1), the upper Eocene Tobacco Road Sand is exposed in a roadcut. The strata here are lithologically similar to units E through H at the Williston locality (Fig. 2). Moreover, *Ophiomorpha nodosa* burrows are also abundant in this exposure and, as in the basal beds at Williston, indicate a shallow marine, intertidal depositional environment. The principal features that may be indicative of paleoseismicity here are 21 clastic dikes (Fig. 8A and B) associated with a set of north-striking joints and small-scale normal faults that cut west-dipping strata. The term "clastic dike" has commonly been used for several decades in the southeast United States in reference to mottled clay features, which are probably related to weathering along joints, faults, and other features. We emphasize that this is not how we are applying the term herein. The clastic dikes that we describe differ from the enveloping strata both in original clastic composition and in texture, but not in the degree of weathering.

The 21 clastic dikes are filled with massive, medium- to coarse-grained sand and cut across poorly bedded, burrowed, clay-rich strata approximately perpendicular to bedding. These relatively planar clastic dikes are 2–15 cm wide, at least 1–2 m long, and at least 1–2 m high (Fig. 8A). The clastic dikes each have subparallel walls, and the contacts with the enveloping strata are sharp (Fig. 8, A and B); therefore, each clastic dike is very uniform in width over its observed length and height of several meters. Furthermore, *Ophiomorpha nodosa* burrows both predate and postdate the clastic dikes, indicating that dike injection occurred near the underwater ground surface (< 5 m) in a marine environment.

The planar geometry and massive, uniform sand fillings of the clastic dikes are consistent with injection along preexisting joints and derivation from an underlying sand source bed. However, the source of the sand is unknown because the bases of the clastic dikes are not exposed in the road ditch and the tops are typically truncated by soil at the ground surface. Thus, the distinction of these clastic dikes as tectonic features rather than mud cracks or desiccation cracks must be inferred indirectly. Features such as very deep mud cracks, desiccation cracks, or paleosols, however, can probably be precluded, inasmuch as *Ophiomorpha nodosa* burrows both predate and postdate the dikes, suggesting that these beds were not exposed to subareal weathering.

At the Hancock Landing exposure, the largest normal fault, with about 0.5 m of offset, is oriented ~010°, ~075°, ~090° (Fig. 9C). No features suggestive of syndepositional displacement (e.g., colluvial wedge, fold, unconformities) were observed either adjacent to this fault or near small reverse faults (Fig. 9D) with less offset. The reverse faults here are parallel to reverse faults at the other localities (Figs. 3E and 7D). The 21 clastic dikes here are all within 2–10 m laterally of this small normal fault and are parallel to the dominant, north-south–striking joint set in this exposure (Fig. 9B). Regionally, Brodie and Bartholomew (1997)

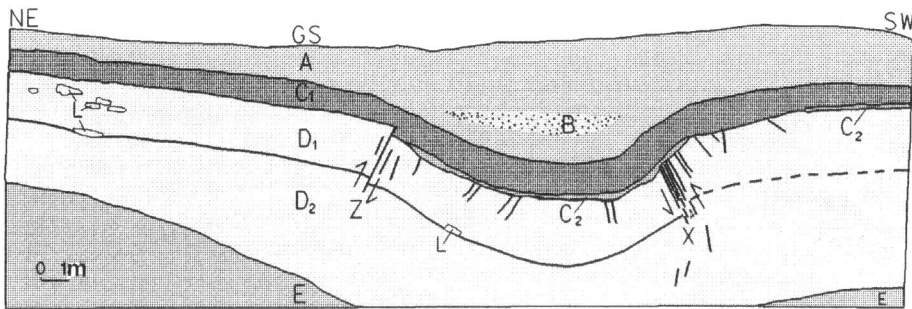

Figure 6. Sketch of exposure near Martin, South Carolina (M in Fig. 1B), showing syncline associated with reverse faults in upper Eocene marine strata of the Tobacco Road Sand overlain by nonmarine, middle to upper Miocene Altamaha Formation. Unit A—light reddish gray, clayey sandstone. Unit B—yellowish gray, sandy pebble conglomerate. Unit C_1—light reddish gray, sandy claystone. Unit C_2—greenish gray claystone; unit D_1—brownish red, well-bedded medium-grained sandstone with lenses (L) of limestone. Unit D_2—brownish red, well-bedded, medium-grained sandstone. Unit E—reddish brown, well-bedded medium-grained sandstone.

70 M.J. Bartholomew et al.

commonly observed strike-parallel joints near normal faults of this scale; thus, we infer that the clastic dikes filled some joints here.

Tectonically induced clastic dikes and associated sand blows are commonly associated with large earthquakes such as the 1886 Charleston earthquake (e.g., Amick and Gelinas, 1991; Dutton, 1889; Obermeier et al., 1990); therefore, movement on this small normal fault at this exposure did not cause injection of these clastic dikes. However, the clastic dikes, the dominant joint set, and the small normal faults are all oriented approximately north-south (Fig. 9, A, B, and C), consistent with these structures forming in response to east-west extension.

Bartholomew et al. (2000) suggested that orthogonal sets of late Quaternary joints in the lower Coastal Plain are consistent with their formation across a slightly arched ground surface above a concealed reverse fault. Furthermore, a similar relationship of orthogonal sets of normal faults developed across a fold arc above a reverse fault is documented herein at the Williston locality. Thus, extensional features at the Hancock Landing locality are inferred to be related to arching above a nearby reverse fault. This locality is within 0.1 km of the projected trace of the Pen Branch fault, which bounds the Triassic Dunbarton basin (Fig. 1) beneath the Savannah River Site, and which was reactivated during Cretaceous to Tertiary time (e.g., Cumbest et al., 1992; Snipes et al., 1993b). Uplift or arching above this fault could have been generated with either reverse or oblique-reverse displacement on the Pen Branch fault; this is consistent with apparent relative reverse offsets of upper Eocene strata along this fault on seismic lines (e.g., Aadland et al., 1995; Domoracki, 1995).

Abutting joint relationships show that the north-south–striking joint set at the Hancock Landing locality is older than northwest-southeast–striking joints here (Fig. 9B), which are strike parallel to normal faults associated with the syndepositional fold at Williston (Fig. 1). This relationship suggests that the clastic dike event at Hancock Landing predated the reverse faulting events at Williston.

POSTDEPOSITIONAL STRUCTURES

The Tobacco Road Sand is also exposed along the railroad at the Savannah River Site (SRS), South Carolina, locality (RR in Fig. 1; Fig. 10). However, the stratigraphic units here display only broad lateral changes in thickness and lithology and lack any evidence that tectonic events affected deposition of these beds, although reverse faults here are subparallel to both the Williston and Martin fault sets. The unusual feature, however, is a set of north-northeast–striking reverse faults (Fig. 11), which are subparallel to those at Hancock Landing (Fig. 9D) but offset older reverse faults parallel to those at both Williston (Figs. 2 and 3D) and Martin (Figs. 6 and 7E). Furthermore, the north-south–striking reverse faults at Hancock Landing (Fig. 9D) are predominantly near vertical and thus are interpreted as fault-reactivated joints that originally formed coeval with the late Eocene clastic dike event. Therefore, the timing of this north-northeast–striking reverse faulting event, which is indicative of

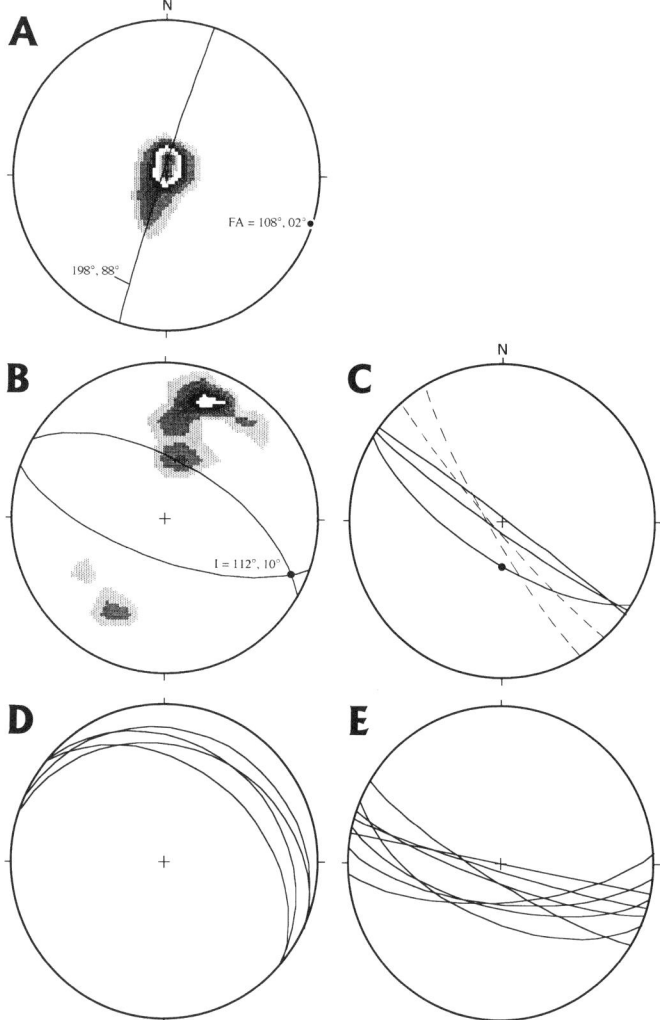

Figure 7. Structural data from bedding and reverse faults in exposure near Martin, South Carolina (M in Fig. 1B), and subparallel reverse faults at other locations. A: 36 poles to bedding (1% area contour) define fold axis (FA) as pole to best fit curve. B: 33 poles to west-northwest–east-southeast–striking reverse faults define intersection (I) of average fault surfaces. C: Three west-northwest–east-southeast–striking reverse faults at Williston that postdate orthogonal normal fault sets associated with main reverse fault there. D: Four west-northwest–striking reverse faults at Hancock Landing. E: Eight east-southeast–striking reverse faults at the SRS railroad locality.

approximately east-west compression, is poorly constrained. It postdates both syndepositional late Eocene and late Miocene faults. The minimum age for the reverse faulting is only constrained by the presence of joints that are related to the modern stress field (σ_1 approximately N65°E, σ_2 approximately vertical, and σ_3 approximately N35°W) in strata of late Quaternary age (Bartholomew et al., 2000). It is consistent with the second oldest event (approximately north-south extension) found in Miocene strata in southeastern Georgia (Bartholomew et al., 1997b). Thus, we consider these crosscutting north-south–striking faults to be Neogene features.

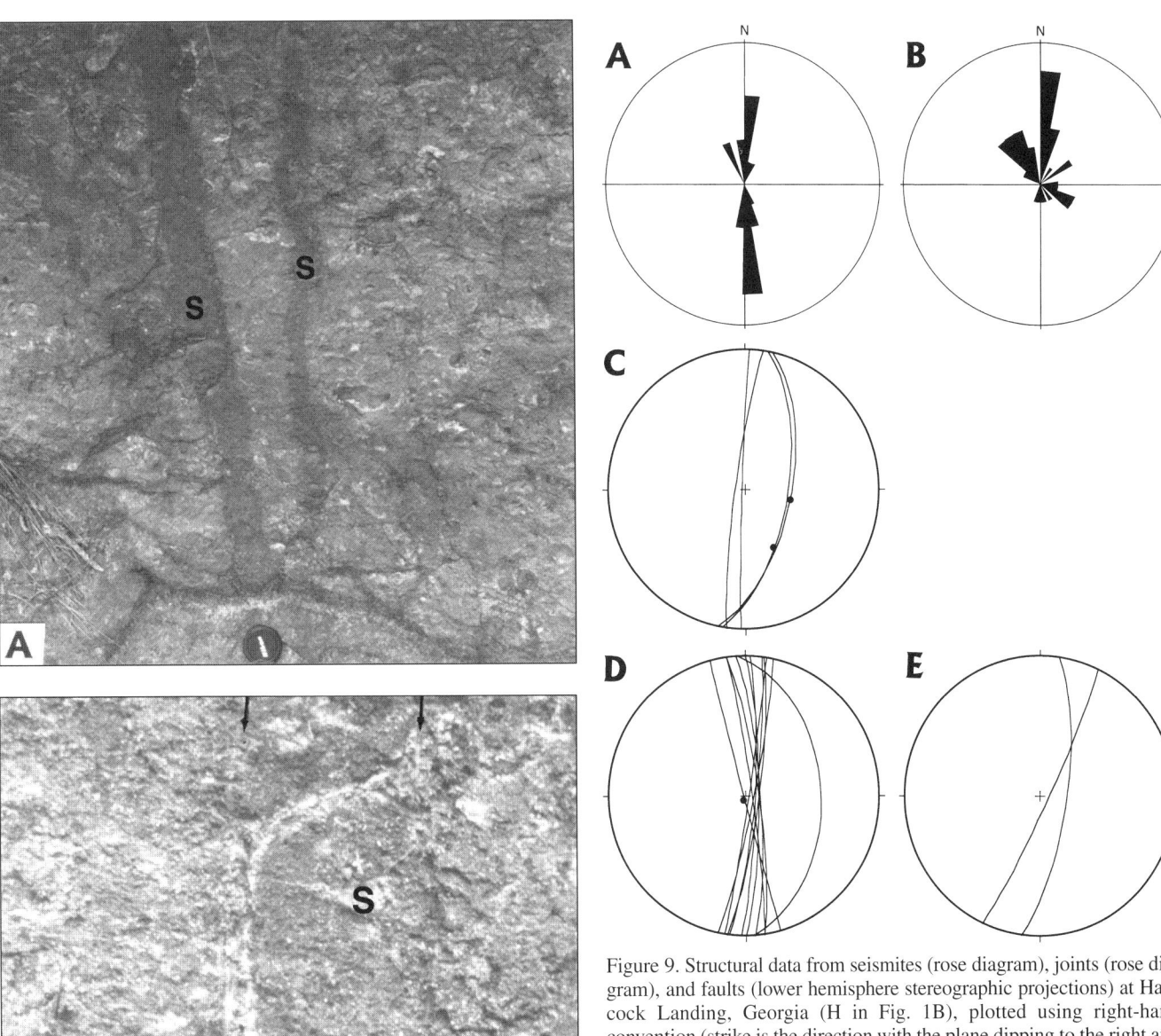

Figure 8. A: Two 1-m-high planar clastic dikes (S, above camera lens cover) that are cut by *Ophiomorpha nodosa* burrows in upper Eocene strata near Hancock Landing, Georgia (H in Fig. 1B). Lens cap is ~6.5 cm in diameter. B: Close-up of clastic dikes showing planar contacts (marked by arrows) and the contrasting massive, coarse-grained sandstone filling of the clastic dike versus the enveloping clay-rich sandstone. Dime is 1.75 cm in diameter.

Figure 9. Structural data from seismites (rose diagram), joints (rose diagram), and faults (lower hemisphere stereographic projections) at Hancock Landing, Georgia (H in Fig. 1B), plotted using right-hand convention (strike is the direction with the plane dipping to the right and pitch is measured from the strike direction). A: 21 north-south–striking, near-vertical clastic dikes; circle is 30%. B: 37 steeply dipping joints; circle is 20%. C: Four normal faults with two slip vectors on a large fault. D: 11 north-striking, steeply dipping reverse faults with slip vector shown on one. E: Two north-striking reverse faults (Fig. 2D) at the SRS railroad locality (RR in Fig. 1) that postdate northwest-striking (Fig. 3D) and west-northwest–striking (Fig. 7E) reverse faults there.

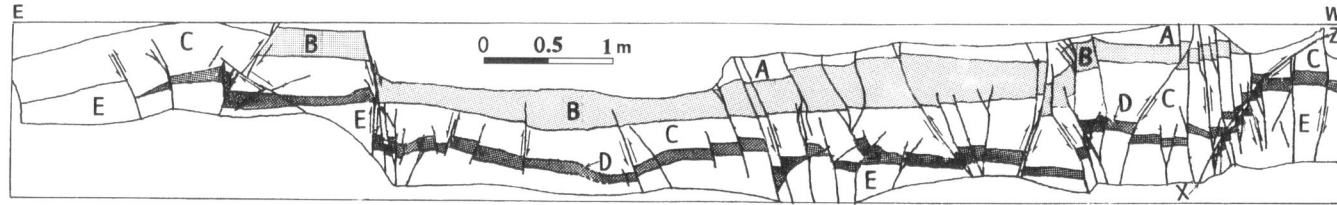

Figure 10. Sketch of exposure along railroad on the Savannah River Site, South Carolina (RR in Fig. 1B), showing younger reverse fault (X–Z) (Figs. 9E and 11) offsetting older reverse faults (Figs. 3D and 7E) in upper Eocene, shallow marine strata of the Tobacco Road Formation. Unit A—mottled light purple to light brown, well-bedded, medium-grained sandstone. Unit B—light gray, massive-bedded, clayey, medium-grained sandstone. Unit C—mottled light purple brown, well-bedded, medium-grained sandstone. Unit D—light brownish gray, laminated, fine-grained sandstone. Unit E—mottled yellowish brown to light gray, massive-bedded, clayey, medium-grained sandstone.

DISCUSSION

The Williston syndepositional features and the Hancock Landing clastic dikes developed nearly contemporaneously, inasmuch as they are contained within and formed during deposition of the upper Eocene Tobacco Road Sand. However, the Hancock Landing event(s) only affected strata similar to the basal beds at Williston. Progressive restoration of the Williston exposure documents that these basal beds preceded the Williston events; thus, we infer that the Hancock Landing event(s) may be slightly older than the Williston events. The Martin event(s) clearly postdate the Williston events and thus provide a chronology of Hancock Landing (oldest), Williston (middle), and Martin (youngest). Furthermore, significant variation (Figs. 3, 7, and 9) of the structural data indicates that either (1) the stress field rotated over time, (2) the stress field varied locally or fluctuated across the area, or (3) two or more stress fields overlapped in this area.

In any case, broad uplift and arching along faults like the Pen Branch were coeval with emergence of the upper Coastal Plain northwest of the Orangeburg scarp. Extension created by arching above the Pen Branch fault likely produced the small normal faults and dominant joints at Hancock Landing. Injection of subparallel clastic dikes along these preexisting joints is an expression of the same period of extension over the uplift, and is indicative of one or more major paleoseismic events during late Eocene time. Syndepositional folding, concomitant normal faulting, scarp-forming reverse faulting, a slump or colluvial wedge, and unconformities at Williston most clearly document several nearby major paleoseismic events during late Eocene time. Syndepositional folding and reverse faulting at Martin also support the interpretation of nearby large paleoseismic events during the middle to late Miocene.

The widespread termination of marine deposition in the upper Coastal Plain of South Carolina and Georgia by the late Miocene (Colquhoun et al., 1991) indicates a large-scale crustal deformation process. The documented mid-Tertiary interaction of the Caribbean and North American plates was such a large-scale crustal deformational process, and it may be the cause of the stress field(s) that produced strong paleoseismicity and surface ruptures on then-active major faults like the reactivated Pen Branch fault. Donnelly (1989) showed a progressive counter-

Figure 11. North-northeast–striking reverse fault (Figs. 9E and 10), marked by arrows (X-Z), offsetting older northwest- and west-northwest–trending reverse faults (Figs. 3D and 7E) at railroad-cut exposure at the Savannah River Site (RR in Fig. 1B). Area within photograph is about 1 m square near the western end of Fig. 10.

clockwise rotation of the North American plate relative to the Caribbean plate that is consistent with a progressive counterclockwise rotation of the stress field in the South Carolina–Georgia area as suggested by the relative ages of the Hancock Landing (oldest), Williston (middle), and Martin (youngest) events. Thus, we suggest that complex interaction between the Caribbean and North American plates resulted in a rotating stress field from late Eocene to late Miocene time.

CONCLUSIONS

1. Near Hancock Landing, orientations of clastic dikes are consistent with both the dominant joint set and small normal faults in a roadside exposure of shallow marine strata. These

near-surface marine features, which are probably related to a major paleoseismic event on the nearby reactivated Pen Branch fault, are interpreted as paleoseismites. The dominant joint set is older than northwest-southeast–striking joints.

2. Near Williston, a syncline formed adjacent to a small southeast-striking reverse fault and served as a depositional trough for material eroded from the flanks of the fold. A slump or colluvial wedge formed on the subjacent side of the fault. Near-surface faulting is indicated by some faults that are truncated by unconformities and by an *Ophiomorpha nodosa* burrow that cuts the trace of a reverse fault. West-northwest–striking reverse faults, parallel to the Martin set, postdate orthogonal sets of normal faults related to the fold and main reverse fault. Both are found within and below the channel deposits, indicating that faulting occurred both during and after deposition of the observable channel strata along the fault scarp. Collectively, these features are interpreted as late Eocene paleoseismites.

3. Relationships in a small depositional syncline near Martin indicate that, at the time of deposition, the ground surface of the trough lens was less than 2 m above the reverse faults that flank the syncline and that formed contemporaneously with the syncline. The thickness of the channel strata provides an approximation of the offset on these zones of small reverse faults. The faulting coincided with emergence of these marine strata and initiation of nonmarine sedimentation. Collectively, these features are interpreted as middle to late Miocene paleoseismites.

4. Crosscutting relationships at the SRS railroad locality indicate that the north-south–striking reverse faults at Hancock Landing are younger than the Williston and Martin syndepositional faults. This fault reactivation of late Eocene joints probably occurred during late Miocene time.

5. Collectively, our data indicate that mid-Tertiary seismites are represented by clastic dikes, syndepositional folds with erosion on uplifted areas, a colluvial wedge or slump, faults deformed by other faults and truncated by unconformities, faults cut by burrow, and faults with different displacement in different units.

6. Our orientation data suggest that all of these mid-Tertiary, near-surface, brittle-deformation features resulted from reverse, left-lateral oblique motion on east-northeast–striking faults, like the reactivated (e.g., Cumbest et al., 1992; Snipes et al., 1993b) Pen Branch fault or the Martin fault (Snipes et al., 1993a). This motion during late Eocene to late Miocene time can account for most of the apparent reverse offsets of Eocene strata noted from drill holes and on seismic lines in previous papers (e.g., Aadland et al., 1995; Domoracki, 1995). Thus, isopach thickness and structural contour changes in upper Eocene strata can now be evaluated using the appropriate displacement vectors across faults with various orientations, to determine if such changes are due to syndepositional faulting.

ACKNOWLEDGMENTS

This study grew out of two larger studies aimed at understanding the effects of geology at the U.S. Department of Energy, Savannah River Site: 1:24 000-scale mapping by the South Carolina Geological Survey (SCGS) Department of Natural Resources (Willoughby); and regional fracture analysis (Bartholomew and associates), which was part of the large project done by the South Carolina University Research and Educational Foundation (SCUREF) for Westinghouse Savannah River Company. Partial support for this study comes from SCUREF task 169 (Bartholomew) and from SCGS contract AB08514N (Nystrom and Willoughby). This study is part of the hazards subprogram of the SCGS (Willoughby). We thank Paul A. Thayer, Charles H. Trupe, Donald T. Secor, Jr., Fredrick J. Rich, and the late Donald J. Colquhoun for their helpful reviews and comments.

REFERENCES CITED

Aadland, R.K., Gellici, J.A., and Thayer, P.A., 1995, Hydrogeologic framework of west-central South Carolina: South Carolina Department of Natural Resources, Water Resources Division Report 5, 200 p.

Amick, D.C., and Gelinas, R., 1991, The search for evidence of large prehistoric earthquakes along the Atlantic Seaboard: Science, v. 251, p. 655–658.

Bartholomew, M.J., Heath, R.D., Brodie, B.M., and Lewis, S.E., 1997a, Post-Alleghanian deformation of Alleghanian granites (Appalachian Piedmont) and the Atlantic Coastal Plain: Geological Society of America Abstracts with Programs, v. 29, no. 3, p. 4.

Bartholomew, M.J., Rich, F.J., Lewis, S.E., and Brodie, B.M., 1997b, Neogene/Quaternary deformational sequence, Atlantic Coastal Plain: Geological Society of America Abstracts with Programs, v. 29, no. 6, p. A231.

Bartholomew, M.J., Rich, F.J., Whitaker, A.E., Lewis, S.E., Brodie, B.M., and Hill, A.A., 2000, Preliminary interpretation of fracture sets in Upper Pleistocene and Tertiary strata of the lower Coastal Plain in Georgia and South Carolina: A compendium of field trips of South Carolina geology with emphasis on the Charleston, South Carolina area: South Carolina Geological Survey, p. 17–27.

Bishop, G.A., and Brannen, N.A., 1993, Ecology and paleoecology of Georgia ghost shrimp, *in* Farrell, K.A., Hoffman, C.W., and Henry, V.J., Jr., Geomorphology and facies relationships of Quaternary Barrier Island complexes near St. Marys, Georgia: Georgia Geological Society Guidebook, v. 13, no. 1, p. 19–29.

Bollinger, G.A., 1977, Reinterpretation of the intensity data for the 1886 Charleston, South Carolina, earthquake, *in* Rankin, D.W., ed., Studies related to the Charleston, South Carolina, earthquake of 1886: A preliminary report: U.S. Geological Survey Professional Paper 1028-B, p. 17–32.

Bramlett, K.W., Secor, D.T., Jr., and Prowell, D.C., 1982, The Belair fault: A Cenozoic reactivation structure in the eastern Piedmont: Geological Society of America Bulletin, v. 93, p. 1109–1117.

Brodie, B.M., and Bartholomew, M.J., 1997, Late Cretaceous-Paleogene phase of deformation in the upper Atlantic Coastal Plain: Geological Society of America Abstracts with Programs, v. 29, no. 3, p. 7.

Colquhoun, D.J., Oldham, R.W., Bishop, J.W., and Howell, P.D., 1982, Updip delineation of the Tertiary Limestone Aquifer, South Carolina: Clemson, South Carolina, Clemson University, Water Resources Research Institute, 93 p.

Colquhoun, D.J., Woollen, I.D., Van Nieuwenhuise, D.S., Padgett, G.G., Oldham, R.W., Boylan, D.C., Bishop, J.W., and Howell, P.D., 1983, Surface and subsurface stratigraphy, structure, and aquifers of the South Carolina Coastal Plain (folio): Columbia, South Carolina, University of South Carolina, 78 p.

Colquhoun, D.J., Johnson, G.H., Peebles, P.C., Huddlestun, P.F., and Scott, T., 1991, Quaternary geology of the Atlantic Coastal Plain, *in* Nonglacial geology, Conterminous U.S.: Boulder, Colorado, Geological Society of America, Geology of North America, v. K-2, p. 620–650.

Cumbest, R.J., Price, V., and Anderson, E.E., 1992, Gravity and magnetic modeling of the Dunbarton Triassic basin, South Carolina: Southeastern Geology, v. 33, p. 37–51.

Domoracki, W.J., 1995, A geophysical investigation of geologic structure and regional tectonic setting at the Savannah River site, South Carolina [Ph.D. thesis]: Blacksburg, Virginia Polytechnic and State University, 236 p.

Domoracki, W.J., Stephenson, D.E., Coruh, C., and Costain, J.K., 1999, Seismotectonic structures along the Savannah River corridor, South Carolina, U.S.A.: Geodynamics, v. 27, p. 97–118.

Donnelly, T.W., 1989, Geologic history of the Caribbean and central America, in Bally, A.W., and Palmer, A.R., eds., The geology of North America: An overview: Boulder, Colorado, Geological Society of America, Geology of North America, v. A, p. 299–321.

Dutton, C.E., 1889, The Charleston earthquake of August 31, 1886: U.S. Geological Survey, Ninth Annual Report, p. 203–528.

Fallaw, W.C., and Price, V., 1995, Stratigraphy of the Savannah River site and vicinity: Southeastern Geology, v. 35, p. 21–58.

Galloway, W.E., Bebout, D.G., Fisher, W.L., Dunlap, J.B., Jr., Cabrera-Castro, R., Lugo-Rivera, J.E., and Scott, T.M., 1991, Cenozoic, in Salvador, A., ed., The Gulf of Mexico: Boulder, Colorado, Geological Society of America, Geology of North America, v. J, p. 245–324.

Howell, D.E., and Zupan, A.–J.W., 1975, Evidence for post-Cretaceous tectonic activity in the area of Westfield Creek, north of Cheraw, South Carolina: South Carolina Development Board, Geologic Notes, v. 18, p. 98–105.

Huddlestun, P.F., 1988, A revision of the lithostratigraphic units of the Coastal Plain of Georgia: The Miocene through Holocene: Georgia Geological Survey, Bulletin 104, 162 p.

Inden, R.F., and Zupan, A.–J.W., 1975, Normal faulting of Upper Coastal Plain sediments, Ideal kaolin mine, Langley, South Carolina: South Carolina Development Board, Geologic Notes, v. 19, p. 161–165.

Johnston, A.C., 1996, Seismic moment assessment of earthquakes in stable continental regions. 3. New Madrid 1811–1812, Charleston 1886 and Lisbon 1755: Geophysical Journal International, v. 126, p. 314–344.

Madabhushi, S., and Talwani, P., 1993, Fault plane solutions and relocations of recent earthquakes in Middleton Place Summerville seismic zone near Charleston, South Carolina: Bulletin of the Seismological Society of America, v. 83, p. 1442–1466.

McDowell, R.C., and Houser, B.B., 1983, Map showing distribution of small-scale deformation structures in a part of the Upper Coastal Plain of South Carolina: U.S. Geological Survey Miscellaneous Field Investigations Map M-1538, scale 1:250 000, 1 sheet.

Nystrom, P.G., Jr., 1998, Middle Eocene to Quaternary surface stratigraphy of Girard Northwest and Shell Bluff Landing 7.5-minute quadrangles, Savannah River site, Barnwell and Aiken counties, South Carolina: South Carolina Geological Survey Open-File Report 107, 9 p. with 1:24 000-scale maps.

Nystrom, P.G., Jr., and Willoughby, R.H., 1992, Field guide to the Cretaceous and Tertiary stratigraphy of the Savannah River Site and vicinity, South Carolina: South Carolina Geological Survey, Field Guide 21, 51 p.

Obermeier, S.F., Jacobson, R.B., Smoot, R.E., Weems, R.E., Gohn, G.S., Monroe, J.E., and Powers, D.S., 1990, Earthquake-induced liquefaction features in the coastal setting of South Carolina and in the fluvial setting of the New Madrid seismic zone: U.S. Geological Survey Professional Paper 1504, 44 p.

Prowell, D.C., 1980, Ductile and brittle faulting in the southern Kiokee belt and Belair belt of eastern Georgia and South Carolina, in Frey, R.W., ed., Excursion in southeastern geology: Leesburg, Virginia, American Geological Institute, p. 80–89.

Prowell, D.C., 1983, Index of faults of Cretaceous and Cenozoic age in the eastern United States: U.S. Geological Survey Miscellaneous Field Studies Map MF-1269, scale 1:2 500 000, 1 sheet.

Prowell, D.C., 1988, Cretaceous and Cenozoic tectonism on the Atlantic coastal margin, in Sheridan, R.E., and Grow, J.A., eds., The Atlantic Continental Margin, U.S.: Boulder, Colorado, Geological Society of America, Geology of North America, v. I-2, p. 557–564.

Prowell, D.C., 1989a, Cretaceous and Cenozoic tectonism in the Appalachians of eastern United States, in Hatcher, R.D., Thomas, W.A., and Viele, G.W., eds., The Appalachian-Oachita orogen in the United States: Boulder, Colorado, Geological Society of America, Geology of North America, v. F-2, p. 362–366.

Prowell, D.C., 1989b, Faults of Cretaceous and Cenozoic age in the eastern United States, in Hatcher, R.D., Thomas, W.A., and Viele, G.W., eds., The Appalachian-Oachita orogen in the United States: Boulder, Colorado, Geological Society of America, Geology of North America, v. F-2, plate 5A, scale 1:5 000 000.

Prowell, D.C., and O'Connor, B.J., 1978, Belair fault zone: Evidence of Tertiary fault displacement in eastern Georgia: Geology, v. 6, p. 681–684.

Secor, D.T., Jr., Barker, C.A., Gillon, K.A., Mitchell, T.L., Bartholomew, M.J., Hatcher, R.D., and Balinsky, M.G., 1998, A field guide to the geology of the Ridgeway-Camden area, South Carolina Piedmont: South Carolina Geology, v. 40, p. 71–83.

Seilacher, A., 1969, Fault-graded beds interpreted as seismites: Sedimentology, v. 13, p. 155–159.

Seilacher, A., 1984, Sedimentary structures tentatively attributed to seismic events: Marine Geology, v. 55, p. 1–12.

Sheridan, R.E., 1989, The Atlantic passive margin, in Bally, A.W., and Palmer, A.R., eds., The geology of North America: An overview: Boulder, Colorado, Geological Society of America, Geology of North America, v. A, p. 81–96.

Snipes, D.S., Hodges, R.A., Warner, R.D., Fallow, W.C., Price, V., Jr., Cumbest, R.J., and Logan, W.R., 1993a, The Martin fault: Southeastern boundary of the Mesozoic Dunbarton basin: Geological Society of America Abstracts with Programs, v. 25, no. 6, p. 468.

Snipes, D.S., Fallow, W.C., Price, V., Jr., and Cumbest, R.J., 1993b, The Pen Branch fault: Documentation of Late Cretaceous-Tertiary faulting in the Coastal Plain of South Carolina: Southeastern Geology, v. 33, p. 195–218.

Snipes, D.S., Hodges, R.A., Price, S.W., Mazur, S.L., Kidd, N.B., Price, V., Jr., and Temples, T.L., 1995, Documentation of late Eocene movement on the Martin fault: Geological Society of America Abstracts with Programs, v. 27, no. 2, p. 88.

Stieve, A., and Stephenson, D., 1995, Geophysical evidence for post Late Cretaceous reactivation of basement structures in the central Savannah River area: Southeastern Geology, v. 35, p. 1–20 and 121.

Talwani, P., 1982, Internally consistent pattern of seismicity near Charleston, South Carolina: Geology, v. 10, p. 654–658.

Talwani, P., and Rajendran, K., 1991, The January 4, 1989, earthquake in Bluffton, South Carolina, and its tectonic implications: Seismological Research Letters, v. 62, p. 139–142.

Tarr, A.C., 1977, Recent seismicity near Charleston, South Carolina, and its relationship to the August 31, 1886, earthquake, in Rankin, D.W., ed., Studies related to the Charleston, South Carolina, earthquake of 1886: A preliminary report: U.S. Geological Survey Professional Paper 1028-D, p. 43–57.

Tarr, A.C., and Rhea, S., 1983, Seismicity near Charleston, South Carolina, March 1973 to December 1979, in Gohn, G.S., ed., Studies related to the Charleston, South Carolina, earthquake of 1886: Tectonics and seismicity: U.S. Geological Survey Professional Paper 1313, p. R1–R17.

Willoughby, R.H., 2000, An overview of Miocene, Pliocene and Pleistocene terraces in the upper, middle, and lower Coastal Plains of South Carolina: Geological Society of America Abstracts with Programs, v. 32, no. 2, p. A83.

Willoughby, R.H., and Clendenin, C.W., 1995, Observations of faulting in an exposure along State Road 62 in Barnwell County, South Carolina: South Carolina Geological Survey Notes, no. 2, 2 p.

Willoughby, R.H., and Nystrom, P.G., Jr., 1994, Field guide to the surface stratigraphy of the area east of the Savannah River Site: Williston, Long Branch, and Snelling 7.5-minute quadrangles: South Carolina Geological Survey, Field Guide 25, 16 p.

Willoughby, R.H., Nystrom, P.G., Jr., Campbell, L.D., and Katuna, M.P., 1999, Cenozoic stratigraphic column of the Coastal Plain of South Carolina: South Carolina Geological Survey, General Geologic Chart 1.

Zupan, A.–J.W., and Abbott, W.A., 1975, Clastic dikes: Evidence for post-Eocene (?) tectonics in the Upper Coastal Plain of South Carolina: South Carolina Development Board, Geologic Notes, v. 19, p. 14–23.

MANUSCRIPT ACCEPTED BY THE SOCIETY MAY 11, 2001

Printed in the U.S.A.

Late Pleistocene soft-sediment deformation structures interpreted as seismites in paralic deposits in the city of Bari (Apulian foreland, southern Italy)

Massimo Moretti*
Piero Pieri*
Dipartimento di Geologia e Geofisica, Università di Bari, Italy

Marcello Tropeano*
Dipartimento di Scienze Geologiche, Università della Basilicata, Italy

ABSTRACT

Late Pleistocene soft-sediment deformation structures are found in lagoonal and eolian sediments exposed in the city of Bari (Murge, Apulian foreland, southern Italy). Deformation structures are represented by load casts, ball-and-pillow, and flame structures. The mechanism of deformation is related with liquefaction and/or fluidization processes. After considering alternative explanations, we conclude that liquefaction and/or fluidization was most probably triggered by a paleoseismic event originating along the south Gargano fault, an active strike-slip zone about 50 km offshore from Bari.

INTRODUCTION

Several examples of Quaternary soft-sediment deformation structures observed in the Murge area (Apulian foreland, southern Italy, Fig. 1) have been interpreted recently as seismites (Moretti, 1997; Pieri et al., 1997). The late Pleistocene age seismites exposed in the city of Bari are well preserved, allowing a detailed morphologic analysis. These seismites are especially interesting because of the regional context in which they formed.

GEOLOGICAL SETTING

Regional features

The city of Bari is on the Adriatic coast of Murge, a karstic area of the Apulian foreland (Avampaese Apulo of Selli, 1962; D'Argenio et al., 1973; Ciaranfi et al., 1983), also known as the Apulian swell (Auroux et al., 1985; Doglioni et al., 1994) (Fig. 1).

The Apulian foreland is the Pliocene-Pleistocene foreland of the South Apennines orogenic system and corresponds to a wide, buckled lithospheric zone (Biju-Duval et al., 1979; Ricchetti and Mongelli, 1980; Royden et al., 1987; Doglioni et al., 1994). It consists of a Variscan crystalline basement with an approximately 6-km-thick Mesozoic sedimentary cover (the Apulian carbonate platform, see D'Argenio, 1974; Ricchetti, 1975; 1980). The Mesozoic rocks are overlain by thin and discontinuous Tertiary and Quaternary deposits (Ricchetti et al., 1988).

The Apulian foreland is a west-northwest–east-southeast–trending ridge, segmented in three large areas with different degrees of uplift: Gargano, Murge, and Salento (Ricchetti et al., 1988; Funiciello et al., 1991; Doglioni et al., 1994; Gambini and Tozzi, 1996) (Fig. 1). Normal faults in all three ridge segments range from Mesozoic to Pleistocene in age, and trend west-northwest–east-southeast, northwest-southeast, and east-west. There are down faulted blocks toward the Bradanic trough to the west-southwest and toward the Adriatic sea to the east-southeast (Carissimo et al., 1963; Pieri; 1980; Ricchetti, 1980). These faults were active during Pliocene-Pleistocene time along the eastern margin of the Bradanic trough (the South Apennines foredeep), as indicated by

---

*E-mails: Moretti—m.moretti@geo.uniba.it, Pieri—p.pieri@geo.uniba.it, Tropeano—tropeano@unibas.it

Moretti, M., Pieri, P., and Tropeano, M., 2002, Late Pleistocene soft-sediment deformation structures interpreted as seismites in paralic deposits in the city of Bari (Apulian foreland, southern Italy), *in* Ettensohn, F.R., Rast, N., and Brett, C.E., eds., Ancient seismites: Boulder, Colorado, Geological Society of America Special Paper 359, p. 75–85.

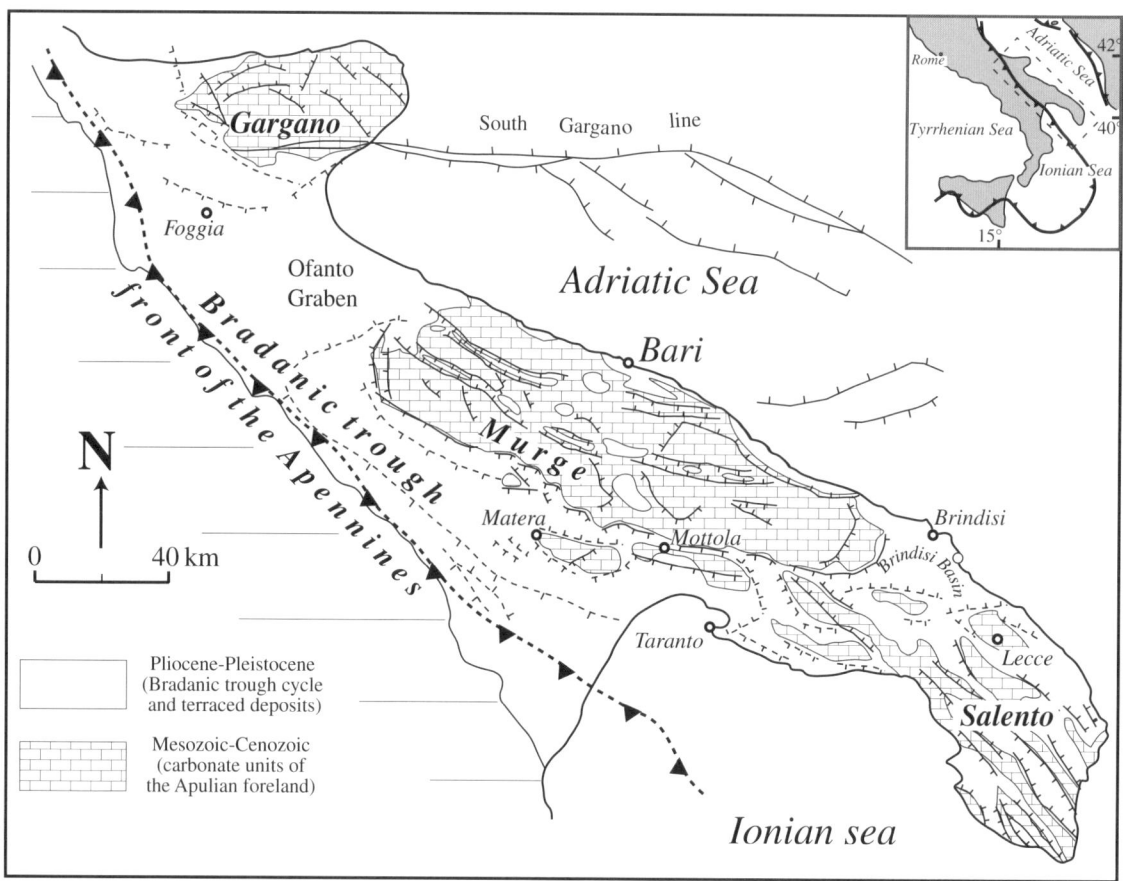

Figure 1. Schematic geologic map of the Apulian foreland.

syntectonic deposits west of Murge (Tropeano et al., 1994) and, in the shelf, west of Salento (Tramutoli et al., 1984).

Gargano is a well-known seismic region, having two main fault zones (North and South Gargano) capable of producing high-energy earthquakes (Postpischl, 1985), and Salento shows evidence of Quaternary tectonic activity, as documented by several normal faults that cut Pleistocene deposits (Martinis, 1962; Palmentola and Vignola, 1980; Moretti, 1996). In contrast, both seismicity and Quaternary faulting in Murge are poorly documented, because few historical earthquakes have been reported. It is also not clear whether the normal faults, which cut the Mesozoic substratum, also cut the thin and very discontinuous Quaternary deposits. For these reasons, Murge has been considered as an aseismic and stable sector of the Apulian foreland (De Vivo et al., 1979; Ciaranfi et al., 1981; Boschi et al., 1995). In spite of this supposed stability, during Pliocene-Pleistocene time the Murge (similar to the Gargano and Salento areas) underwent two distinct vertical movements: subsidence during the Pliocene and early part of the early Pleistocene, and uplift from the upper part of the early Pleistocene (Pieri et al., 1997).

The middle Pliocene–early Pleistocene subsidence of Murge caused flooding of the Mesozoic rocks that had been subaerially exposed since the Late Cretaceous (Pieri, 1980) and the deposition of transgressive and deepening-upward deposits (Calcarenite di Gravina and Argille subappennine Formations) (Iannone and Pieri, 1982). Middle–late Pleistocene–Holocene uplift of Murge (Pieri et al., 1997) is evidenced by 16 orders of uplifted marine terraces showing paleo–sea cliffs, abrasion platforms, and/or thin paralic deposits that disconformably overlie either the Mesozoic limestones or the late Pliocene–early Pleistocene transgressive formations (Ciaranfi et al., 1988). According to late Pleistocene geochronological data, uplift rates are on the order of 0.2–0.3 mm/yr (Cosentino and Gliozzi, 1988; Dai Pra and Hearty, 1989), but stratigraphic data suggest uplift rates of at least 0.5 mm/yr (Ciaranfi et al., 1994; Doglioni et al., 1996; Festa et al., 1999).

Stratigraphy in the city of Bari

In the city of Bari, limestone is overlain by thin terrace deposits (Pieri, 1975). The limestone belongs to two formations: the Calcare di Bari (Cenomanian) and the Calcarenite di Gravina (late Pliocene–early Pleistocene) (Azzaroli and Valduga, 1967; Pieri, 1988). The Calcare di Bari Formation is part of the 5-km-thick Mesozoic succession of the Apulian carbonate platform. In the Bari area, the outcropping Cretaceous limestones are fractured and gently folded. The west-northwest–east-southeast joints and fold axes are parallel to the present-day coastline (Fig. 2B). The Cretaceous rocks are overlain by the late Pliocene–early Pleistocene age Calcarenite di Gravina Formation (Pieri, 1975). The subhorizontal younger limestones partially cover the gentle undulations of the Cretaceous limestones. As a result, the youngest deposits lie disconformably on both limestone formations (Fig. 2, A and B).

One of the late Pleistocene terrace deposits in the city of Bari, 10–15 m above sea level, is the "cordone littorale di Bari" of Pieri (1975), a several-meters-thick and 6-km-long ancient littoral wedge (Fig. 2A). Its stratigraphic and sedimentological features can be seen in a continuous vertical section (a railroad cut) and also in some quarries. In every outcrop, these deposits show soft-sediment deformation structures (Pieri, 1975) (Fig. 2B), and these structures can also be seen in every new foundation trench, characterizing Bari's subsurface.

The late Pleistocene terrace of Bari consists of three units; from bottom to top they are (1) a fine sand unit, (2) a coarse sand unit, and (3) a gravel unit (Fig. 2B).

The fine sand unit (up to 1.5 m) contains abundant ostracodes and characeae oogonia. Where the primary sedimentary features are preserved, fossils or fine sands form very thin laminae. Thin wackestone layers, up to 1–2 cm thick, are present within the fine sands and consist mainly of ostracodes, characeae oogonia, and rare gastropods. The unit has parallel lamination, originally subhorizontal but now intricately folded as a result of the soft-sediment deformation.

The coarse sand unit (2.5 m) overlies the fine sand unit with a sharp (nonerosional) boundary. The well-sorted and well-rounded coarse sand grains are without matrix and without fossils. The sands are layered; subhorizontal beds pass vertically into gently landward-dipping southwest-west cross beds. The lower part of this sand unit shows soft-sediment deformation structures.

The gravel unit cuts both sand units along a 10° seaward-inclined surface and reaches the bedrock, where it forms a gentle antiform (Fig. 2B). The unit consists mainly of discoidal cobbles and pebbles with a coarse-grained sandy matrix. The gravel is well bedded, and lacks soft-sediment deformation structures.

The late Pleistocene age paralic deposits in the city of Bari are one of the younger marine terrace deposits, the fifteenth of those described by Ciaranfi et al. (1988). The fine sand unit can be interpreted as a lagoon or swamp deposit, and the coarse sand unit as an eolian deposit. Thin lagoon depressions were filled by autochthonous deposits (fossiliferous laminae and wackestone layers) and more commonly by terrigenous deposits (laminated fine sands) during flooding of ephemeral streams. The coarse sand unit represents the retrogradation of a backshore dune system that overlapped the lagoon deposits. The gravel unit, which shows beachface features, is not genetically linked to the two preceding units and might be due to a younger marine depositional episode not correlable to that inducing the sand deposition.

SOFT-SEDIMENT DEFORMATION STRUCTURES

Morphology

Deformation involves the fine sand unit (from a few centimeters to 1.5 m, depending on the variable thickness of this unit) and the lower part of the coarse sand unit. The soft-sediment deformation structures are load structures (sensu Allen, 1982), mostly load casts (sensu Reineck and Singh, 1980), varying from 0.20 to 0.60 m in height. They are either subspherical or cylindrical (Fig. 3), but locally, some irregular-shaped load casts were observed (Fig. 4). Internally, some load casts retained primary lamination of the overlying coarse sand unit (Fig. 5), whereas othershave a completely homogenized internal texture (the primary lamination was not preserved; Fig. 6).

Other load structures are sandy ball-and-pillow structures (sensu Reineck and Singh, 1980), found as isolated bodies either within the fine sand unit (Fig. 7) or at its base (Fig. 8), indicating that they dropped through the fine sand deposits. The ball-and-pillow structures show a planar elongation and a completely homogenized internal texture (the primary lamination is not preserved).

Other types of soft-sediment deformation structures in the underlying fine sand unit include folds and some structures that may be of fluid-escape origin (Lowe, 1975; Fig. 6). At the interface between the two units, there are also flame structures (sensu Kelling and Walton, 1957; Fig. 6 here).

Driving force system and mechanism of deformation

The superposition of coarse-grained sands (heavy) on fine-grained sands (light) represents a driving force system (sensu Owen, 1987) related to gravitational instability (Rayleigh-Taylor instability or reverse density gradient system of Anketell et al., 1970). The heavy sands may sink into the light sands, accompanied by a drastic decrease, or complete loss, of the shear strength of the sediments (Allen, 1982; Owen, 1987; Moretti et al., 1999). Some of the observed features in Bari, such as the disappearance of the primary lamination of the sands and the presence of structures with fluid-escape morphologies, suggest that the decrease of shear strength was induced by liquefaction.

Laboratory analyses support this hypothesis, because (1) direct shear tests on the sediments of the layers sampled in the underlying fine sand unit show that these sediments have a high friction angle (33°–34°) and that a drastic reduction of the shear strength is possible only with a sudden increase in the interstitial pressure (liquidization, sensu Allen, 1982) excluding in that way the possibility of anomalous behaviors (sensitivity and/or thixotropy); and (2) sieve analyses of the fine-sand unit (Fig. 9) show that these sediments are silty sands with a high liquefaction potential (sensu Seed and Idriss, 1982).

The formation of load structures (both load cast and ball-and-pillow) was induced by a partial gravitational readjustment that acted simultaneously with the loss of shear strength in the fine sands. The fine-sand unit displays active deformation fea-

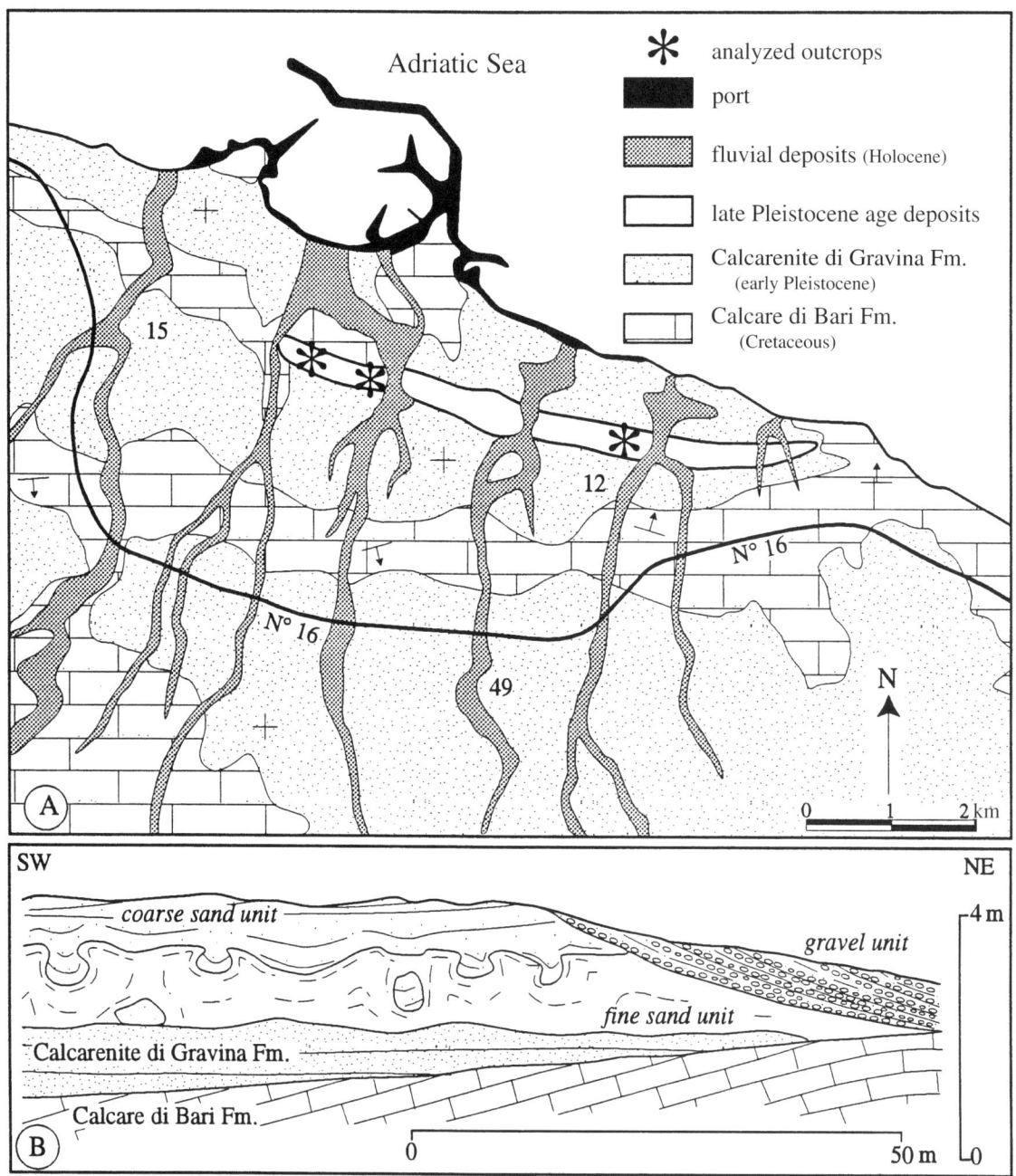

Figure 2. A: Schematic geologic map of the city of Bari. B: Schematic cross section of Bari deposits in the central outcrop shown in A. FM—Formation.

tures (sensu Allen, 1982), while the overlying sands show mainly passive deformation features (sensu Allen, 1982). In fact, the underlying fine sands show evidence of viscous fluid behavior (disappearance of primary lamination and fluid escape structures), whereas the overlying passively deformed sands underwent mainly brittle and/or plastic deformation and subordinately a liquefaction process (as indicated by the lack of lamination in some load structures). These observations suggest that the fine-sand unit was saturated, but the overlying eolian sands were mainly dry or partially saturated (near the bottom of the unit) when the deformation was triggered.

The relationship between all the observed deformations and the loss of shear strength in the fine-sand unit was verified also through morphometric analyses on the soft-sediment deformation structures (Moretti, 1997). For the load casts, we measured the silty unit thickness (S), the load-cast thickness (vertical dimension—D), the thickness of the deformed part of the coarse-sand unit (H), and the distance between the base of the load casts and

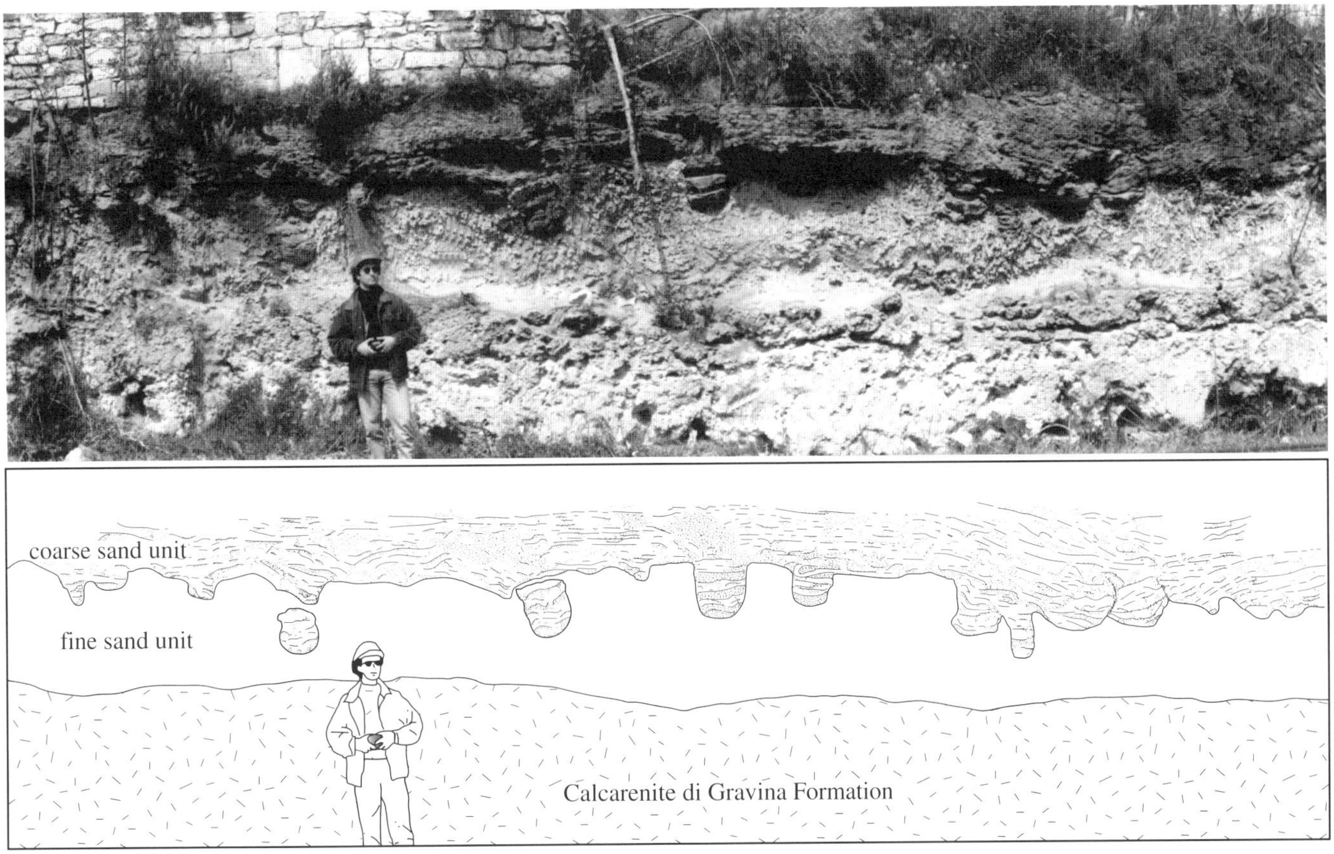

Figure 3. Soft-sediment deformation structures, city of Bari. Note that deformation is restricted mainly to the contact between the silty and sandy units.

Figure 4. Irregular load casts. Primary internal lamination is absent.

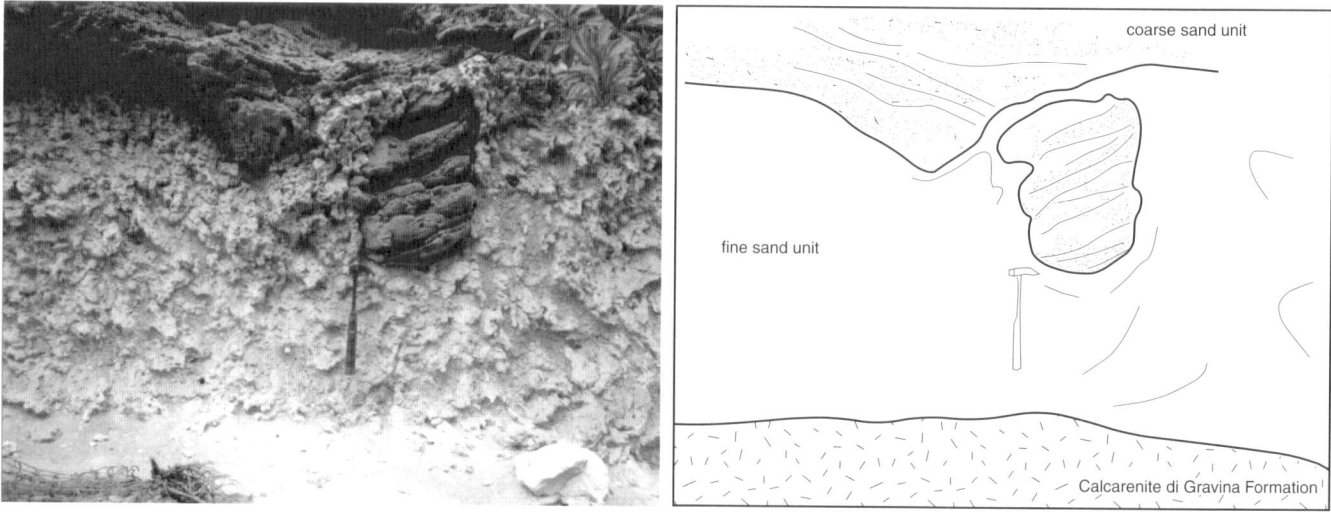

Figure 5. Load cast with preserved primary lamination of the coarse-sand unit; lamination seems partially deformed at the border of the load cast. Hammer is 33 cm long.

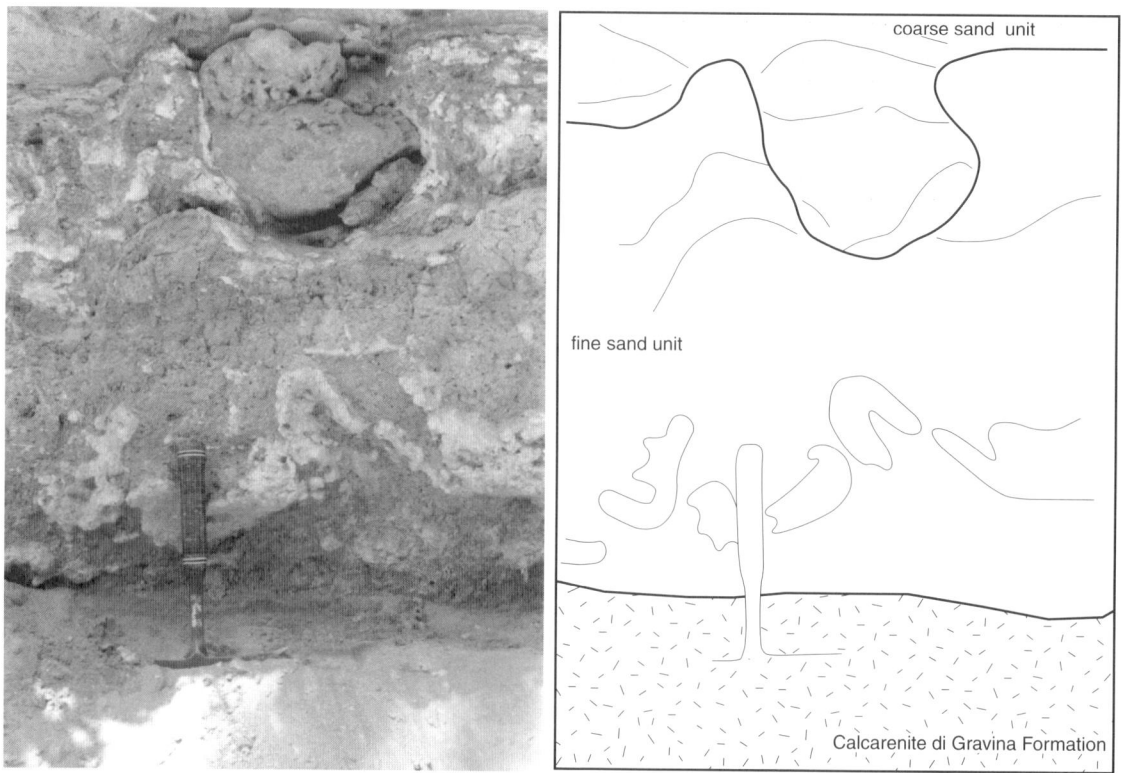

Figure 6. Load cast with totally homogenized internal texture. Note in the underlying fine-sand unit the folded and fragmented calcareous layers. There are sedimentary features with a flame morphology at the contact between the two units. Hammer is 33 cm long.

the base of the fine-sand unit (C) (Fig. 10A). For the ball-and-pillow structures, we measured the fine-sand unit thickness (S), the ball-and-pillow widths (horizontal dimension—T), the depth reached by the pillows in the fine-sand unit, (P) and their distance from the base of the same unit (C) (Fig. 10B). Figure 10C (load casts) indicates a direct relationship between D, H, and S: the thicker the fine-sand unit, the greater both the deformation of the interface between the two units and the thickness of the overlying deformed sands. A direct relation exists also between P and T (Fig. 10D) in the ball-and-pillow structures, and there is a clear transition between ball morphologies near the coarse sand unit top and pillow morphologies up to the base of the same unit.

These observed relationships and data from both field and laboratory analyses demonstrate that (1) the deformation was induced by sudden loss of shear strength in the fine-sand unit, and (2) the deformation degree was directly related to the thickness of the liquefied sediments (fine-grained sand unit).

Trigger mechanism

Liquefaction and fluidization can be induced by several agents, (seismic shaking, overloading, groundwater movement, cyclical and/or impulsive effect of storm waves; Allen, 1982; Owen, 1987, 1996; Molina et al., 1998; Moretti et al., 1999).

What triggered the deformation of the sediments in the city of Bari?

Before considering an external trigger agent (seismic shock or tsunami), we should consider the possibility that an internal agent (sedimentary processes such as overloading or storm waves) induced liquefaction and/or fluidization of the sediments.

The soft-sediment deformation structures of Bari were formed during deposition of the coarse-sand unit in a backshore (eolian) environment. The deformation affected the whole fine-sand unit and the lower half of the coarse-sand unit. Eolian sedimentation continued, and the upper half of the coarse-grained sand unit was deposited on top of the deformed lower half.

An internal trigger mechanism must thus have been related to the back-shore eolian sedimentary environment. In this environment, soft-sediment deformation may have been triggered by (1) gravitational instabilities of dunes during their migration (McKee et al., 1971; McKee and Bigarella, 1972; Collinson and Thompson, 1989), (2) transgression (Glennie and Buller, 1983), (3) various kinds of bioturbation (Plaziat, 1971; Fornos et al., 1986), or (4) cryoturbation (Obermeier, 1996).

We can exclude each of these four internal triggers: 1. Gravitational instabilities of dunes induce deformation only in the eolian sediments (on the lee side of dunes during their migration), and the deformation occurs in the absence of both water and liquefaction-fluidization processes. 2. There is no evidence of transgression within the analiyzed succession (deformed eolian sands are limited at the top by undeformed sands of the same depositional environment) 3. The presence of liquefaction effects indicates that deformation was physically induced, so a biogenic origin is excluded. 4. The Bari deposits are located within latitudes characterized by temperate climate and in which no cryogenic features are regionally recorded in the late Pleistocene deposits. Moreover, cryoturbation produces soft-sediment deformation structures (ice-wedge casts and involutions; see Obermeier, 1996) that are morphologically not comparable to those observed in Bari.

Sudden changes in the water-table level may be another possible trigger mechanism (see Holzer and Clark, 1993, and references therein); this trigger agent produces liquefaction and fluidization structures (sand boils and dikes) whose morphologies are different from those described in this case. Moreover, the effects of fluidization deriving from the bottom (bedrock) are totally absent at the base of the fine-sand unit.

Having excluded every known internal agent, as well as sudden changes in the water level, we conclude that the soft-sediment deformation structures observed in the late Pleistocene age Bari deposits must have been induced by an external agent, such as a seismic shock or a tsunami. A tsunami is unlikely, because

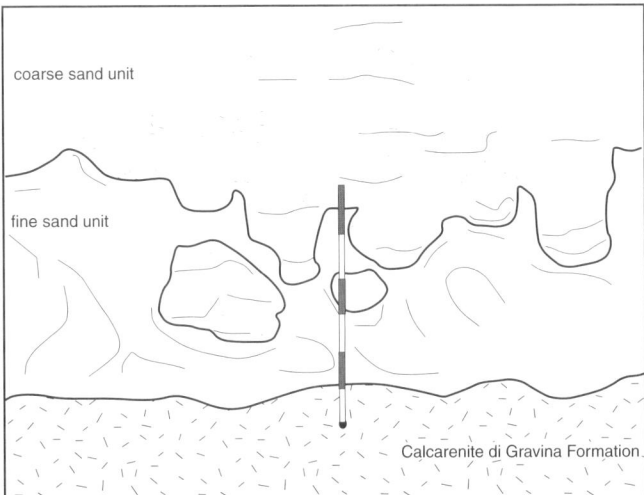

Figure 7. Coarse-grained sandy ball-and-pillow structures isolated within the fine-sand unit. Pole is 1 meter long.

Figure 8. A pillow that has reached the base of the fine-sand unit.

tsunami deposits observed in back-shore environments show specific sedimentologic features (see Benson et al., 1997) that are absent in the deposits we analyzed.

We therefore conclude that the soft-sediment deformation structures are seismically induced. Few examples of soft-sediment deformation structures in an eolian environment interpreted as seismic-induced deformations have been reported (Horowitz, 1982; Plaziat and Poisson, 1992; Alfaro et al., 1999; Moretti, 2000).

CONCLUSIONS

Soft-sediment deformation structures induced by seismic liquefaction and/or fluidization (seismites; sensu Seilacher, 1969) are sedimentary fossil records of moderate to high-magnitude seismic events (M > 5; Ambraseys, 1988) generally occurring within a range of 40 km from the earthquake epicenter (Galli and Meloni, 1993). The Bari seismites must thus have been induced by a seismic event of M > 5 located within the Murge area or the adjacent sector of the Adriatic shelf.

Large deformation zones bounding the Murge from the other foreland areas and horst and graben systems cutting the Murge were mainly produced during pre-Pliocene time (Martinis, 1961; Iannone and Pieri, 1982, 1983), but they were also active during the Quaternary uplift of Murge (Pieri et al., 1997). Baratta (1901) and Kàrnìk (1969) pointed out some strong historical earthquakes inside Murge, but their exact location and actual intensity are considered uncertain. In contrast, low-energy earthquakes (M ~ 3.2) have been instrumentally recorded (Del Gaudio et al., 1997). There is therefore contrast between the palaeoseismic data (both seismites and ancient historical seismic data) and the instrumentally recorded earthquakes. The latter indicate low-energy events (M ~ 3.2) whereas the seismites observed in the Bari deposits require a paleoevent due to an earthquake of M > 5 (Moretti and Tropeano, 1996; Moretti, 1997).

A paleoevent may have been associated with the South Gargano fault zone (Fig. 1), a high-energy seismogenetic fault on shore (Postpischl, 1985; Piccardi, 1998) having present-day seismicity (Favali et al., 1990) and present-day deformation of the sea floor in the offshore continuation of the fault (de Alteriis and Aiello, 1993). This fault zone is the most probable source of high-energy historical and paleoseismic events reported in the eastern part of Murge, including the city of Bari.

ACKNOWLEDGMENTS

The manuscript was greatly improved by the comments and amendments suggested by DeJong and Pope. This chapter is in memory of S. Dzulynski, who patiently explained to us the genetic mechanisms of soft-sediment deformation structures.

The financial support of CNR (Grants 99.00700.CT05 to P. Pieri and G00503F to M. Moretti), and Murst (Grants 60% 1999 and 2000 to L. Sabato; Grants PRIN 2000 to L. Simone and A. Laviano; Grants "Giovani Ricercatori 2000" to M. Moretti) are gratefully acknowledged.

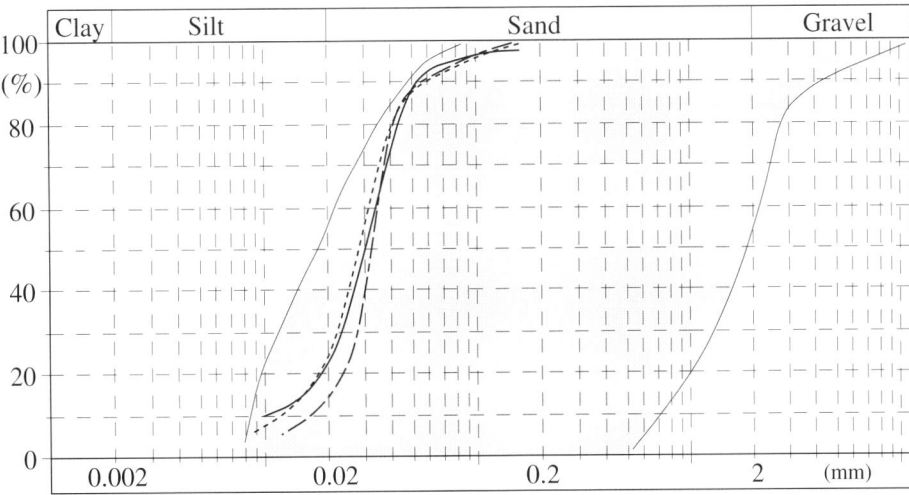

Figure 9. Sieve analysis results of three different layers in the fine-sand unit in the domain of sediments susceptible to liquefaction (after Tsuchida and Hayashi, 1971).

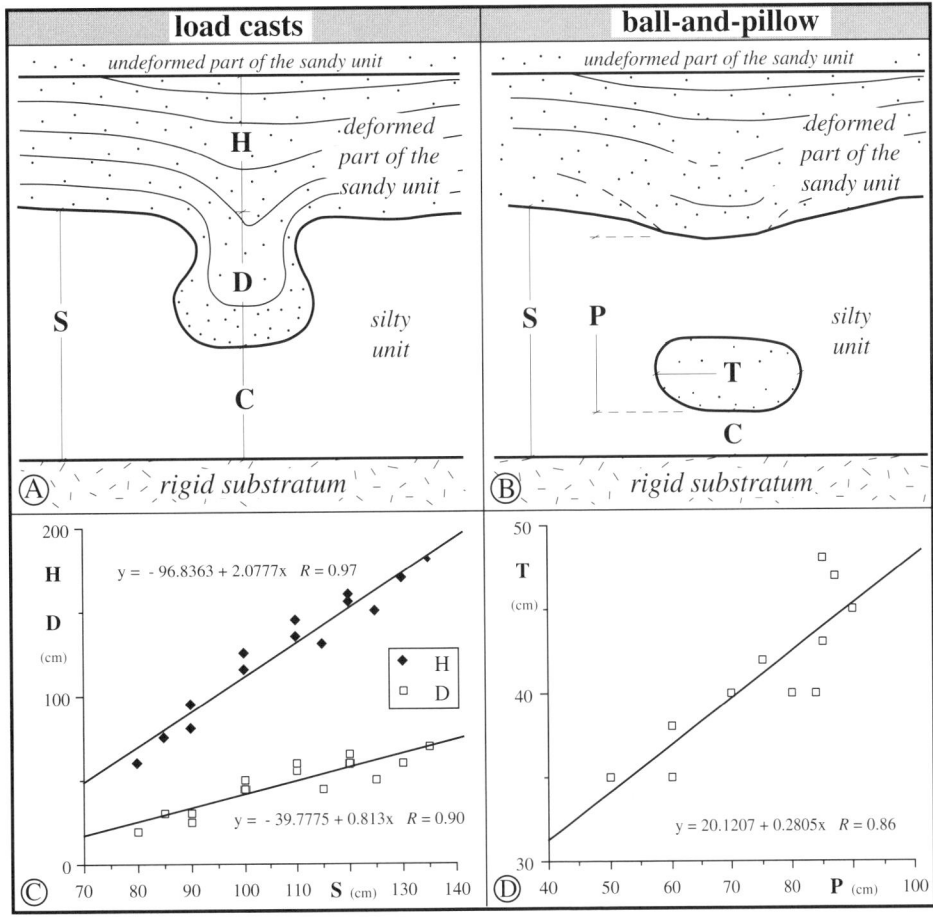

Figure 10. Morphometric features measured in the field on the load casts (A) and on the ball-and-pillow sturctures (B). C, D: Scatter diagrams of main relations between the measured morphometric elements (see text for explanation). S—silty unit in A, fine-sand unit in B; H—coarse-sand unit; D—load cast; C—distance between load casts (in A) and ball-and-pillow structures (in B) and the base of the fine-sand unit; T—ball-and-pillow structures; P—pillows.

REFERENCES CITED

Alfaro, P., Estévez, A., Moretti, M., and Soria, J.-M., 1999, Structures sédimentaires de déformation interprétées comme séismites dans le Quaternaire du Bassin du Bas Segura (Cordillère bétique orientale): Comptes Rendus de l'Academie des Sciences, v. 328, p. 17–22.

Allen, J.R.L., 1982, Sedimentary structures: Their character and physical basis, Volume 2: New York, Elsevier, 663 p.

Ambraseys, N., 1988, Engineering seismology: Earthquake Engineering & Structural Dynamics, v. 17, p. 1–105.

Anketell, J.M., Cegla, J., and Dzulinsky, S., 1970, On the deformational structures in systems with reversed density gradients: Annales de la Société Géologique de Pologne, v. 40, p. 3–30.

Auroux, C., Mascle, J., Campredon, R., Mascle, G., and Rossi, S., 1985, Cadre géodynamique et évolution récente de la Dorsale Apulienne et de ser bordures: Giornale di Geologia, v. 47, p. 101–127.

Azzaroli, A., and Valduga, A., 1967, Note illustrative della Carta Geologica d'Italia, F.177 e 178 "Bari e Mola di Bari": Servizio Geologieo Italiano, 26 p.

Baratta, M., 1901, I terremoti in Italia [1979 edition]: Torino, Italy, A. Forni Ed.

Benson, B.E., Grimm, K.A., and Clague, J.J., 1997, Tsunami deposits beneath tidal marches on northwestern Vancouver Island, British Columbia: Quaternary Research, v. 48, p. 192–204.

Biju-Duvall, B., Letouzey, J., and Montadert, L., 1979, Variety of margins and deep basins in the Mediterranean: American Assoication of Petroleum Geologists Memoir 29, p. 293–317.

Boschi, E., Favali, P., Frugoni, F., Scalera, G., and Smriglio, G., 1995, Mappa della massima intensità macrosismica risentita in Italia: Istituto Nazionale di Geofisica.

Carissimo, L., D'Agostino, O., Loddo, C., and Pieri, M., 1963, Petroleum exploration by AGIP Mineraria and new geological information in central and southern Italy, from the Abruzzi to the Taranto gulf: 6th Petroleum International Congress Section 1, p. 267–292.

Ciaranfi, N., Cinque, A., Lambiase, S., Pieri, P., Rapisardi, L., Ricchetti, G., Sgrosso, I., and Tortorici, L., 1981, Proposta di zonazione sismotettonica dell'Italia meridionale: Rendiconti della Societa Geologica Italiana, v. 4, p. 493–496.

Ciaranfi, N., Ghisetti, F., Guida, M., Iaccarino, G., Lambiase, S., Pieri, P., Rapisardi, L., Ricchetti, G., Torre, M., Tortorici, L., and Vezzani, L., 1983, Carta neotettonica dell'Italia meridionale: Progetto Fininalizzato Geodinamica del Consiglio Nazionale delle Ricerche, no. 515, p. 62.

Ciaranfi, N., Pieri, P., and Ricchetti, G., 1988, Note alla carta geologica delle Murge e del Salento (Puglia centromeridionale): Memorie della Società Geologica Italiana, v. 41, p. 449–460.

Ciaranfi, N., Pieri, P., and Ricchetti, G., 1994, Linee di costa e terrazzi marini pleistocenici nelle Murge e nel Salento: Implicazioni neotettoniche: Riassunti 77° Congresso Società Geologica Italiana, Bari, p. 170–172.

Collinson, J.D., and Thompson, D.B., 1989, Sedimentary structures (second edition): London, Allen and Unwin, 194 p.

Cosentino, D., and Gliozzi, E., 1988, Cosniderazioni sulle velocità di sollevamento di depositi eutirreniani dell'Italia meridionale e della Sicilia: Memorie della Società Geologica Italiana, v. 41, p. 653–665.

Dai Pra, G., and Hearty, P.J., 1989, Variazioni del livello del mare sulla costa ionica salentina durante l'Olocene: Epimerizzazione dell'isoleucina in Helix sp.: Memorie della Società Geologica Italiana, v. 42, p. 311–320.

D'Argenio, B., 1974, Le piattaforme carbonatiche periadriatiche: Una rassegna di problemi nel quadro geodinamico mesozoico dell'area mediterranea: Memorie della Società Geologica Italiana, v. 13, p. 137–159.

D'Argenio, B., Pescatore, T., and Scandone, P., 1973, Schema geologico dell'Appennino meridionale (Campania e Lucania): Accademia Nazionale del Lincei, v. 182, p. 49–72.

de Alteriis, G., and Aiello, G., 1993, Stratigraphy and tectonics offshore of Puglia (Italy, southern Adriatic sea): Marine Geology, v. 113, p. 233–253.

Del Gaudio, V., Iurilli, V., Pierri, P., Ruina, G., Calcagnile, G., Canziani, R., Moretti, M., Pieri, P., and Tropeano, M., 1997, Sismicità di bassa energia e caratteristiche strutturali delle Murge nord-orientali: Atti 15 Convegno Gruppo Nazionale di Geofisica della Terra Solida, p. 325–330.

de Vivo, B., Dietrich, D., Guerra, I., Iannaccone, G., Luongo, G., Scandone, P., Scarpa, R., and Turco, E., 1979, Carta sismotettonica preliminare dell'Appennino meridionale: Progetto Finalizzato Geodinamica, no. 166, 64 p.

Doglioni, C., Mongelli, F., and Pieri, P., 1994, The Puglia uplift (SE Italy): An anomaly in the foreland of the Apenninic subduction due to buckling of a thick continental lithosphere: Tectonics, v. 13, no. 5, p. 1309–1321.

Doglioni, C., Tropeano, M., Mongelli, F., and Pieri, P., 1996, Middle-Late Pleistocene uplift of Puglia: An "anomaly" in the Apenninic foreland: Memorie della Società Geologica Italiana, v. 51, p. 101–117.

Favali, P., Mele, G., and Mattietti, G., 1990, Contribution to the study of the Apulian microplate geodynamics: Memorie della Società Geologica Italiana, v. 44, p. 71–80.

Festa, V., Gueguen, E., Moretti, M., Pieri, P., Sabato, L., and Tropeano, M., 1999, The middle-late Pleistocene uplift of the Murge area (Apulian foreland, southern Italy): An integrated approach: International Workshop on "Large-scale vertical movements and related gravitational processes": Rome-Camerino, Abstracts, p. 4–5.

Fornos, J.J., Pomar, L., and Rodriguez-Perea, R., 1986, Deformation structures on eolian calcarenites recognized as mammal footprints: Kraków, 7th Regional Meeting on Sedimentology, International Association of Sedimentologists, p. 63.

Funiciello, R., Montone, P., Parotto, M., Salvini, F., and Tozzi, M., 1991, Geodynamic evolution of an intra-orogenic foreland: The Apulia case history (Italy): Bollettino Società Geologica Italiana, v. 110, p. 419–425.

Galli, P., and Meloni, F., 1993, Nuovo catalogo dei processi di liquefazione avvenuti in occasione dei terremoti storici in Italia: Il Quaternario, v. 6, no. 2, p. 271–292.

Gambini, R., and Tozzi, M., 1996, Tertiary geodynamic evolution of Southern Adria microplate: Terra Nova, v. 8, p. 593–602.

Glennie, K.W., and Buller, A.T., 1983, The Permian Weissliegend of NW Europe: The partial deformation of aeolian dune sands caused by the Zechstein transgression: Sedimentary Geology, v. 35, p. 43–81.

Holzer, T.M., and Clark, M.M., 1993, Sand boils without earthquakes: Geology, v. 21, p. 873–876.

Horowitz, D.H., 1982, Geometry and origin of large-scale deformation structures in some ancient wind-blown sand deposits: Sedimentology, v. 29, p. 155–180.

Iannone, A., and Pieri, P., 1982, Caratteri neotettonici delle Murge: Geologia Applicata ed Idrogeologia, v. 17, p. 147–159.

Iannone, A., and Pieri, P., 1983, Rapporti fra i prodotti residuali del carsismo e la sedimentazione quaternaria nell'area delle Murge: Rivista Italiana di Paleontologia, v. 88, no. 2, p. 319–330.

Kàrnìk, V., 1969, Seismicity of the European area: Amsterdam, Reidel.

Kelling, G., and Walton, E.K., 1957, Load-cast structures: Their relationship to upper-surface structures and mode of formation: Geological Magazine, v. 94, p. 481–490.

Lowe, D.R., 1975, Water escape structures in coarse-grained sediments: Sedimentology, v. 22, p. 157–204.

Martinis, B., 1961, Sulla tettonica delle Murge nord-occidentali: Accademia Nazionale dei Lincei, Rend.Sc. fis. mat. nat., ser. 8, v. 31, no. 5, p. 299–305.

Martinis, B., 1962, Lineamenti strutturali della parte meridionale della Penisola Salentina: Geologica Romana, v. 1, p. 11–23.

McKee, E.D., and Bigarella, J.J., 1972, Deformational structures in Brazilian coastal dunes: Journal of Sedimentary Petrology, v. 42, no. 3, p. 670–681.

McKee, E.D., Douglass, J.R., and Rittenhouse, S., 1971, Deformation of lee-side laminae in eolian dunes: Geological Society of America Bulletin, v. 82, p. 359–378.

Molina, J.M., Alfaro, P., and Moretti, M., 1998, Soft-sediment deformation structures induced by cyclic stress of storm-waves in tempestites (Miocene, Guadalquivir Basin, Spain): Terra Nova, v. 10, no. 3, p. 145–150.

Moretti, M., 1996, Tettonica sinsedimentaria e strutture sedimentarie deformative nei depositi pleistocenici di Santa Cesarea Terme (LE): Atti Riunione

Gruppo Informale di Sedimentologia del Consiglio Nazionale delle Ricerche, Università di Catania, p. 196–199.

Moretti, M., 1997, Le strutture sedimentarie deformative: Studio delle modalità di deformazione e dell'origine attraverso esempi fossili e modellizzazione in laboratorio [Ph.D. thesis]: Bari, Italy, Università di Bari, 232 p.

Moretti, M., 2000, Soft-sediment deformation structures interpreted as seismites in middle-late Pleistocene aeolian deposits (Apulian foreland, southern Italy): Sedimentary Geology, v. 135, p. 167–179.

Moretti, M., and Tropeano, M., 1996, Strutture sedimentarie deformative (sismiti) nei depositi tirreniani di Bari: Memorie della Società Geologica Italiana, v. 51, p. 485–500.

Moretti, M., Alfaro, P., Caselles, O., and Canas, J.A., 1999, Modelling seismites with a digital shaking table: Tectonophysics, v. 304, p. 369–383.

Piccardi, L., 1998, Cinematica attuale, comportamento sismico e sismologia della Faglia di Monte Sant'Angelo (Gargano, Italia): La possibile rottura superficiale del "leggendario" terremoto del 493 D.C.: Geografia Fisica Dinamica Quaternaria, v. 21, p. 155–166.

Obermeier, S.F., 1996, Use of liquefaction-induced features for paleoseismic analysis: An overview of how liquefaction features can be distinguished from other features and how their distribution and properties of source sediment can be used to infer the location and strength of Holocene paleo-earthquakes: Engineering Geology, v. 44, p. 1–76.

Owen, G., 1987, Deformation processes in unconsolidated sands, *in* Jones, M.E., and Preston, R.M.F., eds., Deformation of sediments and sedimentary rocks: Geological Society [London] Special Publication 29, p. 11–24.

Owen, G., 1996, Experimental soft-sediment deformation: Structures formed by the liquefaction of unconsolidated sands and some ancient examples: Sedimentology, v. 43, p. 279–293.

Palmentola, G., and Vignola, N., 1980, Dati di neotettonica sulla penisola Salentina (Fogli 204 "Lecce", 213 "Maruggio", 214 " Gallipoli", 215 "Otranto" e 223 "S. Maria di Leuca": Progetto Finalizzato Geodinamica, no. 356, p. 173–202.

Pieri, P., 1975, Geologia della Città di Bari: Memorie della Società Geologica Italiana, v. 14, p. 379–407.

Pieri, P., 1980, Principali caratteri geologici e morfologici delle Murge: Murgia Sotterranea, Bollettino Speleologico Martinese, v. 2, no. 2, p. 13–19.

Pieri, P., 1988, Evoluzione geologica e morfologica dell'area di Bari, *in* "Archeologia di una città: Bari dalle origini al X secolo": Bari, Italy, Edipuglia, p. 7–14.

Pieri, P., Festa, V., Moretti, M., and Tropeano, M., 1997, Quaternary tectonic activity of the Murge area (Apulian foreland–Southern Italy): Annali di Geofisica, v. 40, no. 5, p. 1395–1404.

Plaziat, J.C., 1971, Racines ou terriers? Critères de distinction à partir de quelques exemples du Tertiare continental et littoral du Bassin de Paris et du midi de la France: Conséquences paléogéographiques: Géologie de France, v. 7, no. 13, p. 195–203.

Plaziat, J.-C., and Poisson, A.M., 1992, Mise en évidence de plusieurs séismes majeurs dans le Stampien Supérieure continental au Sud de Paris: Enregistrements sédimentaires de la tectonique oligocène: Bulletin de la Société Géologique de France, v. 5, p. 541–551.

Postpischl, D., ed., 1985, Catalogo dei terremoti italiani dall'anno 1000 al 1980: Quaderni Ricerca Scientifica del Consiglio Nazionale delle Ricerche, v. 114, no. 2b, 239 p.

Reineck, H.E., and Singh, I.B., 1980, Depositional sedimentary environments (2nd edition): Berlin, Springer-Verlag, 549 p.

Ricchetti, G., 1975, Nuovi dati stratigrafici sul Cretaceo delle Murge emersi da indagini nel sottosuolo: Bollettino Società Geologica Italiana, v. 94, p. 1083–1108.

Ricchetti, G., 1980, Contributo alla conoscenza strutturale della Fossa bradanica e delle Murge: Bollettino Società Geologica Italiana, v. 49, no. 4, p. 421–430.

Ricchetti, G., and Mongelli, F., 1980, Flessione e campo gravimetrico della micropiastra apula: Bollettino Società Geologica Italiana, v. 99, p. 431–436.

Ricchetti, G., Ciaranfi, N., Luperto Sinni, E., Mongelli, F., and Pieri, P., 1988, Geodinamica ed evoluzione sedimentaria e tettonica dell'Avampaese Apulo: Memorie della Società Geologica Italiana, v. 41, p. 57–82.

Royden, L., Patacca, E., and Scandone, P., 1987, Segmentation and configuration of subducted lithosphere in Italy: An important control on thrust-belt and foredeep-basin evolution: Geology, v. 15, p. 714–717.

Seed, H.B., and Idriss, I.M., 1982, Ground motions and soil liquefaction during earthquakes: Berkeley, California, Earthquake Engineering Research Institute, 134 p.

Seilacher, A., 1969, Fault-graded beds interpreted as seismites: Sedimentology, v. 13, p. 155–159.

Selli, R., 1962, Il Paleogene nel quadro della geologia dell'Italia meridionale: Memorie della Società Geologica Italiana, v. 41, p. 87–107.

Tramutoli, M., Pescatore, T., Senatore, M.R., and Mirabile, L., 1984, Interpretation of reflection high resolution seismic profiles through the gulf of Taranto (Ionian sea, eastern Mediterranean): The structure of Apennine and Apulia deposits: Bollettino di Oceanologia Teorica ed Applicata, v. 2, no. 1, p. 33–52.

Tsuchida, H., and Hayashi S., 1971, Estimation of liquefaction potential of sandy soils, *in* Proceedings of the 3rd Joint Meeting, US–Japan Panel on Wind and Seismic Effects: Tokyo, p. 91–109.

Tropeano, M., Marino, M., and Pieri, P., 1994, Evidenze di tettonica distensiva plio-pleistocenica al margine orientale della Fossa bradanica: L'Horst di Zagarella: Il Quaternario, v. 7, no. 2, p. 597–606.

MANUSCRIPT ACCEPTED BY THE SOCIETY MAY 11, 2001

Indicators of paleoseismicity in the lower to middle Miocene Guadagnolo Formation, central Apennines, Italy

Goffredo Mariotti
Laura Corda
Marco Brandano
Giacomo Civitelli
Dipartimento di Scienze della Terra, Università degli Studi di Roma "La Sapienza," 00185 Rome, Italy

ABSTRACT

Convolute bedding—pillow horizons—of likely seismic origin are identified in a bioclastic carbonate succession, the Guadagnolo Formation, in the central Apennine Mountains of Italy. These sediments, which were deposited in a carbonate-ramp environment, are from Burdigalian to Langhian (Relizian) in age. The lower part, about 500 m thick, consists of marlstones, marly limestones, and calcarenites, representing cyclic, shallow-water, coarsening-upward sequences. The second part, about 100 m thick, is dominated by prograding bodies of calcarenites. The horizons containing the pillow beds are in the topmost of the lower part, about 30 m below the base of the overlying calcarenites. They are present at the same stratigraphic position from the Prenestini to the Ruffi Mountains across a distance of about 20 km. Pillows, 20 cm to more than 1 m thick, are present in all the deformed layers and consist of marly calcarenites, which differ texturally from the enclosing matrix. They are regarded as the product of deformation ensuing from liquefaction of a denser layer overlying a lighter, silty layer that is richer in clay.

These structures are interpreted to reflect liquefaction processes induced by seismic shocks, and they correlate well with coeval Miocene tectonism in this sector of the Apennines.

INTRODUCTION

Liquefaction, the temporary loss of shear strength in granular sediments, is usually regarded as an indicator of paleoseismicity, although it may have other causes, such as lithostatic loading, cyclic wave-induced (storm) loading, and slumping, which can be difficult to identify with certainty (Ricci Lucchi, 1995). Nonetheless, establishing a causal link between the deformation and paleoseismic events is useful in understanding the paleotectonic regime of ancient basins.

Among others, Sims (1973, 1975) specified criteria for correlating deformation with earthquakes: The basin should be located in a tectonically active region; the sediments should be susceptible to liquefaction; the deformation structures should be confined to a specific stratigraphic interval; the structures should occur over a wide area of the sedimentary basin; and, there should be no slopes or slope failures.

This paper describes pillow horizons in the Guadagnolo Formation, a calcareous, marly succession of Miocene age (Burdigalian to Langhian), in the transition area between the Latium-Abruzzi carbonate shelf and the nearby Sabina basin of central Italy (Fig. 1). Satisfying the aforementioned criteria, these pillows are interpreted to be seismically induced deformation structures related to liquefaction. In the following discussion, we use

Mariotti, G., Corda, L., Brandano, M., and Civitelli, G., 2002, Indicators of paleoseismicity in the lower to middle Miocene Guadagnolo Formation, central Apennines, Italy, *in* Ettensohn, F.R., Rast, N., and Brett, C.E., eds., Ancient seismites: Boulder, Colorado, Geological Society of America Special Paper 359, p. 87–98.

the geometric and textural characterization of sedimentary structures to support interpretation of them as likely indicators of paleoseismicity.

GEOLOGICAL SETTING

The Guadagnolo Formation (Fig. 2) is a lower to middle Miocene calcareous, marly succession deposited in shelf-to-basin transition areas along the western edge of the Latium-Abruzzi carbonate shelf (Civitelli et al., 1986a, 1986b). The formation has a variable thickness, which may reach 1000 m as it approaches the basin to the north. Toward the shelf, however, its thickness decreases as some beds disappear, incomplete and/or condensed successions develop, and major erosion results in the uppermost packstone-grainstone resting directly on Upper Cretaceous shallow-water carbonates. The chief feature of the formation is a pattern of cyclic deposition that occurred in slope and carbonate-ramp environments. The basal part is characterized by gravity-flow deposits (Fig. 3A), whereas the bulk of the unit, of Burdigalian and Langhian (Relizian) age, consists of a thick, monotonous stack of calcareous, marly deposits (Fig. 3B) that exhibit repeated shallowing-upward sequences (average thickness about 0.5 to 10 m) with a complex internal organization and numerous unconformities (Civitelli et al., 1986a, 1986b). These unconformities are thought to be mainly related to high-frequency eustatic cycles and, subordinately, to tectonics, the latter being especially important in the topmost part of the succession (Madonna, 1996; Civitelli et al., 1996). The shallowing- and coarsening-upward sequences consist of argillaceous wackestone, commonly with a bioturbated base, gradually passing upward to wackestone-packstone and to bioclastic packstone-grainstone. These facies are organized into sequences (Fig. 4), which have been interpreted to be subtidal cycles deposited on the deeper part of a mid-ramp (sensu Burchette and Wright, 1992; James et al., 1999). In this part of the Guadagnolo Formation, we have observed rare 15- to 20-cm-thick coarser layers with a major bioclastic component (abundant bryozoans). They have a wide lateral extension, are normally graded, and have a sharp, planar base; undulating, hummocky laminae are the main primary structures. We have interpreted these layers as tempestites.

The uppermost part of the Guadagnolo Formation consists of coarse-grained bioclastic grainstone-packstone and rests on the underlying calcareous, marly part; a major unconformity is in between (Fig. 2). In the lower part, these bioclastic carbonates include a 10-m-thick horizon with abundant chert nodules. This is an important correlation horizon, which recurs throughout the Prenestini and Ruffi mountains. This uppermost part forms a rather homogeneous unit, which caps the Guadagnolo Formation throughout the study area.

The pillow structures are located approximately 30 m below the cherty horizon in the uppermost part of the Guadagnolo Formation, and are located at the same stratigraphic position throughout, from the Prenestini to the Ruffi Mountains in the southern Sabina basin (Fig. 5).

PILLOW HORIZONS

The best exposures with pillow structures are in the northwesternmost sector of the Ruffi Mountains, along the road linking the Tiburtina state road with the village of Saracinesco (Fig. 1). There, at least four superimposed horizons can be recognized (Fig. 5).

The lower horizon (Fig. 6; A in Fig. 5) has a thickness of about 1 m. Within it, spheroidal structures can be seen; they are either isolated or aggregated and consist of two or more laterally or vertically packed spheroids (Fig. 7). In some cases, they are not completely separated, because the marl wedges, which appear to be injected between them from below, do not cross the entire layer. The pillows are more coarse grained and much more indurated than the surrounding marl. In the same horizon, an elongate body extends for about 6 m and is bounded by a flat upper surface and by a more undulating lower one. This body is composed of multiple spheroidal structures, with diameters ranging from 30 to 40 cm, which are separated by areas of calcareous marl with weak lamination that follow the contour of the pillows (Fig. 8). This horizon overlies a layer of highly bioturbated calcareous marl and is topped by about 40 cm of marl with both horizontal and vertical traces of equally intense bioturbation.

Above this horizon, the succession continues with 1 m of more calcareous, bioturbated marl, followed by about 2 m of marly limestone. Above this is another 6 m of marlstone, bearing burrows with a basically subhorizontal pattern, mainly represented by *Planolites* and *Palaeophycus*. The subsequent pillow layer (B in Fig. 5) is approximately 80 cm thick and is composed of ovoid bodies in marlstones. The width of these bodies varies from 10 to 60 cm, and their lower and upper surfaces are usually flat. The pillows are separated by marls injected from below.

The next level of deformation (C in Fig. 5) is above 2.5 m of marlstone and is very similar to B; it is less than 1 m thick, and the 20- to 50-cm-thick pillows, consisting of marly limestones, are embedded in marl. Bioturbation is observed in both the pillows and the marl, but never across the marl-pillow boundaries.

The uppermost pillow bed (D in Fig. 5) overlies 3.5 m of calcareous marlstone and intensely bioturbated marlstone. It is composed of two large composite bodies consisting of two or more pillows (Fig. 9). The first is about 2 m long, and the second is 3.5 m long. Both bodies are about 1.5 m high. The pillows inside these bodies generally have a spheroidal shape, but their lower and upper surfaces are fairly flat. They are made up of marly limestone, but their host material has a dominant marly component with cuspate geometry, apparently due to escaping fluids. This stratigraphically younger interval occurs about 30 m below the unconformity shown in Figure 2.

In the same stratigraphic position, about 20 km to the south in the Prenestini Mountains, similar pillow horizons were identified. The best outcrops are near the villages of Guadagnolo and of Capranica (Fig. 1). Below the rock on which Guadagnolo village rests, the topmost pillowed interval can be seen (D in Fig. 5). This interval is about 2 m thick and has simple and composite pil-

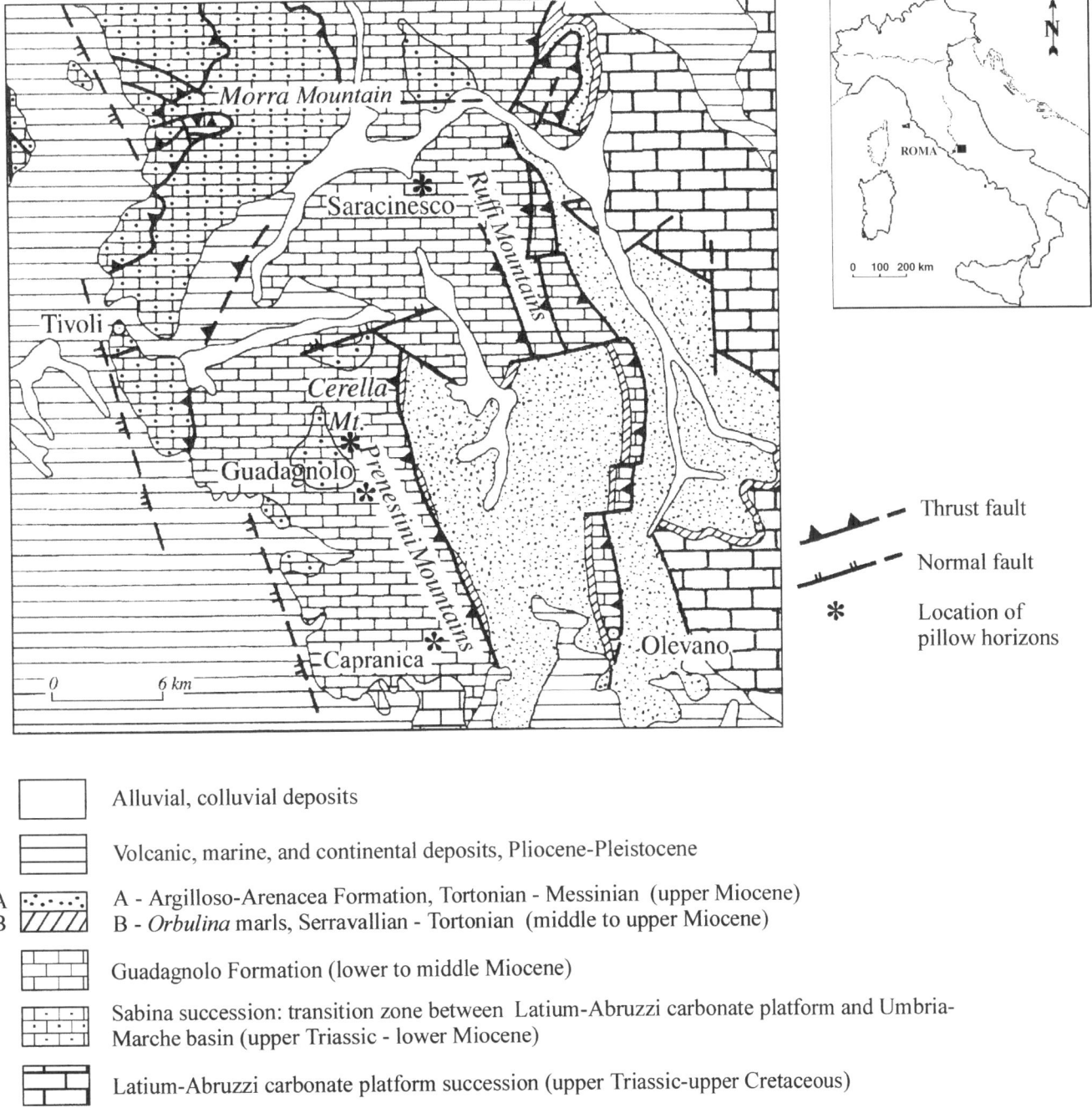

Figure 1. Location of the study area in central Italy.

lows that range from 30 to 60 cm in width. At the northern entrance of Capranica the same deformed horizon (D in Fig. 5) crops out for a few tens of meters. The horizon, more than 1.5 m thick, displays spheroidal- to elliptical-shaped pillows with a main diameter of more than 1 m.

Finally, other exposures with similar pillows at a similar stratigraphic position were identified in the Cicolani Mountains, about 50 km to the north. These are still being investigated. Because of the discontinuous exposure, only the youngest pillow horizon can be correlated from exposure to exposure.

MICROANALYSIS

In the studied horizons, both the pillows and the enclosing material were analyzed with scanning electron microscopy and energy-dispersive spectrometry (SEM-EDS) to better characterize

Figure 2. Outcrop of the Guadagnolo Formation in the Prenestini Mountains. The topmost part consists of coarse-grained bioclastic grainstones overlying a predominantly marly, calcareous deposit separated by a major third-order unconformity (arrow). The arrow also indicates a cliff about 20 m high.

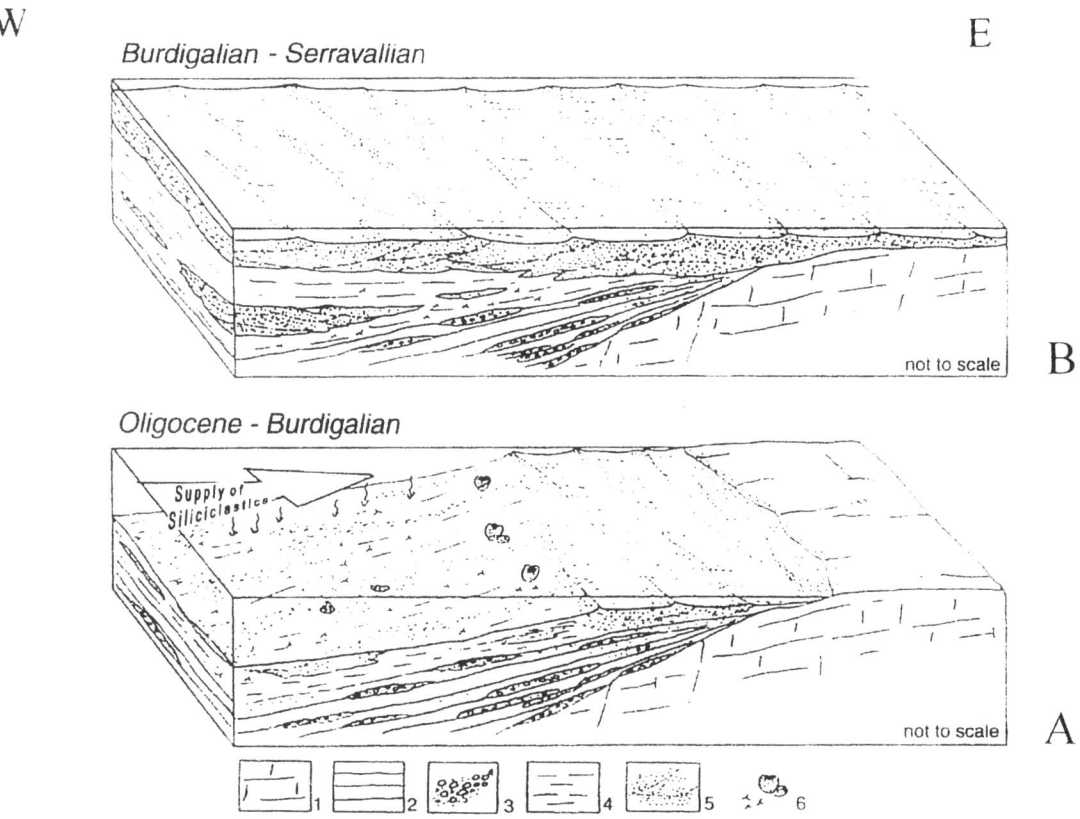

Figure 3. Interpretative diagrams illustrating the evolution of the platform-basin zone in the Sabina area during the Oligocene to mid-Miocene interval. 1—shallow-water carbonates of Latium Abruzzi platform; 2—pelagites; 3—gravity flow deposits in basal part of Guadagnolo Formation; 4—calcareous, spiculitic marly deposits of middle-upper part of Guadagnolo Formation; 5—coarse-grained bioclastic limestones of uppermost part of Guadagnolo Formation; 6—siliceous sponges and spicules (after Civitelli et al., 1986b).

their texture and composition. The pillows consist of a grain-supported fabric with scarce intergranular spaces occupied by small amounts of clay. The individual grain sizes range from 50 to 200 µm. Rounding and sorting are poorly developed (Fig. 10A). Large bioclasts are rarely present and are almost exclusively represented by fragments of echinoids, bivalves, bryozoans, benthic and planktic foraminifers, and siliceous sponge spicules (Fig. 11).

In contrast, the marl squeezed between pillows is matrix-supported, in that bioclasts are not in contact with each other. The individual grains are between 10 and 150 µm in diameter (Fig. 10B), and sorting is even less developed. The biogenic composition, however, is not significantly different from that on the pillow sediments (Fig. 12).

The different density of packing and thus the different matrices are also shown by semi-quantitative chemical analysis (SEM-EDS). The samples taken from the pillows have a high calcium content, whereas silicon is subordinate and aluminum is even scarcer (Fig. 10D). By contrast, the analyses conducted on an equivalent surface area of a rock sample collected from an injected marl show a higher silicon content and the presence of aluminum, potassium, and iron (Fig. 10E). The difference in pillow and nonpillow is due only to the different amount of clay matrix. The elemental chemical composition of the interparticle matrix (Fig. 10C) proves to be consistent with the presence of clay minerals (Fig. 10F). Insoluble-residue analysis, by means of X-ray diffraction, indicates an abundance of clay minerals of the smectite group and subordinate quartz. Carbonate percentages were also determined by analysis of calcium carbonate. Both pillows and injected marl from different deformed horizons were analyzed. The percentage of the $CaCO_3$ is consistently about 75% within the pillows versus 70% in the surrounding marl. These values show a weak but constant difference in all analyzed horizons.

DISCUSSION

The analyzed pillows consist of bioclastic packstone-grainstone with fine-sand and coarse-silt grain sizes. They are surrounded by injected marls represented by wackestone-packstone with medium to fine silt grain sizes. The presence of two main granulometric classes, medium- to fine-grained sands and medium to fine silt (with 30% clay), increases the potential of selective liquefaction and better preserves the resulting deformation (Ricci Lucchi, 1995).

The pillows are characterized by upward-cuspate contacts at the base and by an internal lamination that, if present, follows the external structure. In our opinion, the pillow beds of the Guadagnolo Formation can be interpreted as soft-sediment deformation structures related to the liquefaction processes in an unconsolidated two-layer system. Furthermore, the grain sizes involved in this phenomenon are those that have a high liquefaction potential (Seed and Idriss, 1982). Moretti et al. (1999a) reported some experiments on earthquake-induced liquidization processes in

Figure 4. A shallowing, coarsening-upward, subtidal cycle consisting of argillaceous wackestone passing upward to wackestone-packstone and to bioclastic packstone-grainstone.

unstable, two-layer, density-gradient systems. Liquefaction caused the loss of shear resistance in both strata; the triggered deformation consists of large load casts. Only in the experiment where Moretti et al. (page 377) used "a very soft silty clay with very low shear strength" as the underlying layer, did the overlying sands collapse inside the thixotropic silty clay. The cuspate form of the clay-rich silt squeezed between the pillows was probably due to a lesser effective viscosity of the upper layer compared with the lower one (after Alfaro et al., 1997).

Our analyzed structures formed in a reverse-density gradient system in which a medium fine sand layer, showing greater bulk density as a function of particle concentration, overlies a light silty layer. After liquefaction, the denser layer was completely disrupted, having been reduced to a series of isolated pillows embedded in the lower, less dense underlying layer. The overlying, denser layer, although composed of angular bioclastic grains, was capable of liquefaction because of its water saturation.

Liquefaction processes, as we mentioned above, may have different causes; thus, it is important to review the main causes or processes capable of generating structures similar to the ones we investigated.

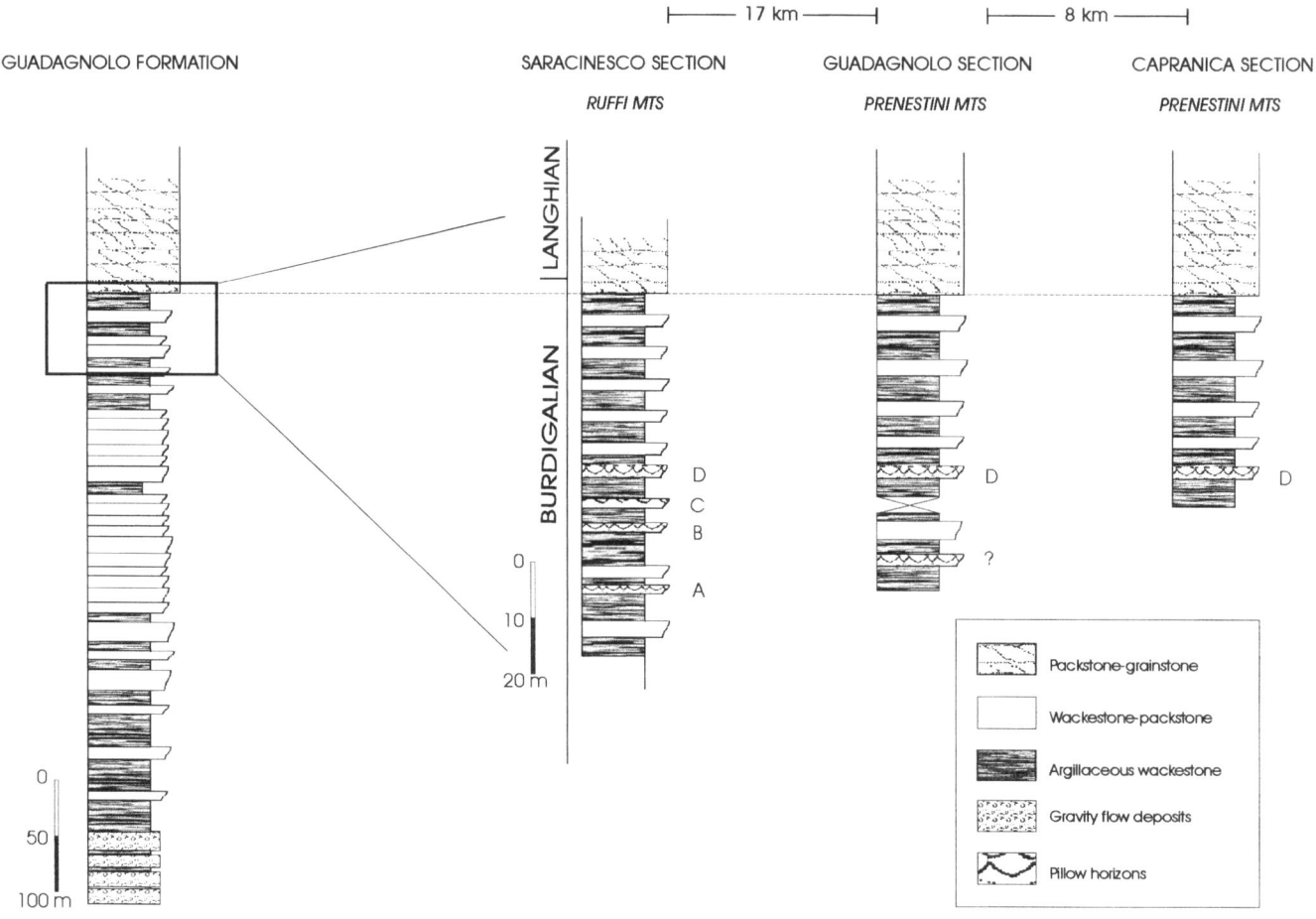

Figure 5. Measured lithostratigraphic sections. Locations shown in Figure 1.

Slumping

The possibility of slumping can be discounted for at least two reasons. First, the pillowed horizons do not show folds, and the geometry of the pillows fails to indicate vergence, which would suggest re-sedimentation due to lateral movement of the materials. The geometry of the escape structures does not indicate any lateral translation, and the deformed horizons are never joined with re-sedimented deposits that would indicate an environment conducive to slumping processes.

Second, the paleogeographic setting for the analyzed part of the Guadagnolo Formation, reconstructed on the basis of a physical and sedimentological stratigraphic study (Civitelli et al., 1996), is a middle ramp with bathymetries of some tens of meters, where the sediments were remobilized by waves and significant tidal currents. The low angle of the ramp was such that it probably did not permit, without an external trigger, slumping and gravity flows. In this connection, we note that in contrast, in the lower part of the succession, where gravity-flow deposits are common and unequivocally indicate sedimentation over a shelf-basin slope pillow structures are never found.

Loading

Unequal loading by density contrast (Dzulynski, 1996) can be responsible for deformational structures similar to the ones we studied. Depending on sedimentation rate, some lithologies may develop nodules floating within more muddy materials. Moretti et al. (1999b) described pillow structures from Miocene (Tortonian) turbidite deposits of the Guadix basin in southern Spain. These structures, cyclically repeated in the succession, are elliptical in shape (long axes between 20 cm and 2 m) and commonly parallel to the paleoslope. According to Moretti et al., the probable triggering mechanism for the studied pillows was the process of overloading induced by rapid mass sedimentation. In the succession that we have analyzed for this study, however, such structures occur only in limited stratigraphic intervals within a succession many hundreds of meters thick that consists of hundreds of cyclically repeated sequences with the same lithologic characters and thicknesses. It seems unlikely that the pillows we studied derive from intrinsic processes or from the mere density contrast of different lithologies. Otherwise, given the lithological homogeneity of the succession, we would expect a greater fre-

Figure 6. Lowermost pillow horizon (A in Fig. 5) at the Saracinesco site (Fig. 1). Note the numerous spheroidal structures across a distance of about 6 m. These structures consist of marly, bioclastic limestones embedded in calcareous marls. Note the flat surface that bounds the top of the layer. Hammer is 33 cm long.

Figure 7. Composite pillow, consisting of two vertically stacked spheroids, inside the pillow horizon shown in Figure 6. Hammer is 33 cm long.

quency in the occurrence of such deformation. Because such a frequency is not present, we assume an extrinsic cause, external to the depositional system and capable of triggering a process that is latent in successions consisting of materials with different density, packing, and fabric.

Storm waves

Many authors have ascribed convolute bedding to storm waves, whose passage can liquefy cohesive and noncohesive sediments, giving rise to pillow structures. Liquefaction due to

Figure 8. Two pillows separated by calcareous marls with laminar structures that tend to parallel the shape of the pillows. Saracinesco site (A in Fig. 5). Hammer is 33 cm long.

Figure 9. The uppermost pillow horizon (D in Fig. 5, locality Saracinesco) consists of pillows that may exceed 1 m in thickness and which are commonly ovoid, with flat bases. The first pillow on the left side is about 120 cm wide.

Figure 10. Scanning Electron Microscopy (SEM) micrographs and plots. A: Sample collected from the inside of a pillow; texturally, it is a bioclastic packstone-grainstone, with grains varying in diameter from 50 to 200 μm and with poor matrix. B: Sample of wackestone-packstone surrounding the pillows; note the larger amount of clay matrix, the smaller average size of the grains, and the lower degree of sorting. C: Enlarged image of the sample in B, with abundant matrix; Energy Dispersive Spectrometry and Scanning Electron Microscopy (EDS-SEM) analyses on the same samples also reveal differences between the pillows and the surrounding material. D: Sample collected from a pillow. E: Sample from marl surrounding the pillows. F: Sample from the interparticle matrix of the injected marl. The high amount of silica in all the samples is also probably related to the numerous sponge spicules both in the pillows and in the embedding marl. EDS spectra represent a 30-s, counting-time acquisition.

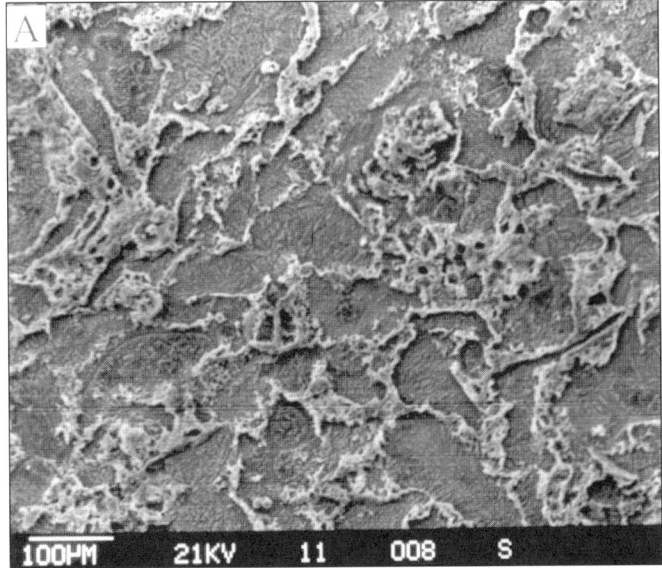
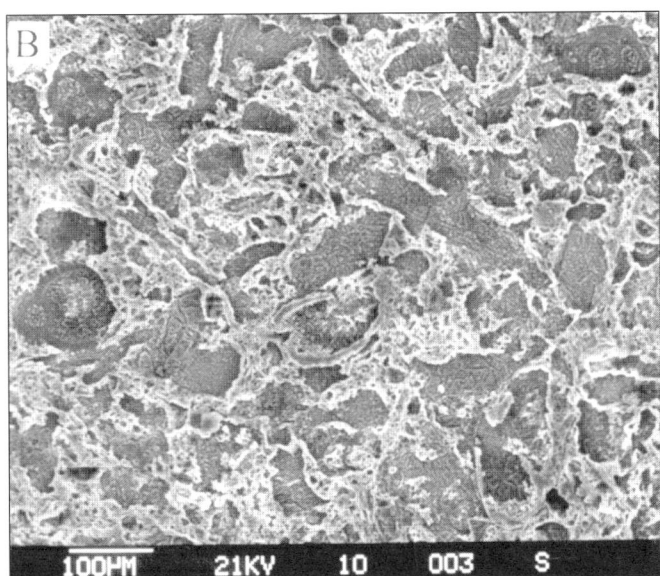
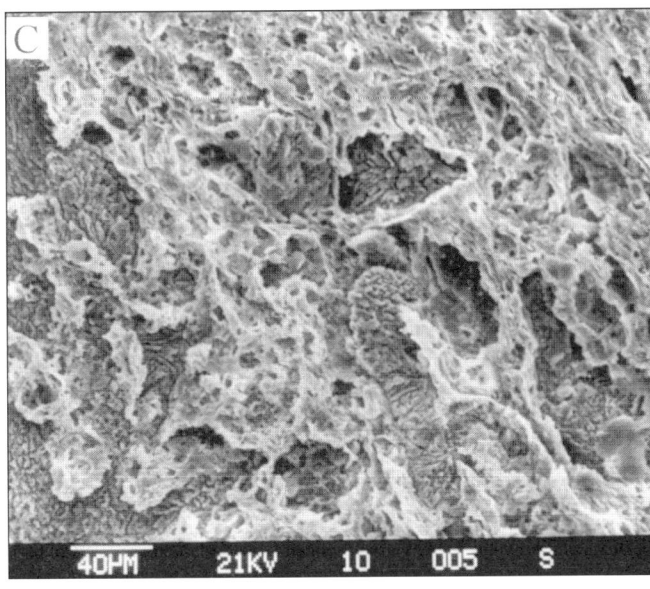
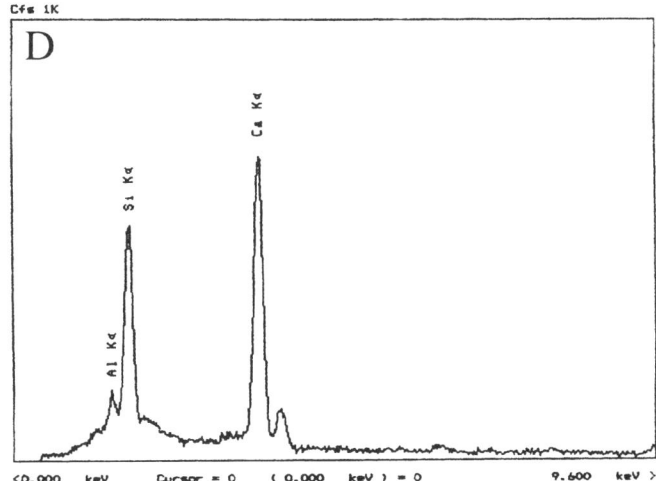
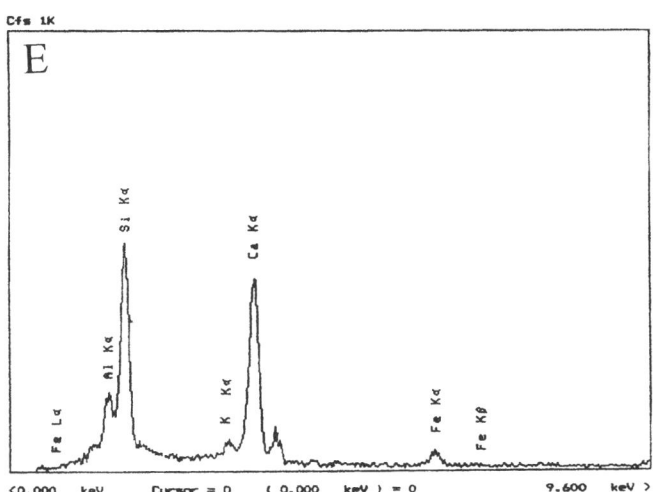
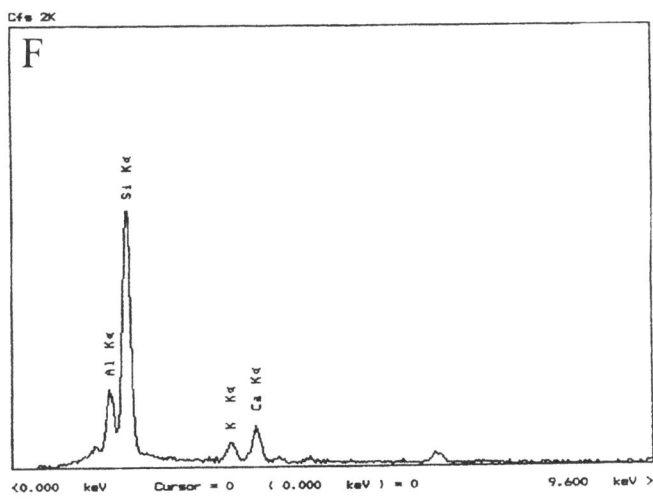

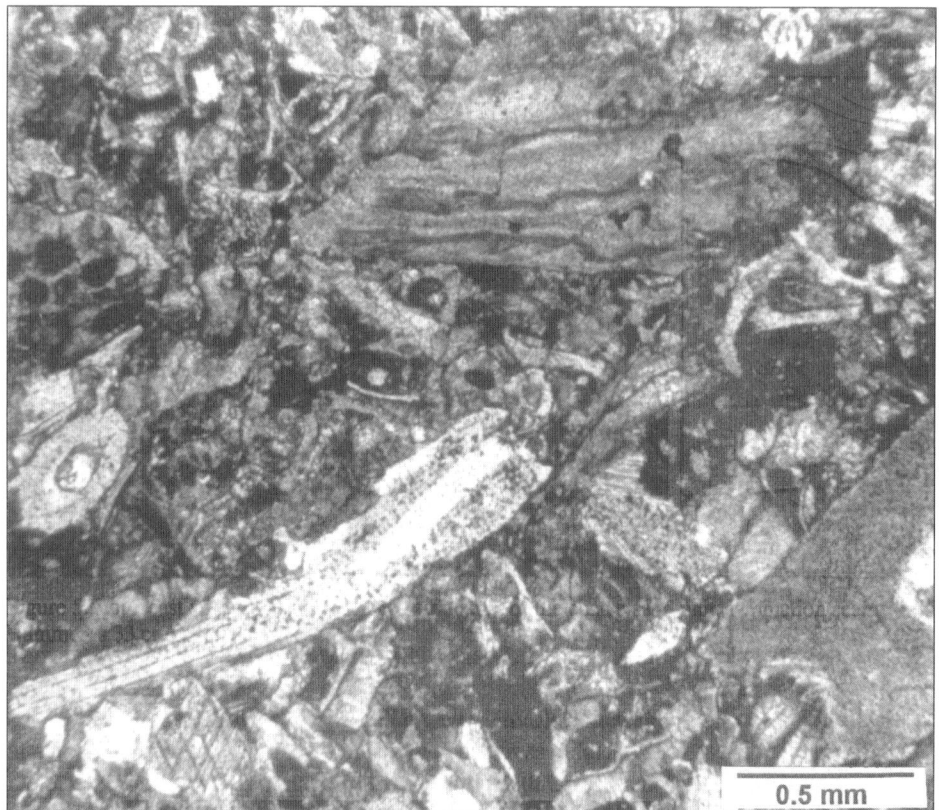

Figure 11. Photomicrograph showing medium-grained packstone, collected from a pillow, with bryozoan and echinoid fragments and benthic foraminifers. It shows poor sorting and a small percentage of matrix.

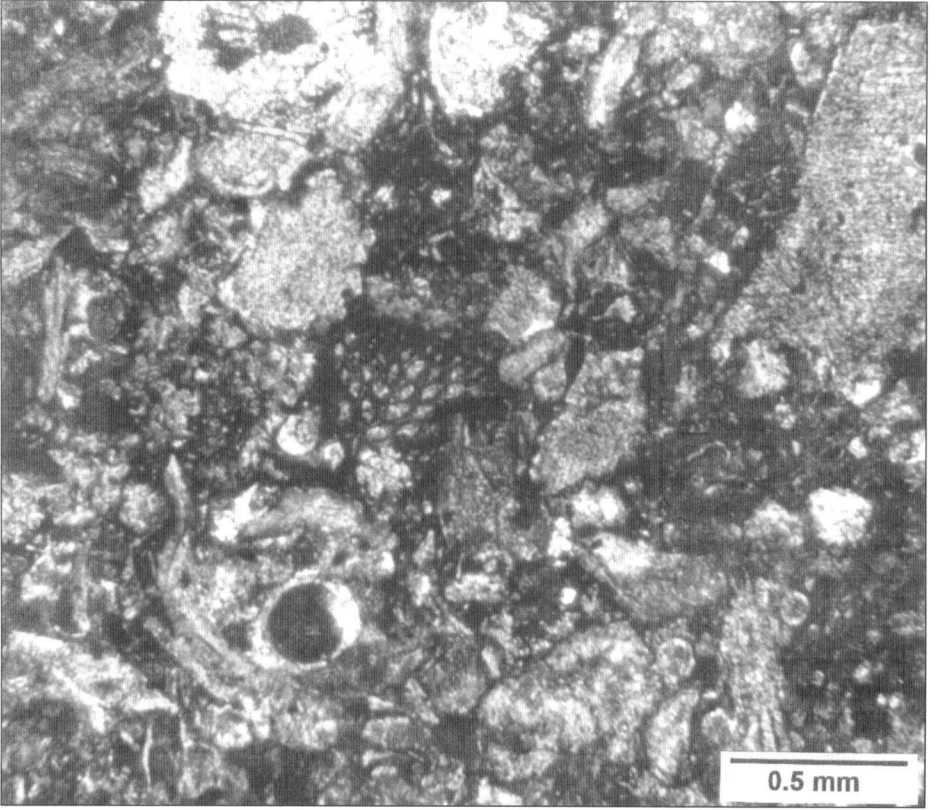

Figure 12. Photomicrograph showing fine-grained packstone-wackestone, collected from marl injected between pillows, containing planktic and benthic foraminifers together with bryozoan and echinoid fragments. Sorting is poor, and the matrix is abundant.

storm waves has been analyzed in modern sediments (Henkel, 1970; Okusa, 1985; Dormieux and Delage, 1988), but few examples have been identified in ancient storm deposits. Nevertheless, an association of soft-sediment deformation structures with hummocky cross stratification is frequently reported (Molina et al., 1998).

In the upper part of the succession we studied, where the pillow horizons are, we did not find any coarser layer with distinctive features we could attribute to storm waves. Thus, on the basis of our findings, the assumption of a possible relation between our deformation structures and storm-wave action is untenable.

Seismicity

In view of the above assumptions, a seismic-shock origin for the pillows is most probable. One of the best arguments supporting this hypothesis is that, in many hundreds of meters of a monotonous succession of cyclically repeated lithologies, the pillows are widespread, but only in the same well-defined stratigraphic interval, between 50 and 30 m below the base of the uppermost calcarenitic part of the Guadagnolo Formation. At the same stratigraphic level, in more northern sectors of the Sabina basin, about 50 km north of the Prenestini Mountains in the Cicolani Mountains, we identified pillow structures very similar to the analyzed ones.

Pillow features comparable to the ones from this study have been found in Spain by Roep and Everts (1992), who indicated these liquefaction features were of seismic origin, on the basis of their observed structures, grain sizes, and presence of dewatering features. Hempton and Dewey (1993) ascribed the deformational structures that they found in Lake Hazar (Turkey) to seismic events. There, the pillow horizons, which are separated by undeformed intervals, are distributed over a wide area and limited to a specific stratigraphic level. Rosskopf et al. (1995) also referred the genesis of the pillow structures encountered in the littoral sediments of Piana del Sele (Campania, Italy) to seismic shocks, given the extent of the affected area and their environment, so far from the shoreline as to exclude lithostatic loading or storm-wave action.

CONCLUSIONS

Given the textural and geometric characters, the well-defined stratigraphical interval, the distance between the outcrops, and the paleogeographic and environmental reconstruction of the Guadagnolo Formation during the pillow-forming time (late Burdigalian), we suggest that these pillows are seismically induced deformational structures. The lithostratigraphic framework and the stratigraphic-physical architecture of the succession containing the deformed horizons is controlled not only by sea-level changes, but also by local and regional tectonics. The Guadagnolo carbonate ramp was part of the foreland basin of the Apennine chain during early Miocene (Burdigalian) time and evolved upward to a foredeep filled by siliciclastic deposits of late Miocene (Tortonian-Messinian) age (Cipollari et al., 1995). On the basis of stratigraphic data from synorogenic sediments deposited during the Miocene in foreland depositioncenters in front of the eastward-migrating Apennine thrust belt, Cipollari et al. (1995) pointed out a Burdigalian thrust phase centered on the northern Sabini Mountains, about 100 km northwestward. This consideration and the widespread areal distribution of the seismically induced pillow horizons suggest that these paleoseismites may reflect the distal effect of tectonics that affected the Apennine accretionary wedge, even if the studied area does not clearly record the presence of faults.

We maintain that the presence of seismically induced deformation structures could help in identifying distal tectonic processes in areas that do not have other important tectonic indicators. In addition, singling out and correctly interpreting such soft-sediment deformation structures could provide useful criteria for timing the evolution of major tectonic stages. Analyses of the microstructure of the rocks indicate the nature of variations in lithology and fabric that could increase the potential of liquefaction.

ACKNOWLEDGMENTS

We are grateful to B.R. Pratt and R. Van Arsdale for their rigorous reviews of this paper and their helpful suggestions, and to F.R. Ettensohn, who provided insightful scientific and editorial comments that improved the manuscript greatly. We also thank Faith Rogers and Salma Monani for their suggestions on the final copyediting.

REFERENCES CITED

Alfaro, P., Moretti, M., and Soria, J.M., 1997, Soft-sediment deformation structures induced by earthquakes (seismites) in pliocene lacustrine deposits (Guadix-Baza Basin, Central Betic Cordillera): Eclogae Geologicae Helvetiae, v. 90, p. 531–540.

Burchette, T.P., and Wright, V.P., 1992, Carbonate ramp depositional system: Sedimentary Geology, v. 79, p. 3–57.

Cipollari, P., Cosentino, D., and Parotto, M., 1995, Modello cinematico-strutturale dell'Italia centrale: Studi Geologici Camerti, v. 1995/2, p. 135–143.

Civitelli, G., Corda, L., and Mariotti, G., 1986a, Il Bacino Sabino. 2. Sedimentologia e stratigrafia della serie calcarea e marnoso-spongolitica (Paleogene-Miocene): Memorie Società Geologica Italiana, v. 35, p. 33–47.

Civitelli, G., Corda, L., and Mariotti, G., 1986b, Il Bacino Sabino. 3. Evoluzione sedimentaria ed inquadramento regionale dall'Oligocene al Serravalliano: Memorie Società Geologica Italiana, v. 35, p. 399–406.

Civitelli, G., Corda, L., Madonna, S., Mariotti, G., Milli, S., Barbieri, M., and Castorina, F., 1996, I depositi miocenici di rampa carbonatica dei Monti Prenestini (Appennino centrale): processi deposizionali e stratigrafia sequenziale: Atti riunione del gruppo di sedimentologia del Consiglio Nazionale delle Ricerche (Catania), p. 123–125.

Dormieux, L., and Delage, P., 1988, Effective stress response of a plane sea-bed under wave loading: Géotechnique, v. 38, p. 445–450.

Dzulynski, S., 1996, Erosional and deformational structures in single sedimentary beds: A genetic commentary: Annales Societatis Geologorum Poloniae, v. 66, p. 101–189.

Hempton, M.R., and Dewey, J.F., 1983, Earthquake-induced deformational structures in young lacustrine sediments, East Anatolia Fault, southeast Turkey: Tectonophysics, v. 98, Letter Section, p. T7–T14.

Henkel, D.J., 1970, The role of waves in causing submarine landslides: Géotechnique, v. 20, p. 75–80.

James, N.P., Collins, L.B., Bone, Y., and Hallock, P., 1999, Subtropical carbonates in a temperate realm: Modern sediments on the southwest Australian shelf: Journal of Sedimentary Research, v. 69, p. 1297–1321.

Madonna, S., 1996, Analisi stratigrafico-sequenziale della successione miocenica marnoso-calcarea di Guadagnolo (M.ti Prenestini, Appennino centrale) [Ph.D. thesis]: Roma, Università degli Studi di Roma "La Sapienza", 178 p.

Molina, J.M., Alfaro, P., Moretti, M., and Soria, J.M., 1998, Soft-sediment deformation structures induced by cyclic stress of storm waves in tempestites (Miocene, Guadalquivir Basin, Spain): Terra Nova, v. 10, p. 145–150.

Moretti, M., Alfaro, P., Caselles, O., and Canas, J.A., 1999a, Modelling seismites with a digital shaking table: Tectonophysics, v. 304, p. 369–383.

Moretti, M., Soria, J., Alfaro, P., and Walsh, N., 1999b, Soft-sediment deformation structures in Tortonian turbiditic deposits (Guadix basin, southern Spain): Atti Geoitalia 1999, Bellaria, p. 124–126.

Okusa, S., 1985, Wave-induced stresses in unsaturated submarine sediments: Géotechnique, v. 35, p. 517–532.

Ricci Lucchi, F., 1995, Sedimentological indicators of paleoseismicity, *in* Serva, L., and Slemmons, D.B., eds., Perspectives in paleoseismology: Seattle, Washington, Association of Engineering Geologists Special Publication 6, p. 7–17.

Roep, T.B., and Everts, A.J., 1992, Pillow-beds: A new type of seismites? An example from a Oligocene turbidite fan complex Alicante, Spain: Sedimentology, v. 39, p. 711–724.

Rosskopf, C., Cinque, A., Ferreli, L., Michetti, A.M., and Vittori, E., 1995, Strutture da liquefazione Interpretabili come sismiti in sedimenti litorali storici della Piana del Sele (Campania): Studi Geologici Camerti, Volume Speciale 1995/2, p. 387–395.

Seed, N.B., and Idriss, I.M., 1982, Ground motions and liquefaction during earthquakes: Berkeley, California, Earthquake Engineering Research Institute, p. 227–251.

Sims, J.D., 1973, Earthquake-induced structures in sediments of Van Norman Lake, San Fernando, California: Science, v. 182, p. 161–163.

Sims, J.D., 1975, Determining earthquake recurrence intervals from deformational structures in young lacustrine sediments: Tectonophysics, v. 29, p. 141–152.

Manuscript Accepted by the Society May 11, 2001

Stratigraphic and sedimentological evidence for late Paleozoic earthquakes and recurrent structural movement in the U.S. Midcontinent

Daniel F. Merriam
Kansas Geological Survey, University of Kansas, Lawrence, Kansas 66047, USA

Andrea Förster
GeoForschungZentrum Potsdam, Telegrafenberg, D-14473 Potsdam, Germany

"Hallo, Eeyore," said Christopher Robin, as he
opened the door and came out. "How are you?"
"It's snowing still," said Eeyore gloomily.
"So it is."
"And freezing."
"Is it?"
"Yes," said Eeyore. "However," he said, brightening
up a little, "we haven't had an earthquake lately."
—Benjamin Hoff, *The Te of Piglet*, 1992

ABSTRACT

The present major structures and their subsidiary features in the U.S. Midcontinent, in eastern Kansas and northeastern Oklahoma in particular, were formed near the end of the Mississippian and in the early Pennsylvanian, except for continued minor adjustments and regional tilting.

These major structures overprinted the earlier Paleozoic structure and were controlled by the inherent Precambrian basement fracture-fault pattern. This pattern is reflected in the overlying Paleozoic sedimentary section, and recognition of continued rejuvenation is based on stratigraphic and sedimentological evidence. The local folds, known as "plains-type folds," continued to develop after their formation, as chronicled by differential compaction, mostly in the shaly units, over "buried hills," which for the most part are irregularities created by the differential movement of the basement fault blocks. Recurrent movement on the fault blocks is corroborated by sedimentary convolute bedding (also known as soft-sediment deformation) formed in the Permian-Pennsylvanian sequence, again occurring mostly in the clastic units. The compaction is greatest and convolute features are most abundant in the older shaly units, as would be expected, indicating that the recurrent movement was most intense immediately following and closest to the Ouachita orogeny in southern Kansas and northern Oklahoma near the deformation source. The convolute bedding generally is recognized as having been formed in special dewatering situations in the sediments and triggered by intense mechanisms such as paleo-earthquakes. The paleo-earthquakes and the differential movement of fault blocks are a reflection of changing stress within the basement. Magnitude (M) on some of the paleo-earthquakes is estimated to be between 6.5 and 7, affecting between 16 000 and 26 500 km².

Merriam, D.F., and Förster, A., 2002, Stratigraphic and sedimentological evidence for late Paleozoic earthquakes and recurrent structural movement in the U.S. Midcontinent, *in* Ettensohn, F.R., Rast, N., and Brett, C.E., eds., Ancient seismites: Boulder, Colorado, Geological Society of America Special Paper 359, p. 99–108.

INTRODUCTION

Recognition of paleo-earthquakes is difficult at best. The record in the U.S. Midcontinent as interpreted in the sedimentary sequence is sparse and diffuse, but it can be pieced together with some reliability with stratigraphic and sedimentological evidence. McCalpin and Nelson (1996, p. 10–11) presented a hierarchical classification of paleoseismic features and recognized two categories: primary (created by tectonic deformation) and secondary (created by seismic shaking). Faults and folds are considered primary stratigraphic evidence, whereas convolute bedding is considered secondary. The late Paleozoic deformational history of the area was documented recently by Merriam and Förster (1996). Pettijohn et al. (1972), Blatt et al. (1980), and Collinson (1994), as well as many others, have suggested that at least some convolute bedding was formed by shock (paleo-earthquakes) applied to unconsolidated sediments. Lucchi (1995) used the term *seismite* for these features and attributed the term to Adolf Seilacher.

The stratigraphic evidence consists of the time and place of differential shale compaction over irregularities in the section, specifically over buried hills, during the formation of the Midcontinent plains-type folds. The amount and duration of the compaction over the tilted Precambrian basement fault blocks indicates recurrent movement on these structures in response to changing stresses in the basement.

The sedimentological evidence is in the form of convolute bedding, which is interpreted as probably being triggered by paleo-earthquakes generated by movement along the basement faults. The recurrence of these sedimentological features in the same location corroborates recurrent movement on the faults.

Our purpose here is to document the late Paleozoic structural adjustments that occurred during and after the Ouachita orogeny, which took place south of the stable Midcontinent region, affecting eastern Kansas and northeastern Oklahoma. We have noted the time and place of recurrent movement on plains-type folds and the formation of the convolute bedding (or soft-sediment deformation) caused by paleo-earthquakes. In addition, on the basis of field data, we have made some preliminary estimates of the magnitude of the paleo-earthquakes and the size of the affected area.

STRATIGRAPHY

The stratigraphy is described only briefly here; see Merriam (1963) and Zeller (1968) for details.

Overlying the crystalline Precambrian basement is a series of relatively thin, persistent, and regionally uniform lower Paleozoic units (Fig. 1). A basal Paleozoic sandstone (Upper Cambrian) nonconformably overlies the Precambrian basement and ranges in thickness from 0 to 35 m (Cole, 1975). This sandstone is overlain by the thick carbonate (mostly dolomite) Cambrian-Ordovician Arbuckle Group (125–250 m). Unconformably overlying the Arbuckle is the clastic (mostly sandstone) Simpson Group, 0 to 15 m thick. The Viola Limestone and Maquoketa Shale (both Ordovician) are absent over most of the area, but locally as much as 25 m thick. The black Chattanooga Shale, which is Devonian-Mississippian, is a persistent marker over much of the area and is up to 45 m thick (Merriam, 1963). Overlying the Chattanooga is a Mississippian carbonate section (mainly limestone) ranging in thickness from 75 to 125 m.

A major change in the structural configuration and sedimentary regimen occurred in late Mississippian–early Pennsylvanian time. A large influx of clastic material was deposited across this region by fluctuations of sea level in a rhythmic manner of alternating thin, but relatively persistent, beds of clastic and carbonate deposits. The Pennsylvanian deposits range in thickness from about 175 to 700 m; they unconformably overlie the Mississippian and are conformably overlain by similar-type Permian units ranging from 0 to 225 m thick (Merriam, 1963). The total sedimentary sequence overlying the Precambrian basement in southeastern Kansas, which is representative of the area, ranges from about 500 m in the extreme southeast on the flank of the positive Ozark Uplift to 1800 m thick westward in the deeper parts of the basins.

STRATIGRAPHIC EVIDENCE

The origin and development of plains-type folds have been discussed by many authors (see Merriam, 1999 for a history of the concept). Essentially, the folds are formed by differential compaction over tilted Precambrian basement faults.

The major change in the structural configuration in the Midcontinent at the end of the Mississippian and in the early part of the Pennsylvanian is mirrored in a change in the sedimentary regime near the end of the Mississippian. The Ouachita orogeny resulted in the major structures as recognized today, with only subsequent minor structural changes and regional tilting. The major features are basins separated by positive features (Fig. 2).

Most of the thinning of stratigraphic units in the late Paleozoic took place in the dominantly clastic units composed mostly of shale—for example, in the Cherokee, Pleasanton, Douglas, and Admire groups. This is explained partly by the greater adjustments possible in the weaker shales than in the intervening more rigid limestones that compose much of the Marmaton, Lansing–Kansas City, and Shawnee groups (Merriam, 1963). The thick clastic wedges, marking renewed episodic orogenic activity in the late Paleozoic Ouachita belt to the south, were spread over adjacent areas to the north. In southern Kansas and northeastern Oklahoma, large amounts of these clastics were distributed in a shallow sea interspersed with and periodically overwhelming the carbonate sedimentation.

The clastics were affected most by differential movement of the local fault blocks, which were responding to forces transmitted through the rigid crystalline basement. The response in the sediments was the differential compaction of the soft shales over irregularities resulting from the differential elevation of the basement fault blocks or the nondeposition of sedimentary material over topographic highs on the seafloor.

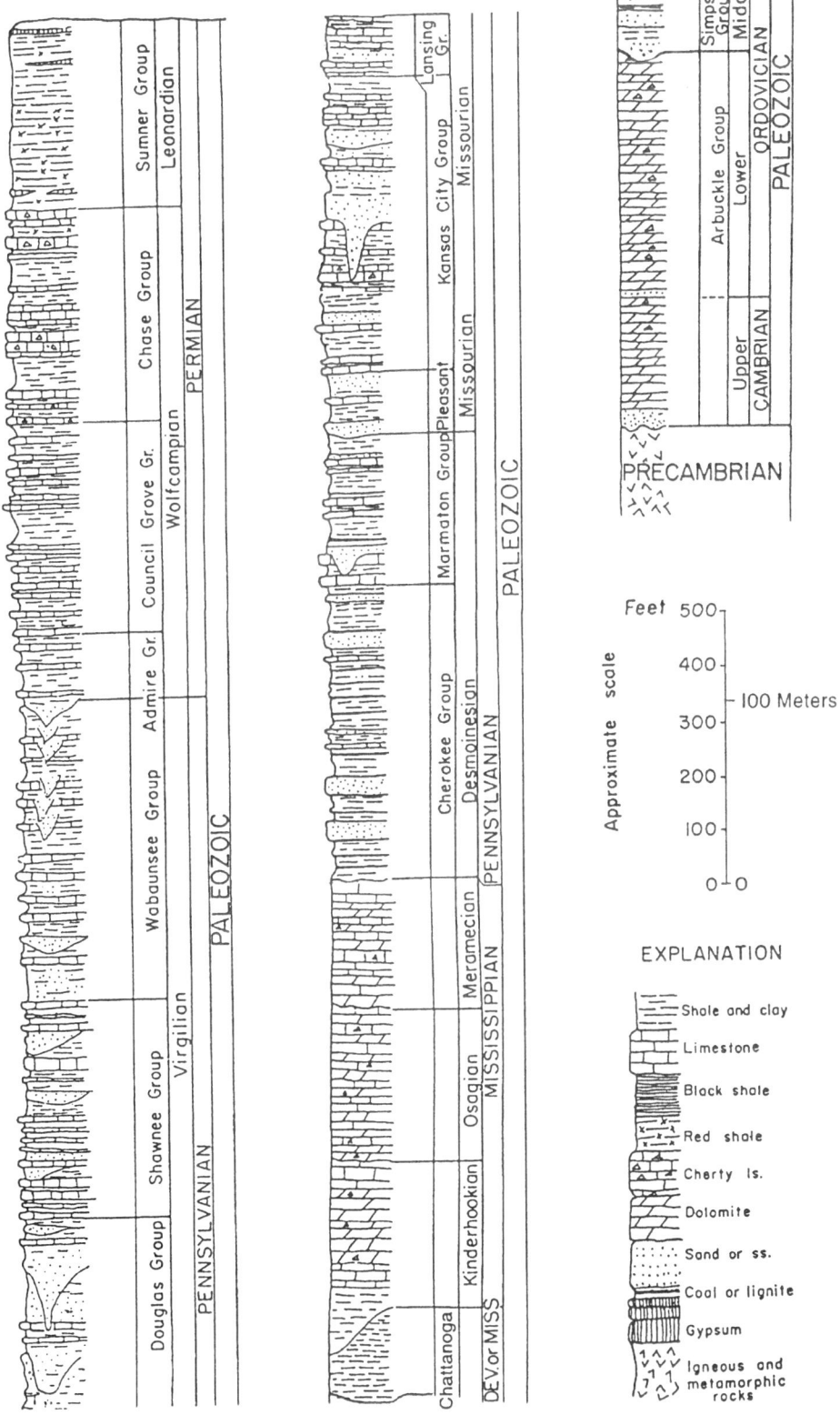

Figure 1. Generalized stratigraphic section of rocks present in southeastern Kansas, which is representative of eastern Kansas and northeastern Oklahoma. Lower Paleozoic units are known in subsurface only (Merriam, 1963).

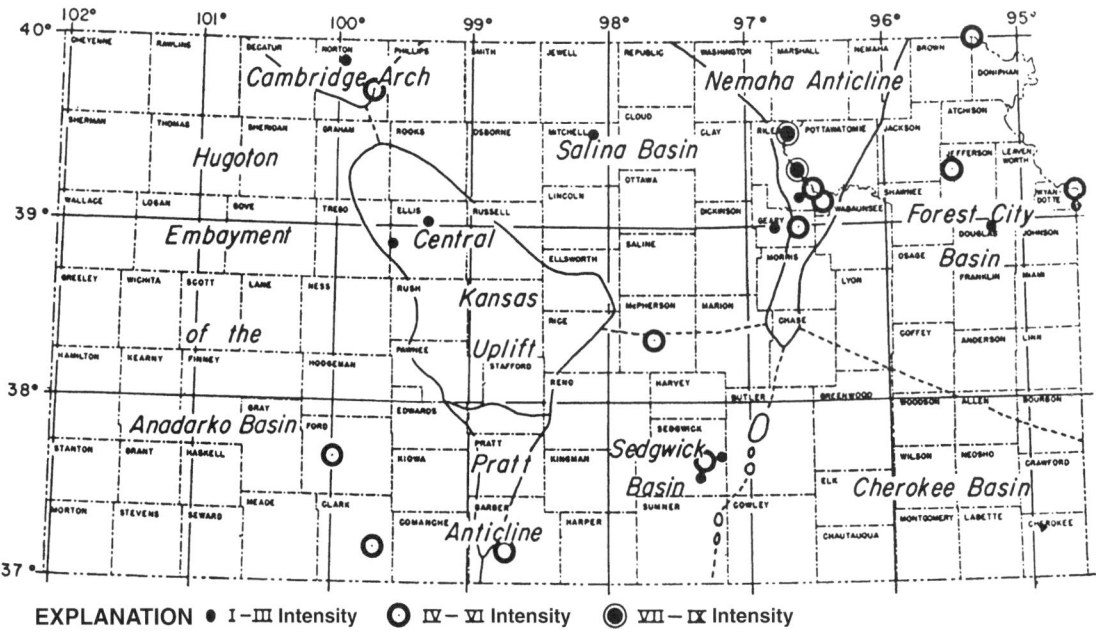

Figure 2. Location and major structures in area discussed here. Symbols show recorded earthquakes in Kansas as of the mid-1950s (from Merriam, 1956).

Visually, as seen in Figure 3, much of the structural activity took place in Desmoinesian and early Missourian time; activity decreased in the Virgilian and Wolfcampian. This early surge of activity reflects the earlier, orogenic activity in the Ouachitas. The activity has decayed through time partly because the structures were being masked by the ever-increasing influx of sediments, and partly because the orogenic disturbance was coming to an end and the area was settling back into a state of semiquietude (Merriam, 1963).

No progressive, systematic change in the spatial arrangement of the block movement can be seen in an arrangement of blocks through time (Fig. 3), but the most continuously active structures are on or adjacent to the Nemaha Anticline, which is the dominant positive structure in eastern Kansas and Oklahoma and seemingly is quiet today. The activity on this major feature controlled, to some extent, the structural development of satellite structures.

SEDIMENTOLOGICAL EVIDENCE

The convolute bedding noted in the Permian-Pennsylvanian section of the area meets the criteria for recognizing paleoseismicity listed by J.D. Sims (1975; reported by Maltman, 1994, p. 288). These criteria are: (1) the features are restricted to single layers separated by undisturbed units; (2) they were formed in essentially flat-lying beds; and (3) other evidence points to the area being seismically active.

It is not surprising, however, that more of the convolute-bedding features have not been noted in surface sections in eastern Kansas and northeastern Oklahoma; this probably is because the clastics are poorly exposed. Also, the number and amount of clastics in the section decrease as the effects of the distant orogeny and of the local readjustment wane, resulting in fewer conditions for forming the features. The seeming paucity of convolute bedding, taking into account all factors, is related to the special circumstances and timing under which it formed (Pettijohn et al., 1972): (1) in thin (2–25 cm) beds, (2) as fine sand on coarse silt, (3) having sharp anticlines and broad synclines, (4) convolutions die out upward and downward and (5) horizontal beds above and below. Most of the convolute structures observed in outcrop in eastern Kansas and northeastern Oklahoma meet these conditions. The features are in thin beds with fine-grained sand on coarse-grained silt. Horizontal beds are above and below, and the convolutions die out upward and downward, but without the pillar and dish structures (Fig. 4). Examples of the convolute-bedding structures in eastern Kansas are shown in Figure 5. Similar features have been reported in the stratigraphic section in units ranging from Ordovician to Tertiary in age and also have been attributed to paleo-earthquakes (Brenchley and Newall, 1977; Keighley and Pickerill, 1998; Keighley et al., 1998; and references in those papers).

Locations of known surface and, in the subsurface, convolute structures are shown in Figure 6. They are more abundant in the older units and to the south, reflecting the distribution in abundance and location of recurrent movement on the anticlinal structures. Microseisms are known today to occur in association with the Precambrian basement block faults, which have continued to adjust, albeit slowly, to tectonic forces outside the area (Merriam, 1956; Steeples and Brosius, 1996). Many of the convolute features are located on or near these local structures. Therefore, it is reasonable to assume that the convolute-bedding

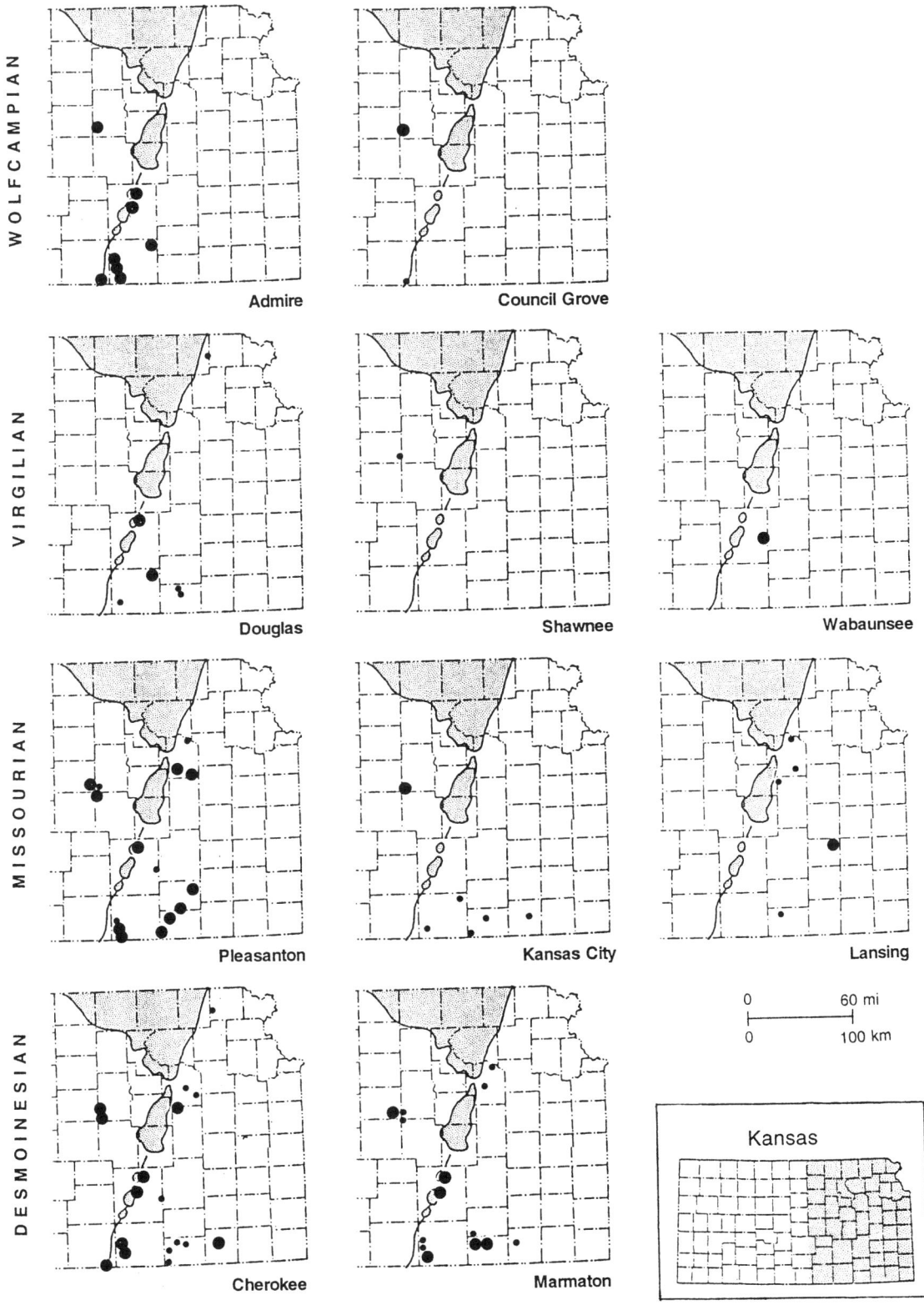

Figure 3. Maps of Permian-Pennsylvanian time slices (represented by groups of Fig. 1), showing structures in eastern Kansas where major structural movement (large dots) and minor movement (small dots) took place. Note that most of the movement took place on structures located nearest Nemaha Anticline (gray shading), in southern Kansas, and movement decreased with time. Left to right is oldest to youngest; stratigraphic sequences are arranged from oldest (bottom), Pennsylvanian Desmoinesian, to youngest (top), Permian Wolfcampian.

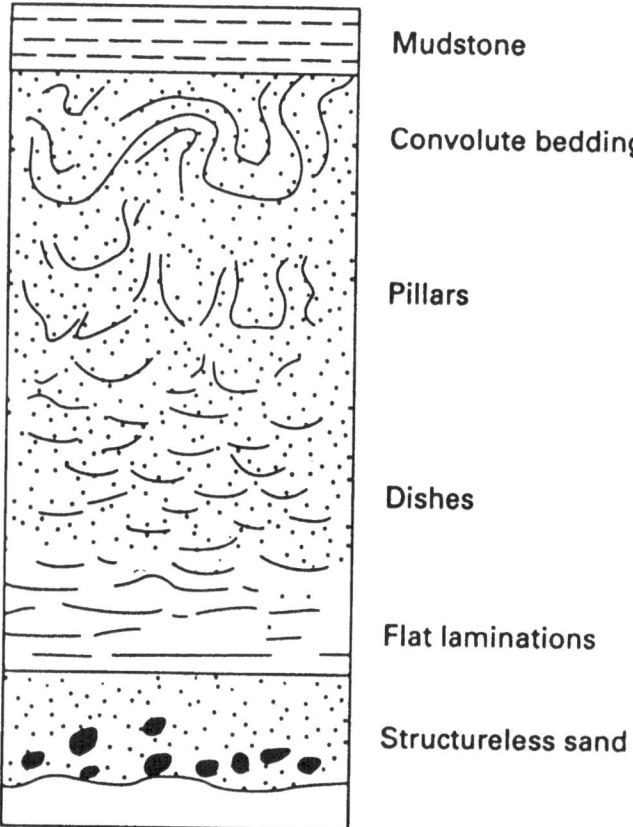

Figure 4. Schematic diagram of observed contorted sedimentary structures attributed to paleo-earthquakes, which are representative of those in the Permian-Pennsylvanian section exposed in eastern Kansas and northeastern Oklahoma (from Obermeier, *in* McCalpin and Nelson, 1996, p. 385).

features in the Permian-Pennsylvanian strata were triggered by, and are associated with, the movements on these fault blocks. Similar findings were reported in eastern Kansas by Fenster et al. (1984) and in northeastern Oklahoma by Gardner (1922), Rascoe (1975), and Visher and Cunningham (1981).

In a detailed subsurface study of cores in the Cherokee Group (Desmoinesian) in southeastern Kansas, Harris (1984) showed convolute bedding in almost every unit in the group. The convolute bedding seemingly is continuous in several units between some of the more closely spaced coreholes. If cores from other units were available, we probably would note convolute structures more abundant and widespread than known heretofore, reflecting the continued unrest of the region in the aftermath of the Ouachita orogeny. Huffman (1991) and Reinholtz (1982) studied areas to the north and west of Harris's (1994) study area in the Cherokee Basin and also noted convolute structures throughout the Cherokee section.

Convolute structures also are abundant in the older Pennsylvanian units exposed in Oklahoma where they have been described and illustrated from several units in the Jackfork Group (Morrowan) and the Atoka Formation (Atokan) (Suneson and Hemish, 1994).

MAGNITUDE AND SIZE OF PALEO-EARTHQUAKES

Using McCalpin and Nelson's (1996) classification scheme, we cannot determine the magnitude and size of the Midcontinent paleo-earthquakes from primary evidence (Fig. 7), but we use secondary evidence (instantaneous or coseismic, off-fault features), which in this instance is convolute bedding to estimate some preliminary magnitude values. McCalpin and Nelson (1996) suggested that a moment magnitude (M_w) threshold of 5 to 5.5 is necessary for the event to be recorded and preserved in the stratigraphic record. A quake of this magnitude probably involves a minimum area approximately 300 km^2 (115 mi^2), and the area expands exponentially as the magnitude increases.

We realize that the stratigraphic sequence is incomplete, that sedimentary features may not be recognized, and that events may not have been recorded and preserved in the record; nevertheless, we make a first attempt to summarize available data on the magnitude and occurrence of paleo-earthquakes in this area. The number of probable paleo-earthquakes recorded by convolute structures in Kansas is low, but we have plotted them on a time scale (Fig. 8). Note that the greatest numbers are mainly in two of the clastic units—the Cherokee group (Desmoinesian) and Douglas group (Virgilian). This substantiates the stratigraphic evidence of differential compaction on plains-type folds. Although the data are incomplete and sparse, they may reflect the actual frequency.

Preliminary calculations, admittedly approximate, on the areal extent of the Cherokee paleoquakes gives an affected area of about 16 500 km^2 (6500 mi^2), which, according to McCalpin and Nelson (1996, p. 16), would be about 6.5 to 7 on the M_w scale. This value is obtained by using the maximum distance between the localities of convolute bedding in the Cherokee shale, as currently known in Kansas, as the diameter (radius might be closer to the truth, because Oklahoma is not considered) of the area affected by any one of the quakes. Similarly, a preliminary estimate of the magnitude for paleoquakes affecting units in the Douglas Group gives approximately 26 500 km^2 (10 200 mi^2) and about magnitude 7 on the McCalpin-Nelson scale. The magnitude values seem reasonable but relatively high, considering the known magnitude of similar historic earthquakes in the Midcontinent; this may be because of the assumptions made about the size of the affected area.

As noted, earthquakes of about the same estimated magnitude and size of affected area have occurred in Kansas in historic time, mostly along the Nemaha Anticline (Fig. 2; Merriam, 1956). These quakes ranged in intensity from III to VIII on the modified Mercalli scale (on a scale of I to XII). An intensity of III is equivalent to about 2.5 on the Richter scale, and VIII is equivalent to 6. The 24 April 1867 earthquake in Kansas, which occurred near Manhattan on the Nemaha Anticline, was felt over an area of 777 000 km^2 (300 000 mi^2), but had an intensity of only VIII on the modified Mercalli scale, whereas the Oklahoma City earthquake of 9 April 1952, also along the Nemaha Anticline, was a VII on the modified Mercalli scale and was felt over an area of 360 000 km^2 (140 000 mi^2) (Merriam, 1956, p. 93).

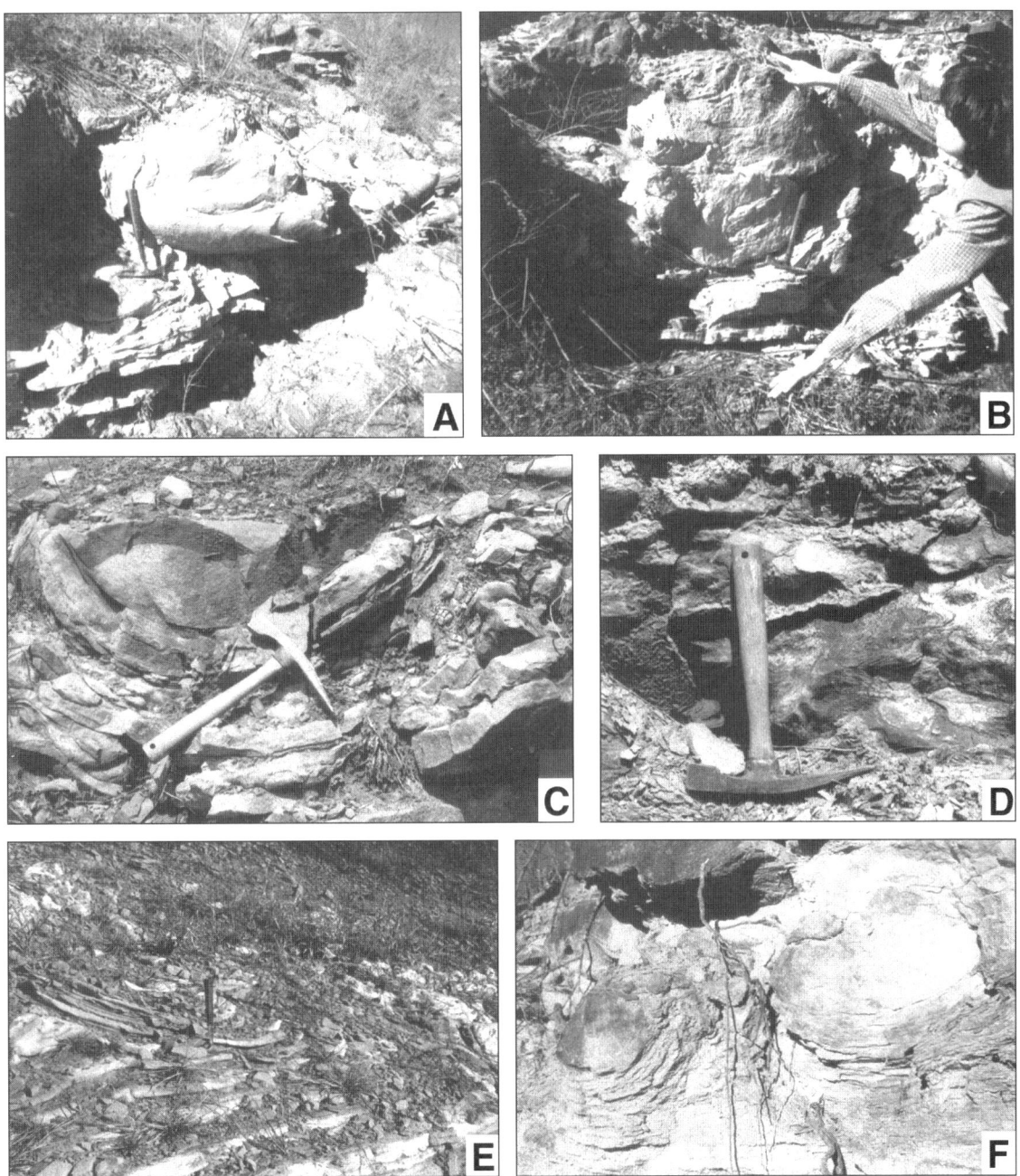

Figure 5. Exposures of convolute structures in eastern Kansas. Hammer is 28 cm long. A: Douglas Group (along north-south county road in SE NE sec. 24, T. 31 S., R. 14 E., Montgomery County). B: Douglas Group (along north-south county road in SE NE sec. 24, T. 31 S., R. 14 E., Montgomery County). C: Douglas Group (along east-west county road in SW SE sec. 7, T. 35 S., R. 12 E., Chautauqua County). D: Bachelor Creek Member of Howard Formation, Wabaunsee Group (on old Kansas State Highway 166 west of Grant Creek in S2 sec. 3, T. 34 S., R. 9 E., Chautauqua County). E: Douglas Group (on U.S. Highway 400 northwest of Fredonia in C sec. 3, T. 28 S., R. 14 E., Wilson County). F: Calhoun Shale, Shawnee Group (at Calhoun Bluffs in SE sec. 16, T. 11 S., R. 16 E., Shawnee County).

Considering the area affected by the historic earthquakes, it seems that the area affected by the paleo-earthquakes is too small. This can be accounted for by assuming that either the known geographic distribution of the convolute features is too small or that the area is the *effective* area of convolute-bedding formation. If, in fact, the paleo-earthquakes were of magnitude 6.5 to 7, then the total area affected would have been larger than estimated.

Gordon (1983) has suggested that these large Midcontinent quakes had foci below the depth to the Precambrian basement. He further noted that 70% of the quakes in the Stable Continental

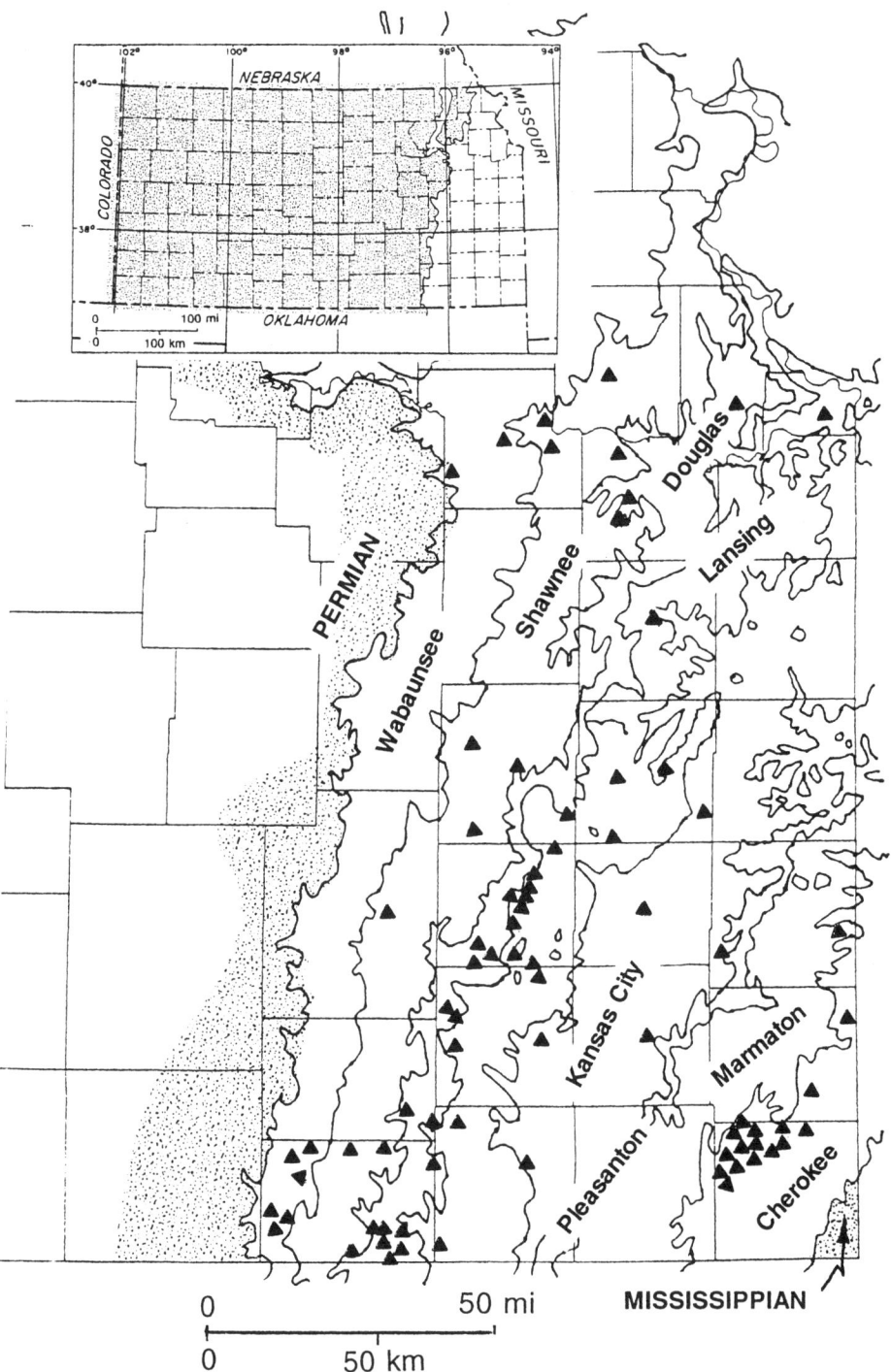

Figure 6. Locations of convolute sedimentary structures by stratigraphic unit in eastern Kansas. Features are more numerous in older units and farther south toward Oklahoma, closer to source of Ouachita orogeny structural disturbance.

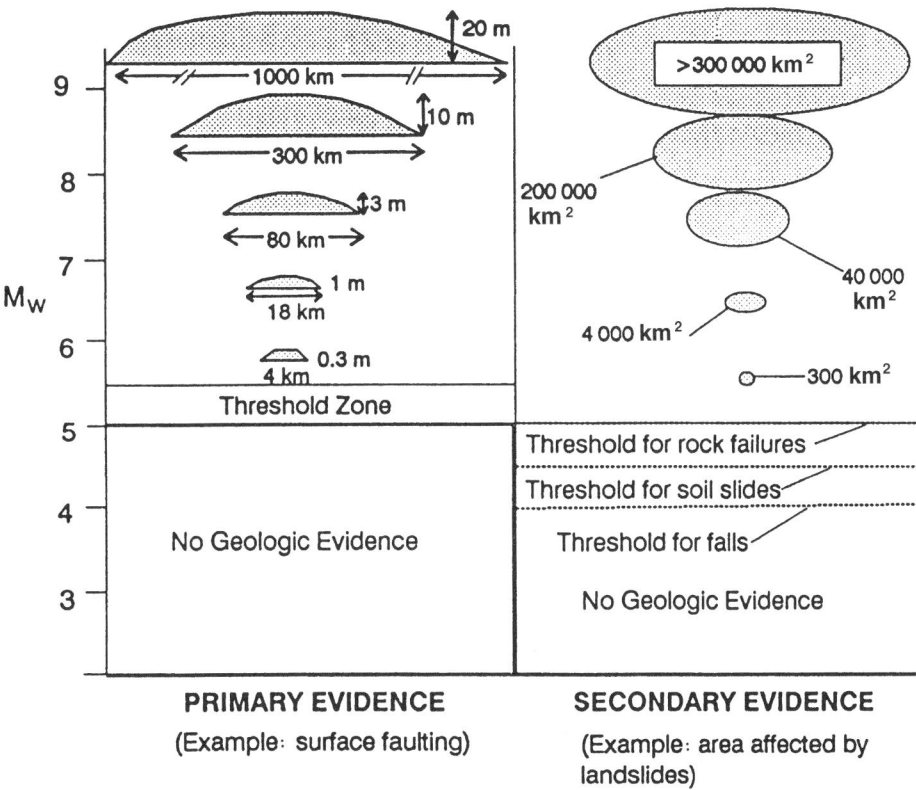

Figure 7. Schematic diagram of size and extent of paleo-earthquakes, based on primary or secondary evidence. Our estimate of size and extent is based on secondary evidence, according to McCalpin and Nelson (1996) classification. M_w is moment magnitude based on observed historical earthquakes (from McCalpin and Nelson, 1996, p. 16).

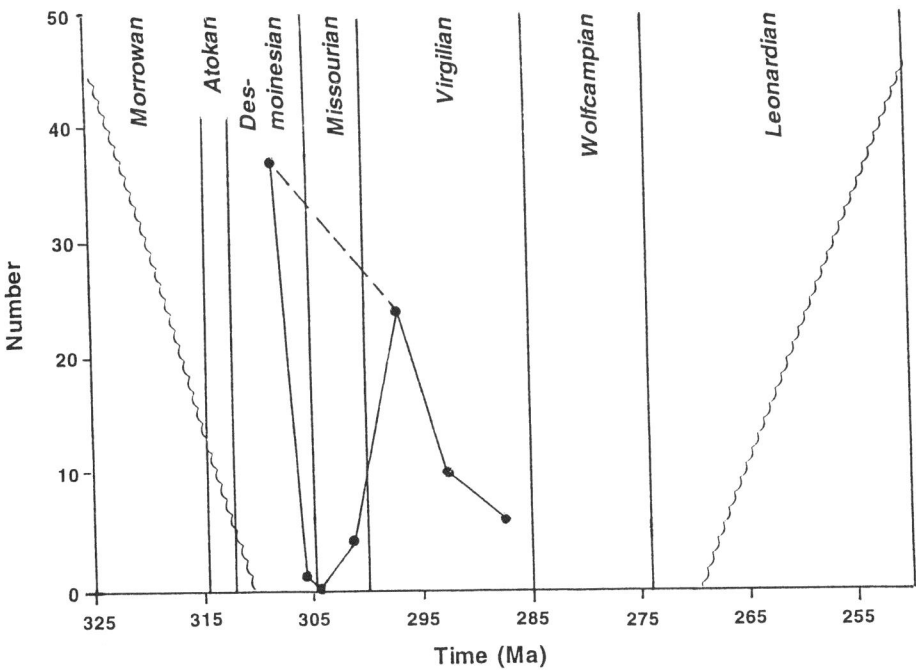

Figure 8. Plot of paleo-earthquakes as interpreted from convolute structures through time. Record incomplete, but evidence of activity is greatest in clastic units of the Desmoinesian and Virgilian. Overall decline in activity through time (shown by dashed line from Desmoinesian to Virgilian) parallels record of structural history. Lower and upper part of Permian-Pennsylvanian section is missing (wavy line).

Interior occur in association with the ancient plate tectonic structures of the Wichita Aulacogen of southern Oklahoma, the northwest-trending Colorado lineament in the northern Great Plains, and the Keweenawan Midcontinent rift system.

A swarm of microseisms was recorded from 1977 to 1989 in eastern Kansas (Steeples and Brosius, 1996). These microseisms seemingly are associated with continued structural movement along the Nemaha Anticline and subsidiary folds.

CONCLUSIONS

Stratigraphic and sedimentological evidence support the contention of recurrent tectonic movement on local structures during late Paleozoic time in parts of the Midcontinent. The movement waned through the remainder of the Paleozoic after the Ouachita orogeny, but seemingly continued intermittently until the present. The intensity of deformation was greatest in the south and decreased progressively to the north, as indicated by the stratigraphic and geographic location of structural events and sedimentological evidence. Paleo-earthquakes suggested by seismites indicate recurrent movement along the inherent Precambrian basement fracture-fault pattern. In eastern Kansas, paleo-earthquakes during Middle and late Pennsylvanian time are estimated to have been in the magnitude 6.5 to 7 range, affecting 16 500 to 26 500 km^2.

ACKNOWLEDGMENTS

We benefited from discussions with Rick Groshong of the University of Alabama and Bryce Hand of Syracuse University about structure, paleoseismicity, and convolute structures. We thank Philip Heckel of the State University of Iowa, Ron Pickerill of the University of New Brunswick, and Frank Ettensohn of the University of Kentucky for suggestions on clarifying the presentation.

REFERENCES CITED

Blatt, H., Middleton, G., and Murray, R., 1980, Origin of sedimentary rocks (2nd edition): Englewood Cliffs, New Jersey, Prentice-Hall, Inc., 782 p.

Brenchley, P.J., and Newall, G., 1977, The significance of contorted bedding in Upper Ordovician sediments of the Oslo region, Norway: Journal of Sedimentary Petrology, v. 47, no. 2, p. 819–833.

Cole, V.B., 1975, Subsurface Ordovician-Cambrian rocks in Kansas: Kansas Geological Survey, Subsurface Geology Series 2, 18 p.

Collinson, J., 1994, Sedimentary deformational structures, in Maltman, A., ed., The geological deformation of sediments: London, Chapman and Hall, p. 95–125.

Fenster, D.F., Trapp, J.S., and Bandoian, C.A., 1984, Late Pennsylvanian penecontemporaneous deformation: Detailed studies of unanticipated features at a nuclear power plant site in the Midcontinent [abs.]: Boston, Massachusetts, Association of Engineering Geologists 27th Annual Meeting, Abstracts and Program, p. 54.

Gardner, J.H., 1922, Rock distortion on local structures in the oil fields of Oklahoma: American Association of Petroleum Geologists Bulletin, v. 6, no. 3, p. 228–243.

Gordon, D.W., 1983, Seismicity and plate tectonic remnants in the central stable region of the United States (abs.): Earthquake Notes, v. 54, no. 3, p. 19.

Harris, J.W., 1984, Stratigraphy and depositional environments of the Krebs Formation lower Cherokee Group (Middle Pennsylvanian) in southeastern Kansas [M.S. thesis]: Lawrence, University of Kansas, 139 p.

Huffman, D.P., 1991, Stratigraphy and depositional environments of the Cherokee Group (Middle Pennsylvanian) Bourbon Arch region, east-central Kansas [M.S. thesis]: Lawrence, University of Kansas, 177 p.

Keighley, D.G., and Pickerill, R.K., 1998, Mudstone-clastiform conglomerates and trough-shaped depressions from the Pennsylvanian lower Port Hood Formation of eastern Canada: Occurrences due to soft-sediment deformation: Journal of Sedimentary Research, v. 68, no. 5, p. 901–912.

Keighley, D.G., Andersson, D., Flint, S., and Howell, J., 1998, Soft-sediment deformation structures as potential indicators of synsedimentary tectonic control in alluvial-lacustrine sequences, Green River Formation, Nine Mile Canyon, Unita Basin, east-central Utah [abs.]: American Association of Petroleum Geologists Annual Convention, Extended Abstracts, v. 2, p. A355.

Lucchi, F.R., 1995, Sedimentological indicators of paleoseismicity, in Serva, L., and Slemmons, D.B., eds., Perspectives in paleoseismology: Association of Engineering Geologists Special Publication 6, p. 7–17.

Maltman, A., 1994, Deformation structures preserved in rocks, in Maltman, A., ed., The geological deformation of sediments: London, Chapman and Hall, p. 261–307.

McCalpin, J.P., and Nelson, A.R., 1996, Introduction to paleoseismology, in McCalpin, J.P., ed., Paleoseismology: San Diego, Academic Press, p. 1–32.

Merriam, D.F., 1956, History of earthquakes in Kansas: Seismological Society of America Bulletin, v. 46, no. 2, p. 87–96.

Merriam, D.F., 1963, The geologic history of Kansas: Kansas Geological Survey Bulletin 162, 317 p.

Merriam, D.F., 1999, "Plains-type folds" in the Midcontinent region (USA): Chronicling a concept: Earth Science History Journal, v. 18, no. 2, p. 262–294.

Merriam, D.F., and Förster, A., 1996, Precambrian basement control on "plains-type folds" (compactional features) in the Midcontinent region, USA, in Oncken, O., and Janssen, C., eds., Basement tectonics 11: Europe and other regions: Dordrecht, Netherlands, Kluwer Academic Publishers, p. 149–166.

Obermeier, S.F., 1996, Using liquefaction-induced features for paleoseismic analysis, in McCalpin, J.P., ed., Paleoseismology: San Diego, Academic Press, p. 331–396.

Pettijohn, F.J., Potter, P.E., and Siever, R., 1972, Sand and sandstone: New York, Springer-Verlag, 618 p.

Rascoe, B., Jr., 1975, Tectonic origin of preconsolidation deformation in Upper Pennsylvanian rocks near Bartlesville, Oklahoma: American Association of Petroleum Geologists Bulletin, v. 59, no. 9, p. 1626–1638.

Reinholtz, P.N., 1982, Distribution, petrology and depositional environment of "Bush City shoestring sandstone" and "Centerville Lagonda sandstone" in Cherokee Group (Middle Pennsylvanian), southeastern Kansas [M.S. thesis]: Iowa City, University of Iowa, 180 p.

Steeples, D.W., and Brosius, L., 1996, Earthquakes: Kansas Geological Survey, Public Information Circular 3, 6 p.

Suneson, N.H., and Hemish, L.A., eds., 1994, Geology and resources of the Eastern Ouachita Mountains Frontal Belt and southeastern Arkoma Basin, Oklahoma: Oklahoma Geological Survey Guidebook 29, 294 p.

Visher, G.S., and Cunningham, R.D., 1981, Convolute laminations: A theoretical analysis: Example of a Pennsylvanian sandstone: Sedimentary Geology, v. 28, no. 3, p. 175–188.

Zeller, D.E., ed., 1968, The stratigraphic succession in Kansas: Kansas Geological Survey Bulletin 189, 81 p.

MANUSCRIPT ACCEPTED BY THE SOCIETY MAY 11, 2001

Printed in the U.S.A.

Critical evaluation of possible seismites: Examples from the Carboniferous of the Appalachian basin

Stephen F. Greb
Garland R. Dever Jr.
Kentucky Geological Survey, 228 MMRB, University of Kentucky, Lexington, Kentucky 40506-0107, USA

ABSTRACT

Numerous Carboniferous depositional features in the central Appalachian basin have previously been interpreted as being structurally controlled. Five features were analyzed to evaluate if they might be seismites based upon (1) deformation style, (2) extent and trend of deformation, and (3) plausibility of a seismic origin.

After analysis, earthquake shaking seemed a likely triggering mechanism for two of the five examples studied. The Big Sinking bed is a fine-grained carbonate that (1) contains a widespread, correlative zone of large flow rolls on the downthrown margin of a fault, (2) exhibits a regional trend of decreasing deformation away from a fault, (3) contains a wide variety of soft-sediment deformation features, which may cut across lateral facies, (4) locally contains unusual, stacked flow rolls between fluidization pipes, and (5) locally exhibits vertical stacking of brachiopod valves in a convex-down position. This style of valve stacking may indicate sudden liquefaction and rapid resettling of the valves, as might follow a seism.

The second example is a ball-and-pillow deposit that occurs at the same horizon as the Poison Honey bed paleo–debris flow. The deposit exhibits opposing orientations of pillows, with axes oriented close to the horizontal. This orientation may record a back-and-forth, side-to-side shear that could have been caused by seismic-wave oscillation, rather than a vertical load or lateral translation. The potential zone of deformation, however, cannot be demonstrated to be widespread.

INTRODUCTION

During the past two decades, the importance of structural influence on Carboniferous sedimentation in the central Appalachian basin has been well documented. Some of the influences are based upon regional trends of increasing stratal thickness from the northern and western margins of the central Appalachian basin toward the deeper basin to the southeast (Rice et al., 1979; Chesnut, 1992b; Greb and Chesnut, 1996), stratigraphic pinchouts and thickness changes along reactivated basement faults and structures (Ettensohn, 1977; Horne and Ferm, 1979; Dever, 1977, 1986, 1990, 1999; Dever et al., 1979), the coincidence of paleovalley trends to faults and other structures (Hester and Taylor, 1977; Greb and Chesnut, 1996), and rectangular trends of coal-thickness distribution (Weisenfluh and Ferm, 1991; Greb et al., 1999). Aside from these regional stratigraphic trends, several types of sedimentary features have also been inferred to be structurally controlled where they occurred in the vicinity of possibly reactivated faults, including paleoslumps and debris flows (Dever et al., 1979; Hester and Brant, 1981; Greb et al., 1990; Greb and Weisenfluh, 1996), and an unusual stratum of ball-and-pillow structures (Greb and Chesnut, 1990; Dever, 1999).

Stratigraphic and tectonic setting

Carboniferous strata in the central Appalachian basin consist of a Mississippian carbonate sequence, informally termed the "Big Lime" and designated as the Slade Formation on the western outcrop margin of the basin (Fig. 1). The Slade is overlain by mixed carbonates and siliciclastics of the Paragon Formation (Fig. 1). These are unconformably overlain and progressively truncated toward the basin margin by Pennsylvanian siliciclastics of the Breathitt Group (Fig. 1).

The central Appalachian basin is a sub-basin of the larger Appalachian basin. The basin formed as a series of foreland basins, initially above a possible Precambrian aulocogen called the Rome Trough (shaded in Fig. 2). It subsequently became enlarged during the Taconian, Acadian, and Alleghenian-Hercynian orogenies. The northern margin of the trough in eastern Kentucky is defined by two subparallel, east–west fault trends: the Irvine–Paint Creek (IPCF in Fig.2) and Kentucky River (KRFZ in Fig. 2) fault zones, which define the northern margin of the Rome Trough (Fig. 2). The Paintsville–Warfield anticline (PWA in Fig. 2) is located along the Irvine–Paint Creek trend in the eastern part of the coal field (Fig. 2). The southern margin of the Rome Trough is less well defined but parallels the axis of the Eastern Kentucky syncline in the eastern part of the coal field (Fig. 2). Westward, on the flank of the Cincinnati Arch, faults are oriented north-south or northeast-southwest and overall movement is down-to-the-east.

As mentioned previously, regional thickness and distribution of Carboniferous units were profoundly influenced by basin tectonics (Tankard, 1986; Quinlan and Beaumont, 1984; Chesnut, 1992b). Structural movement could have produced earthquakes, some of which may have been large enough to leave a record in fine-grained sediments. In this study, examples of several Mississippian (western limit shown with an M in Fig. 2) and Pennsylvanian (western limit shown with an P in Fig. 2) deformational features, which have been inferred to be structurally controlled, are examined to determine if they are seismites.

Seismites

The term *seismite* was introduced by Seilacher (1969) for a postdepositional, "fault-graded" bed exhibiting soft-sediment deformation, inferred to be earthquake-generated. Cita and Lucchi (1984) expanded the term to include deformation induced by seismic events or tsunami (seismically generated) waves, which (1) disturbed sediment but left it in place (as in the original definition), (2) produced an en masse movement with limited internal deformation (e.g., debris flows, slumps, slides), and (3) completely reorganized preexisting sediments into bulk-homogenous deposits, as may follow a tsunami (e.g., homogenites). An increased awareness of the significance of these deposits has recently been sparked by the use of seismites (1) as tools for inferring earthquake recurrence intervals (Seed et al., 1968), (2) as indicators of potential areas of ground failure in advance of earthquakes (Seed et al., 1983), and (3) as criteria for interpreting the magnitude of Holocene and Quaternary earthquakes (Obermeier, 1998).

Many different types of features have been interpreted as seismites, and several overviews of the criteria used to define sedimentary deposits as seismites have been offered (Seilacher, 1984; Vittori et al., 1991; Obermeier, 1996). Herein, various types of landslides and soft-sediment deformation features that can be seismically triggered are summarized for comparison to Carboniferous features analyzed in this study.

Seismogenic soft-sediment deformation features. When near-surface, cohesionless sediment layers are subjected to earthquake-induced cyclic loads, they may be subject to liquefaction (Seed, 1968). Liquefaction features, including clastic dikes and lateral spreads, are the most common types of deformational features described following modern, large-magnitude earthquakes (Seed and Idriss, 1982; Youd and Perkins, 1987). Sand dikes occur when seismic loading causes sedimentary grains to become suspended in the pore water, and pore-fluid pressure is as high as the initial confining stress, forcing water and sand vertically out of the soil-water suspension (Youd, 1973; Obermeier, 1996). The minimum magnitude (M) for soil liquefaction has been reported as approximately M 5.0 (Kuribayashi and Tatsuoka, 1975; Youd, 1973), although the minimum magnitude for sand-dike formation is sometimes considered to be M 6.2 (Obermeier, 1996; 1998). In some cases, widespread trends or swarms of sand dikes have provided conclusive evidence of liquefaction and have been used for determining historical earthquake characteristics such as recurrence and magnitude (Seed and Idriss, 1982; Seed et al., 1983; Youd and Perkins, 1987; Obermeier, 1996, 1998).

Sand dikes attributed to seismic origins have also been noted in ancient strata (Rascoe, 1975; Plint, 1985; Guiraud and Plaziat, 1993). In the rock record, however, a broader suite of sedimentary features inferred to be the product of sudden liquefaction have been interpreted as seismites. Some of these features are more commonly formed through hydroplastic deformation processes, sometimes in combination with local fluidization and liquefaction (see Lowe, 1975, for definitions). Features interpreted as seismites that have formed through some combination of these processes are varied, but they include ball-and-pillow structures (Kelling and Williams, 1966; Sims, 1975; Weaver, 1976; Weaver and Jeffcoat, 1978; Brenchley and Newell, 1977) and flow rolls "encased" or "layered" above and below by undisturbed strata (Rascoe, 1975; Rast and Ettensohn, 1995; Pope et al., 1997).

In some cases, the processes responsible for soft-sediment deformation features could be triggered by mechanisms other than earthquakes. Liquefaction-induced sand dikes can also form as a result of artesian pressure changes (Li et al., 1996). Ball-and-pillow structures and flow rolls can form through (1) a gravitationally unstable gradient of bulk density between stacked layers of different grain and pore sizes, (2) nonuniform confining pressures from overlying strata, (3) slope-induced shear (which can be shallow), (4) fluid flow at the sediment-fluid interface, and (5) fluid flow within the sediment itself (Allen, 1982; Mills, 1983). These can form in response to (1) tidal flux (Bjerrum, 1973), (2) rapid changes in river stage (Li et al., 1996), (3) storm-generated waves (Palmer, 1976), (4) breaking waves (Dalrymple, 1979), (5) slumping (Elliott and Williams, 1988), and (6) rapid sediment loading (Dzulynski and Walton, 1965).

Critical evaluation of possible seismites—Examples from the Carboniferous of the Appalachian basin

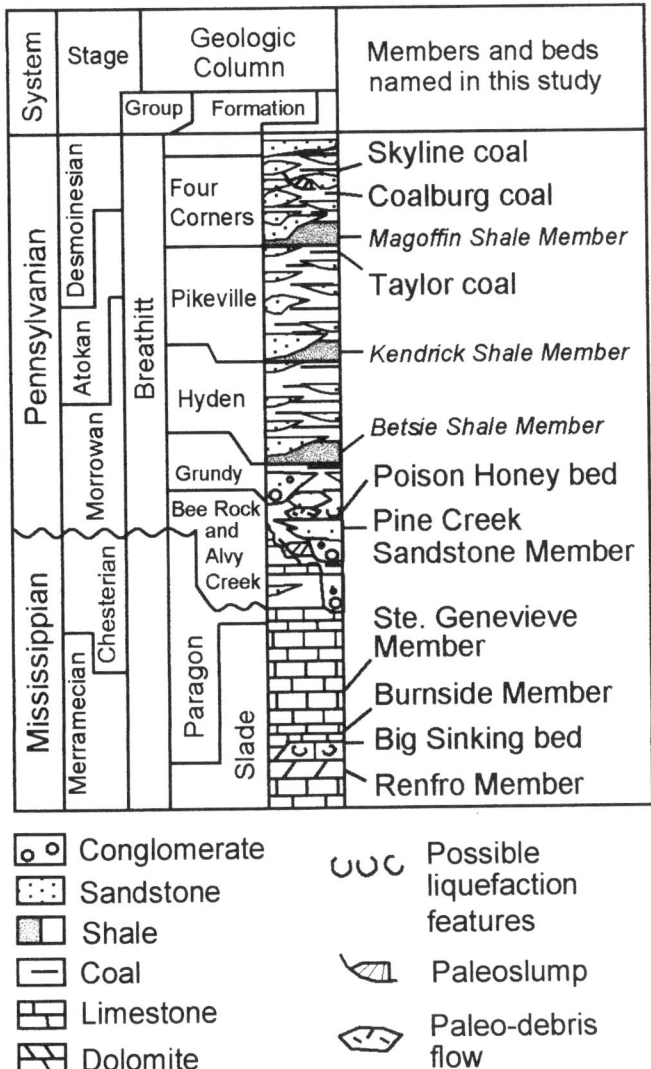

Figure 1. Upper Mississippian and Pennsylvanian stratigraphy of the Appalachian basin in eastern Kentucky.

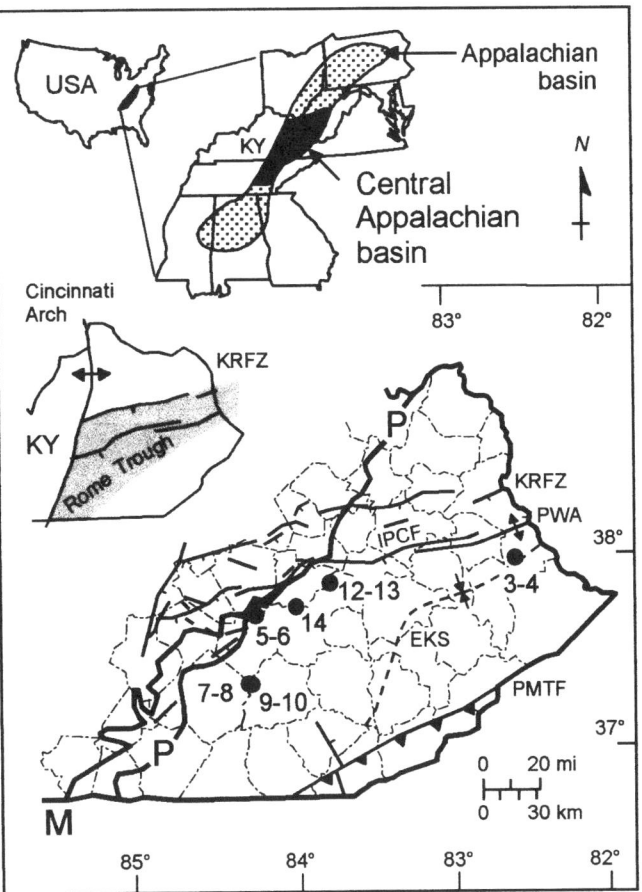

Figure 2. Location of the central Appalachian basin (shaded black), tectonic structures, and strata evaluated in this report as possible seismites. Numbers next to locations (black dots) denote figures in this report. EKS—Eastern Kentucky syncline, KRFZ—Kentucky River fault zone, IPCF—Irvine–Paint Creek fault zone, PMTF—Pine Mountain thrust fault, PWA—Paintsville–Warfield anticline. The western outcrop limit of the Mississippian "Big Lime" sequence (M) and Pennsylvanian strata (P) are indicated to delineate the basin.

This report evaluates two occurrences of soft-sediment-deformation features, (1) ball-and-pillow structures in the lower Breathitt Group, and (2) large flow rolls in the Big Sinking bed, a Meramecian carbonate of the Slade Formation (Fig. 1), to determine whether they are likely to have been generated by paleo-earthquakes.

Seismogenic landslides. Another group of features commonly reported following modern, large-magnitude earthquakes are various types of landslides, including (1) subaerial slumps (Seed, 1967), subaerial slides (Seed and Wilson, 1967), and submarine slumps and slides (Heezen and Ewing, 1952). Analysis of historical earthquake data suggests greater magnitude earthquakes are required to initiate lateral spreads, landslides such as subaqueous slides, and rock slumps than are required to trigger rock slides and rock falls (Keefer, 1984; Jibson and Keefer, 1993); the minimum threshold for soil slumps is M 4.5, and for rock slumps and earth flows it is M 5.0.

Obviously, not all landslides are seismically triggered, and differentiating seismic triggers from nonseismic triggers is problematic (Jibson and Keefer, 1993). Subaerial landslides can be triggered by many mechanisms, including excess water saturation on slopes, which do not have to be steep (Keefer, 1984). Likewise, submarine slides and flows can be triggered by rapid deposition, or oversteepening of sedimentary deposits on slopes (Morgenstern, 1967) and wave pressures on the bed (Henkel, 1970).

We evaluate three examples of paleolandslides to determine if they are likely to have been generated by earthquakes: (1) a large, monocline-bordered, Morrowan paleoslump of the lower Breathitt Group, (2) a Desmoinesian, fault-bordered paleoslump associated with sand dikes in the Four Corners Formation of the

Breathitt Group, and (3) a Morrowan paleo–debris flow called the Poison Honey bed, of the Bee Rock Sandstone of the Breathitt Group (Fig. 1).

Criteria for interpreting seismites

Widespread trends of deformation along correlative horizons can be useful in identifying possible seismites, and differentiating them from nonseismic mechanisms that could produce similar features. Keefer (1984) noted that in studies of modern earthquakes, large-magnitude earthquakes can produce widespread regional trends in the types of landslides as much as 100 km away from the epicenter. Numerous investigations of seismogenic liquefaction features have noted widespread trends of decreasing deformation away from the epicenter, and decreasing deformation with decreasing magnitude (Youd and Perkins, 1987, Obermeier, 1996). These trends have been used to infer potential sources and relative magnitudes of paleoseismites where absolute ages were not available (Obermeier, 1998, for example). What they also suggest is that large earthquakes produce zones of deformation across the landscape that increase in scale or number toward an earthquake source. Also, if widespread deformation is characteristic, then deformation might cut across local paleoenvironments and facies, as long as grain size and depth and water pressure (among other attributes) are in the range in which hydroplastic deformation, liquefaction, or fluidization are possible.

Purpose

To evaluate the probability of seismic versus other origins for five Carboniferous deposits, each was investigated to determine if they (1) contain evidence of liquefaction or of styles of deformation that differ from the surrounding fabric or facies of the deposits' time and place, (2) occur as part of a widespread trend of deformation along a correlative horizon, and (3) contain evidence that the deformation is less likely to have been formed by other causal mechanisms of the deposits' place and time than by a seismic mechanism.

POSSIBLE CARBONIFEROUS SEISMITES

Paleoslump with sand dikes

A paleoslump above the Coalburg coal, Four Corners Formation, Breathitt Group (Fig. 1), in Martin County, Kentucky (location 3-4 in Fig. 2), was previously described by Greb and Weisenfluh (1996), and inferred to be structurally controlled. Slumped sandstones and shales are on the downthrown margin of a listric normal fault, and movement is in the direction of throw (Fig. 3). Several sandstone beds within the slump thicken toward the plane of rotation. The underlying coal is offset, and it thickens on the downthrown margin, splitting and bending downward toward the fault along an apparent hinge line. At the hinge, the coal contains several very fine grained sandstone dikes, which are V-shaped, downward-thickening intrusions into the underlying coal (dv in Fig. 4A), and a broadly horizontal boudinage-like parting that enters and exits the top of the coal (Fig. 4B). Laterally, the paleoslump and upper part of the coal are truncated by additional paleoslumps, and a possible paleochannel deposit (Fig. 3). The overlying Stockton coal (St in Fig. 3) is offset along the fault, but its thickness does not change across the offset.

The paleoslump is south of the east-west–trending Paintsville-Warfield anticline and north of the axis of the Eastern Kentucky syncline (location 3-4 in Fig. 2). The paleoslump area is also within 10 km of the inferred trace of several basement faults, including structures that define the southern edge of the Rome Trough (Ammerman and Keller, 1979; Drahovzal and Noger, 1995). The area was apparently a structural hinge line during much of the Early and Middle Pennsylvanian, with significant accommodation space developed south of the hinge line (Horne and Ferm, 1978). Vertical stacking of sandstones 2 km east of the paleoslump has been used to indicate structural control of a slightly younger stratigraphic interval between the Coalburg and Taylor coal bed (see maps in Andrews et al., 1996).

Interpretation. Thickening of the Coalburg coal and sandstone beds within the paleoslump indicate growth along the slump-bounding fault during deposition (Greb and Weisenfluh, 1996). Sand dikes in the coal and the boudinage-like parting may have been liquefied or fluidized during injection. The sand dikes are connected to the overlying strata and show no connection to underlying strata, so they are inferred to originate from the overlying strata, perhaps filling tension tears in the peat along the hinge of downward bending. This suggests that injection occurred after (or during) the rotation of those strata into their current position, rather than during initial failure. Likewise, headward thickening of strata within the paleoslump suggests growth during rotation, which might be more common in slow, creeplike movement, rather than sudden failure and rapid rotation. Also, the paleoslump is laterally truncated by a paleochannel (Fig. 3). Slope failure being instigated by channel migration and undercutting, or a combination of channel processes and creep movement along the fault, must be considered a likely possibility for this paleoslump. Moreover, sand dikes were reported in an Illinois basin coal along the basal Pennsylvanian unconformity, which was not adjacent to a known fault or structure, or correlated to a widespread trend of deformation (Nelson et al., 1985). Thus, these types of fluidized injection features in coals are not diagnostic of seismic triggering mechanisms.

Few outcrops are available at the Coalburg horizon in the area, to determine if there is a widespread horizon of deformation or if there are any trends along the same horizon. Thus, although it is possible, the lack of additional liquefaction evidence at a correlative horizon and the likelihood of slumps along channel margins or fault scarps in the fluvial-deltaic environment interpreted for this part of the Breathitt Group make the likelihood of a seismic trigger instead of normal channel-margin mechanisms unlikely, or at least, difficult to substantiate.

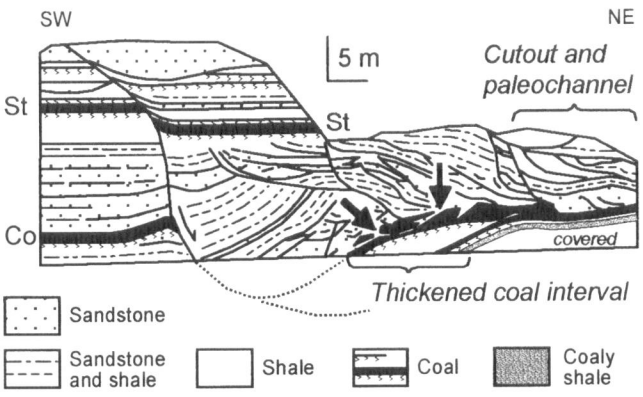

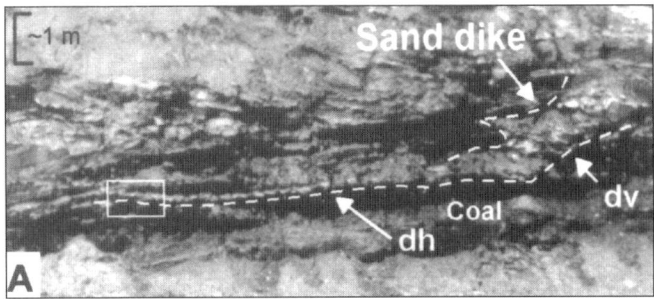

Figure 3. Illustration of a paleoslump and bounding listric normal fault on Kentucky Highway 3 near Inez, Martin County (modified from Greb and Weisenfluh, 1996, Fig. 5B, p. 122). Arrows show location of sand dikes in the coal beneath the slump. Coalburg coal bed (Co) is overthickened beneath the slump. St—Stockton coal bed.

Figure 4. Clastic dikes in the Coalburg coal bed. A: Position of subvertical dikes (dv) connected by horizontal dike (dashed line, dh). B: Detail of near-horizontal, boudinage-like dike from boxed area in A.

Large paleoslump on monocline

Another paleoslump, near Mt. Vernon, Kentucky (location 5-6 in Fig. 2), occurs in basal Pennsylvanian strata at the margin of the Mt. Vernon monocline (Fig. 5). Mid-Carboniferous reactivation of the Mt. Vernon monocline is inferred from stratigraphic thinning across its crest (Chesnut, 1992a, 1992b). This paleoslump has been described previously by Dever et al. (1979), Cobb et al. (1981), Chesnut (1992a), and Greb and Weisenfluh (1996), all of whom noted the possibility that movement on the monocline may have triggered slumping, because the failure is oriented down structural dip. The slump consists of a 10-m-thick sequence of bioturbated sandstone, siltstone, and shale, and a large fold in the toe region of the slump (Fig. 6). The underlying Mississippian carbonates are undeformed. Several other paleoslumps are known along the Mississippian-Pennsylvanian unconformity surface on the western margin of the basin (Hester and Brant, 1981; Greb and Weisenfluh, 1996), but most are less than 6 m thick.

Interpretation. Zones of widespread slumping have been inferred to have a seismic origin in some areas, where other mechanisms seemed unlikely (Seilacher, 1984). The Mt. Vernon paleoslump is only one of many along the Mississippian-Pennsylvanian unconformity surface, a surface in the basin that is partly structurally controlled (Chesnut, 1992b; Greb and Chesnut, 1996). Numerous paleoslumps have been identified along the unconformity in exposures across more than 200 km of the western margin of the basin (Hester and Brant, 1981; Greb and Weisenfluh, 1996). A comparison of the mid-Carboniferous unconformity in the central Appalachian basin with the unconformity in the Illinois basin indicates that slumping is also common there. A report about the southern Illinois basin noted a pervasive area of paleoslumps along the unconformity near the Ste. Genevieve fault zone, which was active during the Carboniferous (Nelson and Devera, 1995).

Figure 5. Cross section along Interstate 75 in Rockcastle County, showing location of large paleoslump on the margin of the Mt. Vernon monocline (modified from Chesnut, 1992a). Pennsylvanian strata: Pb—Breathitt Group. Mississippian strata: Mp—Paragon Formation, Mssg—top of Ste. Genevieve Limestone of Slade Formation, Msr—Renfro Member of Slade Formation, Mb—Borden Formation.

Paleoslumps, however, have also been noted in paleovalleys along the unconformity with no relation to structure (Bristol and Howard, 1971, 1980). Thus, the processes that formed the unconformity itself might be considered as alternative triggering mechanisms. The large scale of some paleoslumps at this horizon might be a function of deeper paleovalley incision during formation of this sub-Pennsylvanian unconformity surface. Downcutting during sea-level fall has been interpreted to have resulted in steep-walled valleys in both basins (Bristol and Howard, 1971; 1980; Greb and Chesnut, 1996). In fact, the unconformity is a time-transgressive surface in the central Appalachian basin (Chesnut, 1992b). If the surface is time-transgressive, however, paleoslumping may also be time-transgressive, in which case features along the unconformity are not likely to have been triggered by a single

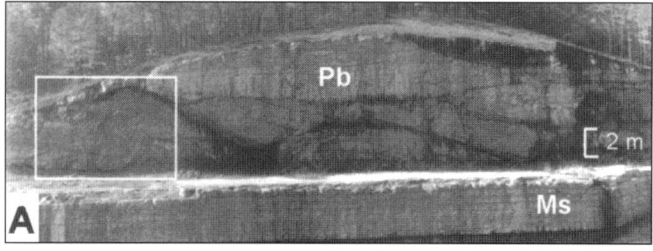

Figure 6. The paleoslump near Mt. Vernon. A: Slump is along contact between Mississippian strata (Ms—Slade Formation) and Pennsylvanian strata (Pb—Breathitt Group). B: Close-up of boxed area in A showing folded and sheared area near toe of slump.

seismic event. This does not mean that individual paleoslumps were not triggered by earthquakes, or that multiple earthquakes might not have occurred, just that a widespread trend from a single large event is difficult to demonstrate. In addition, there is no evidence of liquefaction features along the unconformity in the area, even though fine-grained, quartz arenites, which would probably have been prone to liquefaction, are present.

Bioturbation and sedimentary structures preserved in the slumped strata suggest deposition in a tidal-marine environment (Greb and Weisenfluh, 1996). Tidal flux can instigate slumping, especially in tidal channels (Bjerrum, 1973). The deposits within the paleoslump do not appear to be part of a channel, but the outcrop is not laterally extensive. Smaller slumps in Carboniferous tidal-estuarine facies are not uncommon in the basin (Horne, 1979; Greb and Chesnut, 1992).

Although it is possible that the paleoslump was seismically triggered, it is also possible that it resulted from increased slope on the structure, a combination of increased slope and tidal flux, or increased slope and loading of unconsolidated, rapidly accumulating tidal sediments on consolidated Mississippian carbonates.

Fault-bordered paleo–debris flows

In many basins, debris flows are common and would not be considered evidence of a seism, because they are triggered by rainfall more often than large-magnitude earthquakes. In the Carboniferous of the central Appalachian basin, however, debris flows are rare. One such deposit, the Poison Honey bed (Greb et al., 1990) is located on the western margin of the basin (location 7-8 in Fig. 2). It consists of a complex mosaic of matrix-supported, shale-clast and siderite-clast conglomerates, and laminated sandstones (Fig. 7, bottom). In the conglomerates, angular shale and siderite clasts are supported in a matrix of fine-grained sandstone (Fig. 8A). Shale clasts longer than 30 cm are not uncommon. Clasts exhibit an eastward imbrication in the lower part of the unit (Figs. 7, bottom, 8B). The conglomerates are complexly interbedded with relatively clast-free, fine-grained, massive to laminated sandstone (Figs. 7, bottom, 8C).

The Poison Honey bed is on top of the quartzose Pine Creek Sandstone Member of the Bee Rock Sandstone of the Breathitt Group (Fig. 7, top), previously considered part of the Lee Formation (Chesnut, 1992b). The bed is just east of a fault, and clast imbrication and bed dip show that paleoflow was eastward, coincident with the direction of throw on the fault (Fig. 7). Paleoflow indicators in the underlying sandstone are to the southwest (PC in Fig. 7, bottom), which was the dominant paleoslope for much of the Early Pennsylvanian (e.g., Greb and Chesnut, 1996). Shale and siderite do not occur beneath or lateral to the Poison Honey bed on the downthrown side of the fault, but they are common in lower Breathitt Group deposits on the upthrown block to the west (Fig. 7, top).

Interpretation. Sandstones in the Bee Rock Formation have been interpreted as dominantly bedload-fluvial deposits, and there is local evidence of estuarine sedimentation and tidal channels, which formed as base level rose (Archer and Greb, 1995; Greb and Chesnut, 1996). The Poison Honey shale-clast conglomerates are matrix supported, which suggests deposition in a debris flow or succession of debris flows (Nemec and Steele, 1984). These flows contain siderite nodules and shale clasts, which were not derived from the underlying quartz-arenite, but from Breathitt Group shales on the upthrown fault block (Fig. 7, top). For the siderite pebbles to have been incorporated into the debris flows, they must have been elevated relative to the quartz-arenite across the fault, prior to or concurrent with at least the initial debris flow. Movement of faults and small arches along this margin of the Cincinnati Arch has been inferred for the Early Pennsylvanian on the basis of thinning of Lower Pennsylvanian strata across or along faults and arches, as well as the abrupt thinning of the Bee Rock Formation along its western margin (Chesnut, 1992b; Greb and Chesnut, 1996).

A similar matrix-supported conglomerate in the Illinois basin (rare in that basin as well) occurs on an anticline inferred to have been reactivated during sedimentation (Potter, 1957). Thinning of Lower Pennsylvanian strata across the anticline supports inferences of movement during the time in which the Illinois basin conglomerate was deposited (Nelson et al., 1991). Paleo–debris flows have also been identified within coal-bearing strata of the Triassic Callide coal measures in Australia, along faults that are inferred to have been active during deposition of the flows (Jorgensen and Fielding, 1999). Thus, fault-bounded paleo–debris flows in basins or stratigraphic intervals where they are generally absent might be good candidates for seismite interpretations.

As with paleoslumps, however, the failure of local paleoslopes is difficult to discount when failure orientation corresponds to paleoslope. It is possible that fault movement preceded debris flows. Seasonal rains or a sudden flash flood might have caused the flow to develop on an exposed fault scarp or slope well after movement. Also, stratification within the Poison Honey bed suggests multiple flows (Greb et al., 1990), rather than a single flow from a single coseismic event. The bed may represent (1) two seismites, (2) a seismite followed by a nonseismically triggered debris flow, or (3) two nonseismically triggered flows. A possible correlative horizon of deformation, which is described below, may indicate a horizon of deformation not restricted to a single facies, and possibly a trend of deformation, both of which increase the possibility that at least the initial flow in the Poison Honey bed might be a seismite; but the bed by itself has several possible triggering mechanisms.

Down-dip ball-and-pillow structures

Down structural dip from the Poison Honey bed is a deposit of ball-and-pillow structures (Fig. 7, top; briefly described by Greb and Chesnut, 1990). The ball-and-pillow structures occur within the Pine Creek sandstone, either at the same level as the Poison Honey beds or within approximately 10 m of that level. Isolated and folded "pillows" of very fine grained sandstone in a shale matrix, above a broad scour in underlying, medium-grained, quartzose sandstones of the Bee Rock Formation (Fig. 9A). The shaly interval contains numerous listric glide planes oriented toward the thickest part of the shale lens, as well as glide planes oriented toward the horizontal between and across listric glide planes(dotted lines in Figs. 9A, 10A). Individual pillows of deformed sandstones have preserved internal lamination. The pillows tend to be folded symmetrically across the axis of folding (Fig. 10, A and B), and exhibit a strong east-west orientation (Fig. 9B). In vertical sequence, pillows above successive glide planes commonly show opposing orientations (Fig. 10, A and B). Strikes of elongated rolls, without obvious folding, exhibit modes (Fig. 9C) subperpendicular to the direction of overturning (Fig. 9B). These orientations differ from cross-bed dips in the underlying sandstone (Pc on Fig. 9D). The strike axes of flow rolls are subparallel to the apparent trend of the scour axis (SC on Fig. 9D), which may represent an abandoned scour of the underlying sandstone. Several overturned pillows are stacked at near-vertical angles (Fig. 10C). There are also several near-vertical disruptions in the shales, where bedding is upturned or absent (dashed lines in Fig. 10A).

Interpretation. Ball-and-pillow structures form primarily through hydroplastic deformation, and in some cases they have been interpreted as seismites (e.g., Kelling and Williams, 1966). Because the features in this example are beneath load casts in the overlying deposit, deformation triggered by rapid loading and the

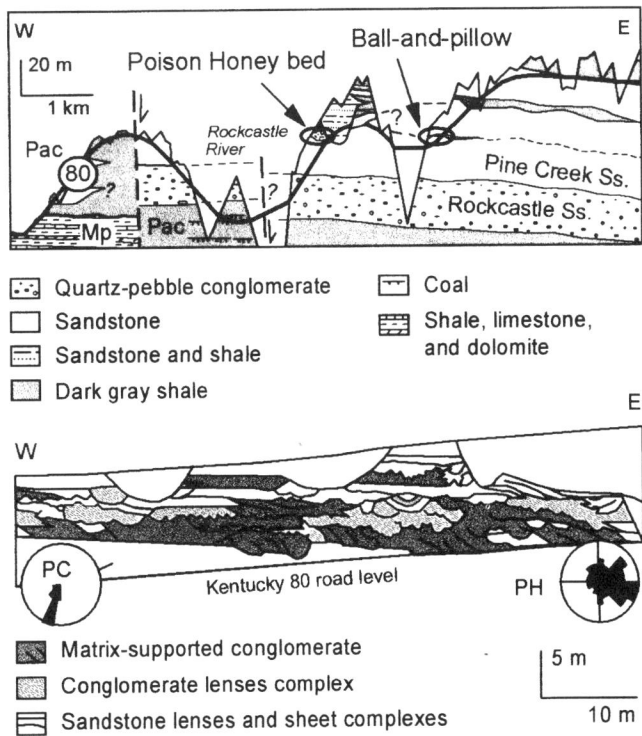

Figure 7. Diagrams showing the Poison Honey bed. Top: Cross section along Kentucky Highway 80, near Billows, Laurel County, showing location of the Poison Honey bed and a correlative zone of ball-and-pillow structures (modified from Greb et al., 1990). Mp—Mississippian, Paragon Formation; Pac—Pennsylvanian, Alvy Creek Formation. Bottom: Facies mosaic of the Poison Honey bed structures. Surrounding strata are not shown in order to highlight the complex facies relationships of the Poison Honey bed (modified from Greb et al., 1990). PC—crossbed dip orientation in the Pine Creek sandstone; PH—chert imbrication orientations in the Poison Honey bed.

density difference between unconsolidated mud and sand (Dzulynski and Walton, 1965; McKee and Goldberg, 1969; Mills, 1983; Allen, 1982) should also be considered. Although load casts are present at the base of the overlying sandstone, the grain size and bedding within pillows is different from the overlying sand, so that the pillows do not appear to be foundered remnants of the capping sandstone.

Another alternative triggering mechanism is slope. The units contain several listric shear planes, but most of the pillows are not oriented in a preferred rotation direction, parallel or perpendicular to the dip of the scour, as might be expected if deformation was initiated by down-slope movement along the limb of the scour. Only the pillows and shear planes on the westernmost margin of the scour show preferred orientation down the apparent dip of the scour limb, and even these show numerous flow rolls oriented against slope.

Laboratory experiments have produced axially symmetrical pillow structures by applying a shock to unconsolidated sediments (Kuenen, 1958; Allen, 1982). In nature, such a shock could

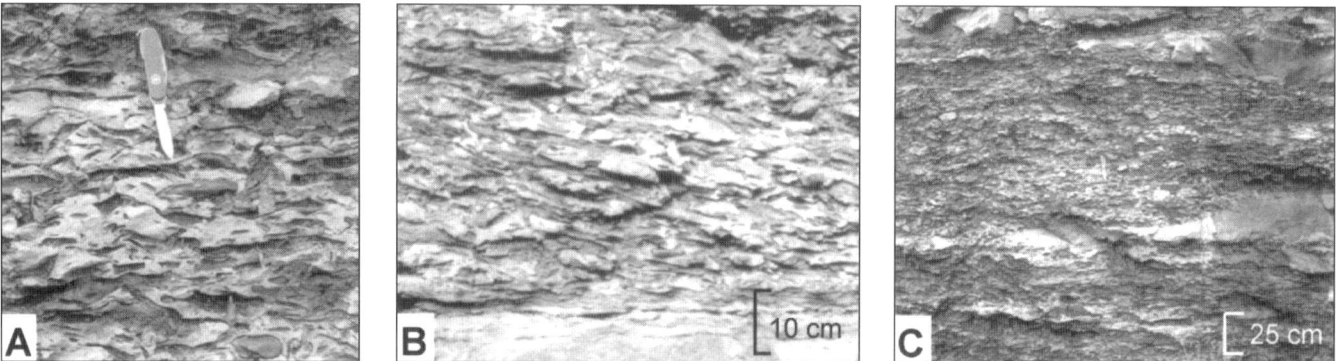

Figure 8. Poison Honey bed features. A: Shale and siderite clasts in a matrix of fine-grained sandstone. Knife is 15 cm long. B: Shale and siderite clasts forming crude imbrication. C: Complex interfingering of clast-rich and clast-poor facies.

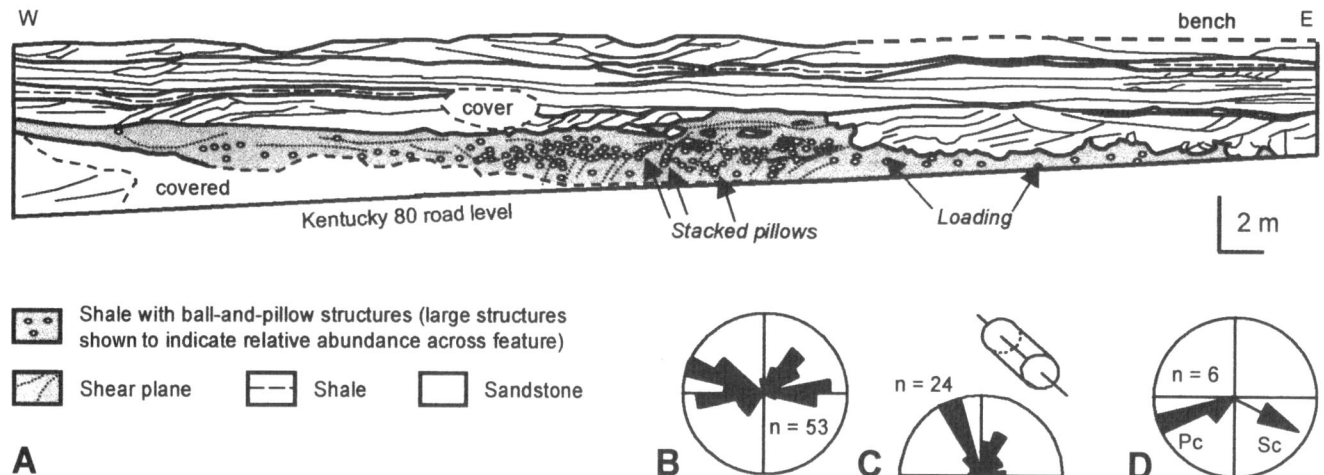

Figure 9. Ball-and-pillow structures within shale-filled scour. A: Line drawing from photomosaic of shale-filled scour with abundant ball-and-pillow structures and load casts in the overlying sandstone. B: Orientation of direction of pillow overturning. C: Strike axes of balls or rolls in which overturning could not be determined. D: Apparent dip directions of other significant sedimentary features at the ball-and-pillow horizon. Pc—Pine Creek sandstone cross-beds; Sc—scour axis.

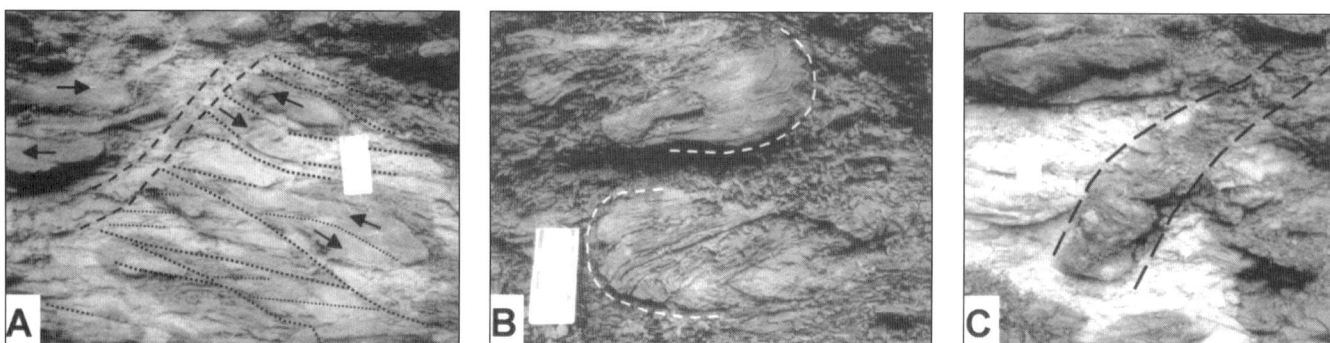

Figure 10. Ball-and-pillow structures. A: Multiple shears in the gray shale (dotted lines), vertical shale dike (dashed lines) that disrupts overlying bedding, and opposing orientations of pillow folding (small arrows) between shear planes. B: Detail of pillow showing lamination and opposing fold orientations. C: Stacked pillows, arranged in a crude vertical pipe (dashed lines), which disrupts overlying bedding.

be produced by an earthquake, although Sims (1975) noted the problems with directly relating these experimentally produced structures to actual seisms. More compelling evidence of a seismic origin for the pillows may be the opposing directions of folding in the pillows (Figs. 9B, 10, A and B). Lignier et al. (1998) noted opposing orientations of pillows within Quaternary glaciolacustrine sediments in the Alps of France. The opposing orientations of these pillows were between shear planes in a type of feature termed fold-graded bedding, inferred to be similar to fault-graded bedding of Seilacher (1969). The opposing fold directions of the pillows between anisotropic slumps were inferred to result from horizontal shear and oscillation of material during seismic shaking. In the ball-and-pillow deposits analyzed here, successive pillows are oriented in opposing directions, even in some cases where the glide planes between pillows are oriented downslope. The opposing orientations might suggest back-and-forth motions that are unlikely to have been caused by a vertical load or downslope movement.

Another mechanism that might theoretically cause opposing orientations of structures is tidal flux. Tidal-estuarine sedimentation is common at the top of the Pine Creek sandstone (Greb and Chesnut, 1996). Tidal current drag by itself does not exert enough shear stress on the bed to initiate liquefaction (Allen and Banks, 1972), but rapid falls in water level during ebb tides can create sufficient pore-fluid pressure change to initiate liquefaction (Terzaghi, 1950; Wunderlich, 1967; Bjerrum, 1973). Also, in the modern Bay of Fundy, pressure differences on the bed formed by relatively shallow (0.1 to 0.3 m) but breaking waves caused hydroplastic deformation and liquefaction, resulting in a horizon of thin ball-and-pillow–like structures along the crest of well-sorted, fine-grained tidal megaripples (Dalrymple, 1979).

The grain sizes and bedding in the Pine Creek example are different from those in the Bay of Fundy, so wave-induced liquefaction in an estuarine setting based on that analog seems unlikely. Similarly, the Pine Creek ball-and-pillow structures differ from flow rolls noted in other estuarine facies in the basin in appearance (Horne, 1979, for example), and in exhibiting opposing orientations of folding. Estuarine facies have been widely studied in the central Appalachian basin, and similar ball-and-pillow structures have not been reported (Greb and Chesnut, 1992; Martino and Sanderson, 1993; Greb and Archer, 1998). Thus, although a possible cause, tidal mechanisms are not diagnostic for the ball-and-pillow structures either.

Because ball-and-pillow structures occur (1) at the same horizon as the Poison Honey bed, another rare lithofacies in the basin, and (2) on the downdip from the downthrown margin of a fault, there is a possible, but weak, trend of deformation that can be tied to a potential seismic source.

Carbonate flow rolls and soft-sediment deformation

The Big Sinking bed of the Slade Formation has long been noted for its flow rolls and contorted bedding (Butts, 1922), and was recently described in detail by Dever (1999), from which the following description is summarized. The Big Sinking bed is present at or near the top of the Renfro Member of the Slade Formation in east-central Kentucky and beneath the Burnside Member of the Slade (Fig. 1), which is partly equivalent to the St. Louis Limestone. The Big Sinking bed is a 3-m-thick, mainly very fine grained, well-sorted, bioclastic, pelletal calcarenite. Carbonate particles principally range from coarse silt to fine sand. The limestone is finely laminated, thin to thick bedded, with thin to very thin interbeds of greenish-gray shale, mainly in the lower part. The unit commonly contains stringers of laminated chert. It is sparsely fossiliferous (brachiopods, bryozoans, gastropods, and crinoid plates), but orthotetid and productid brachiopods are locally abundant.

In the central part of the area in which the bed is found, flow rolls, slumped flow rolls, and other forms of soft-sediment deformation are common. Deformation occurs along and on the upthrown side of the Locust Branch fault trend (Fig. 11); the thickness of the deformed interval and the scale of individual flow rolls generally decrease away from the fault trend toward the Irvine–Paint Creek fault system (Fig. 11C). In addition, the overlying Burnside Member is missing across much of the Locust Branch Fault trend (Fig. 11B). Where the Burnside is missing, the Big Sinking bed is overlain by the Ste. Genevieve Member of the Slade Formation (Fig. 11C).

In the interval of contorted bedding, curvilinear patterns of deformed laminae and chert stringers outline bodies of rolled-up sediment. Bedding ranges from wholly planar to totally deformed, but more commonly it is only partly deformed, and the interval of contorted bedding is either underlain by, overlain by, or within undeformed, planar-bedded, fine-grained limestone.

In the Big Sinking bed (Fig. 12), flow rolls dominate most exposures; they have a wide variance in sizes and orientations (Figs. 12, 13A). In some areas, single flow rolls occupy the entire deformed zone (Fig. 12, A and B). Large flow rolls are commonly separated by vertical "dike-like" or "pipe-like" features generally lacking internal bedding, or showing crude vertical lamination (Figs. 12A, 13C). These dikes commonly contain abundant chert bodies, but they are composed of similar grain size and material as the rocks that they truncate. At one location, flow rolls on either side of large vertical dikes were stacked on top of each other (Figs. 12A, 13D).

Small flow rolls exhibit overall random orientations (Fig. 12C), although locally they occur in small "lows" between "highs" or irregularities on the basal surface. In other areas, the basal contact is sharp. In most exposures, deformation is common toward the base, but in others, the lower part of the unit is undeformed and flow rolls are above the base. Dolomitization is common along the basal surface and may extend upward into the zone of deformation, in many cases masking the lower contact of the unit. Although the basal surface is flat to slightly irregular in most areas, it locally contains distinct scours. Scour-form basal contacts may be overlain by small flow rolls as in other areas or slump-form rolls (Fig. 13, E and F). In some cases, the slump-form flow rolls appear to be onlapped by flat-laminated, unde-

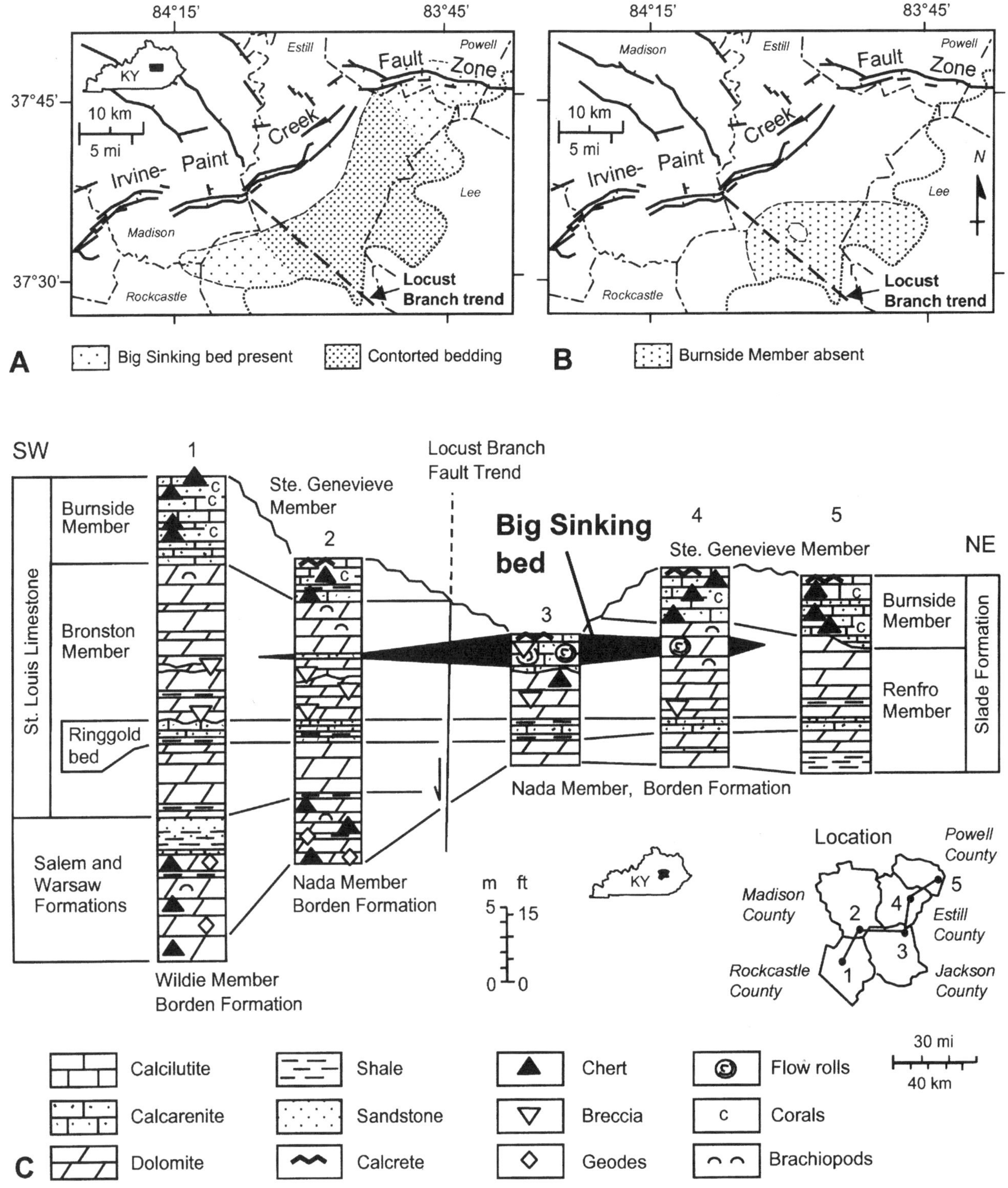

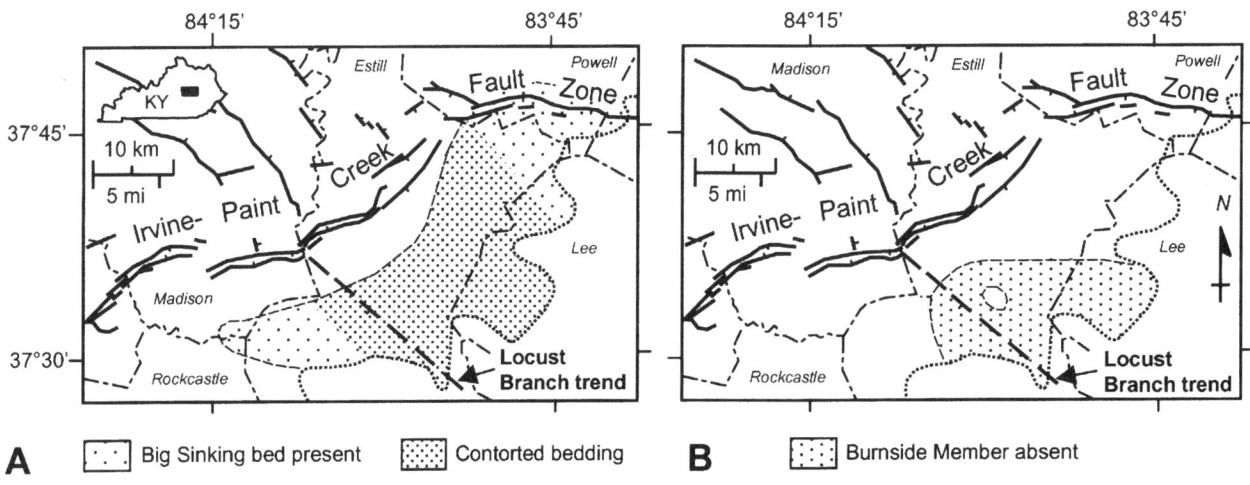

Figure 11. Maps of Big Sinking bed distribution. A: Area of Big Sinking bed, area of contorted bedding, and local fault trends (modified from Dever, 1999, Fig. 35, p. 42). B: Same region, showing area where the Burnside Member is missing (modified from Dever, 1999, Fig. 33, p. 40). Dotted line on right side of both maps is where units dip into the subsurface. C: Cross section of the Big Sinking bed and surrounding strata (modified from Dever, 1999, Fig. 10, p. 17).

formed strata (Fig. 13E). In other cases, the contact between deformed and undeformed strata above scours is gradational (Fig. 13F). Where gradational, undeformed beds may pass laterally into deformed beds or protrude upward into undeformed beds (Fig. 13F).

The upper contact is generally sharp. The Big Sinking is overlain by thin, planar-laminated, relatively poorly sorted, coarser grained, bioclastic calcarenite in the lower part of the Burnside Member (Fig. 11A). Flow rolls may extend up to this contact, or the upper part of the Big Sinking bed may be undeformed beneath it. Locally in Powell, Pulaski, and Rockcastle counties, Big Sinking and Burnside limestones intertongue through intervals less than 1 m thick (Dever, 1999). At one location, a thin bed of limestone capping the zone of flow rolls contains abundant productid brachiopod valves that are stacked vertically in a convex-down position (Fig. 14). This stacking is positioned just above a large flow roll. Laterally, where flow rolls are smaller and farther beneath the limestone bed, preferred brachiopod valve orientations are not discernible or the brachiopod-rich layer is absent.

Interpretation. The Big Sinking bed has been interpreted as being structurally controlled because of its position along and on the upthrown side of the Locust Branch fault trend (Fig. 11A). This fault is probably related to the nearby Irvine–Paint Creek fault system (Figs. 11A), which stratigraphic evidence indicates was reactivated during at least the Mississippian (Dever, 1986). Eastward, the Irvine–Paint Creek fault system also influenced the thickness of Pennsylvanian stratigraphic units (Horne and Ferm, 1978). The absence of the overlying Burnside Member along the Locust Branch fault trend (Fig. 11B) suggests continued movement following deposition of the Big Sinking carbonates. Thickening of the stratigraphic interval beneath the Big Sinking bed southward (Fig. 11C) also suggests pre–Big Sinking bed movement along the structure (Dever, 1999).

The grain size of the rock is within the range in which liquefaction can occur, although much of the deformation in the Big Sinking bed may have resulted from hydroplasticity and fluidization. The large size of some flow rolls at this horizon is analogous to the large size of Bartlesville Sandstone (Oklahoma) flow rolls, which were interpreted as seismites (Rascoe, 1975). Likewise, flow rolls encased by undeformed strata in carbonates have been interpreted as seismites (Weaver and Jeffcoat, 1978; Rast and Ettensohn, 1995; Pope et al., 1997). One of the features used to infer seismite origin of flow rolls in these other carbonate examples is the nonerosive base of the deformed zone (Pope et al., 1997). This is not always true for the Big Sinking bed.

In some places, the base of the Big Sinking is nonerosive. In others, dolomitization masks the nature of the basal contact. In still other places, scours are present along the basal contact. These scours may represent small tidal or storm channels. The scours and irregularities on the basal surface locally appear to have affected the rotation direction of sediment, the orientation of glide planes, and positions of flow rolls. The overall trend of flow roll axes, however, is random (Fig. 12C). If a single slope failure caused deformation, a directional trend would seem more likely. If multiple slope failures occurred off a fault scarp or topographic high along the fault trend, a directional trend would still be expected. No obvious trend has been noted.

The fact that deformed strata above some scours appear to be overlain by undeformed strata (Fig. 13E) could be explained by normal depositional processes burying slumps within the channels. In many areas, however, deformed strata are gradational with overlying undeformed strata. In fact, flow rolls may extend upward into undeformed strata, indicating deformation following burial (Fig. 13F). Also, deformation is not restricted to the channels or channel flanks. Because flow rolls are at different levels within the zone, and the upper contact of the unit is sharp and horizontal across the study area, it would appear that most of the deformation occurred after burial, rather than syndepositionally by slumping.

The Big Sinking bed commonly is present between the Renfro Member, a sparsely fossiliferous, laminated dolomite, generally inferred to have been deposited in upper tidal-flat environments, and the Burnside member of the Slade Formation (partly equivalent to the St. Louis Limestone), a fossiliferous, relatively poorly sorted limestone, inferred to have been deposited in a subtidal environment (Dever, 1990). Stratigraphic position and parallel lamination in undeformed parts of the pelletal Big Sinking bed suggest that it was deposited in a lagoonal environment prior to deformation. Similar scales and extents of flow rolls have not been reported from lagoonal or other pelletal facies in other Mississippian deposits in the basin, although tidal fluctuation is known to create smaller scale soft-sediment deformation features in modern tidal environments (e.g., Wunderlich, 1967).

Another possible triggering mechanism for flow rolls is vertical loading. The Big Sinking flow rolls do not occur within a facies sequence that suggests rapid loading of sand-size particles on muds, nor are there any load structures at the base of the overlying unit. Also, in some areas, laminated strata within the Big Sinking bed occur between the flow rolls and overlying strata, which would be unlikely if vertical loading triggered liquefaction.

In marine carbonates, the possibility that storm waves produced sudden pore-fluid pressure changes in the bed, or shear stress on the bed, which could result in flow rolls, must also be considered. Where the Big Sinking bed is undeformed and in overlying strata, there are no tempestites, hummocky stratification, and large-amplitude wave ripples, as would be typical if storms affected sedimentation.

Within the Big Sinking bed are areas of hydroplastically deformed relatively tabular bedding and areas of hydroplastically deformed bedding above lenticular scours. In Figure 12A, deformation appears to cut across both channel-form and tabular bedding, although with fewer effects in the channel-form facies. This may suggest that deformation cut across facies boundaries, but occurs within facies of similar grain sizes.

The structureless dikes between large flows may have formed by local liquefaction or fluidization. Vertical stacking of flow rolls between dikes is not typical of slumping, sliding, or

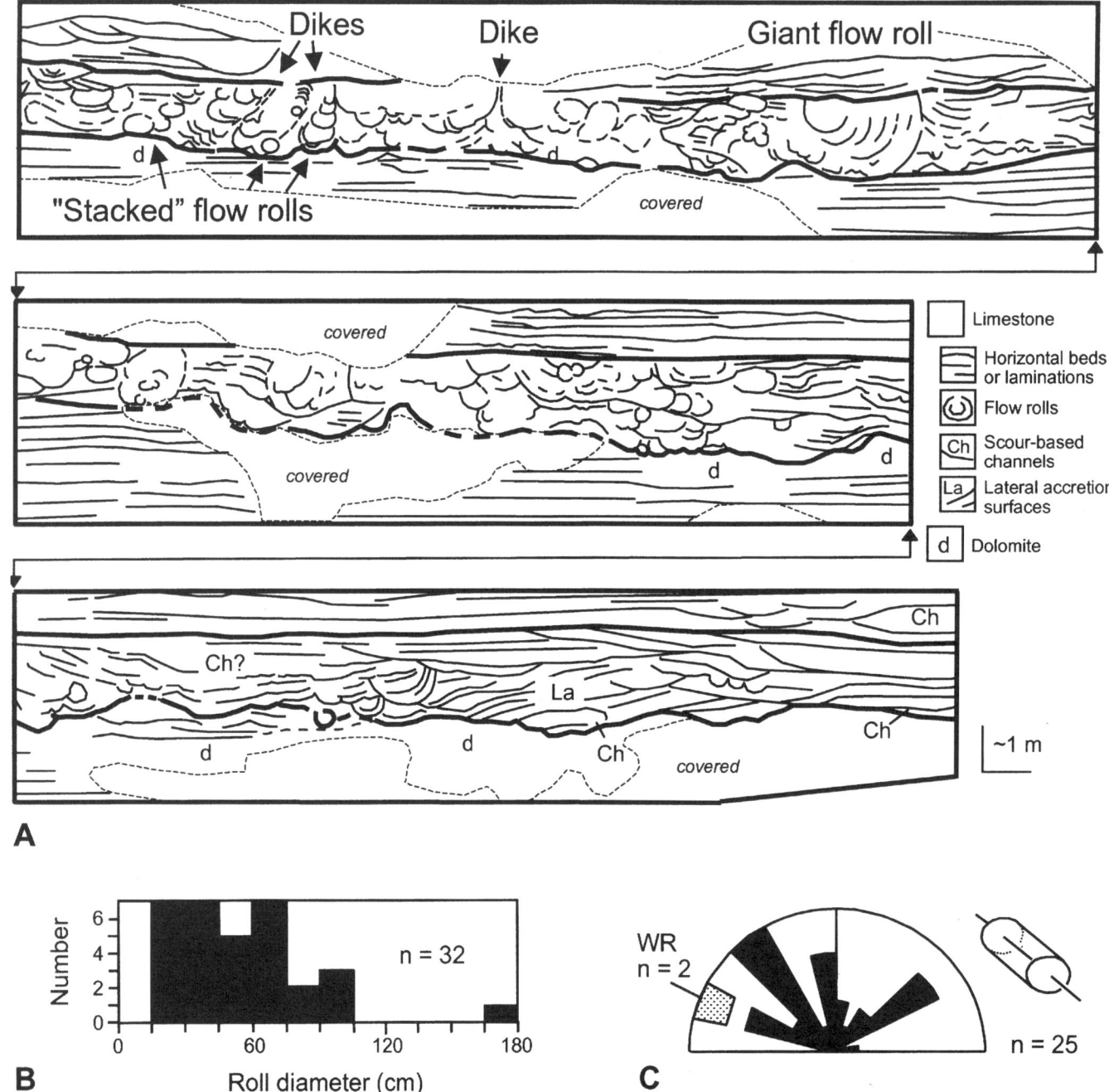

Figure 12. Bedding features of the Big Sinking bed, Hatton Hollow, Lee County, Kentucky. A: Trace of a photomosaic. B: Compilation of flow roll thickness. C: Strike orientations of flow rolls (black) and wave ripples (WR, stippled).

storm-induced deformation, and it may suggest an in situ, near-vertical, or oscillating energy source in the bed, similar to the hydroplastic deformation inferred for the opposing flow rolls in the Pine Creek example. Similar features have not been reported from modern marine carbonate seismites (for which there are few data), so this feature is also inconclusive, although possibly suggestive.

Another feature that hints at a seismic origin for the Big Sinking bed is the local stacking of brachiopod valves above flow rolls (Fig. 14). Similar stacking was noted in studies of biostratinomy by Seilacher (1984), and was interpreted to result from liquefaction triggered by a seism on unconsolidated sediment or by the passage of a tsunami above unconsolidated sediment. Convex-up orientations would be more stable positions if the valves

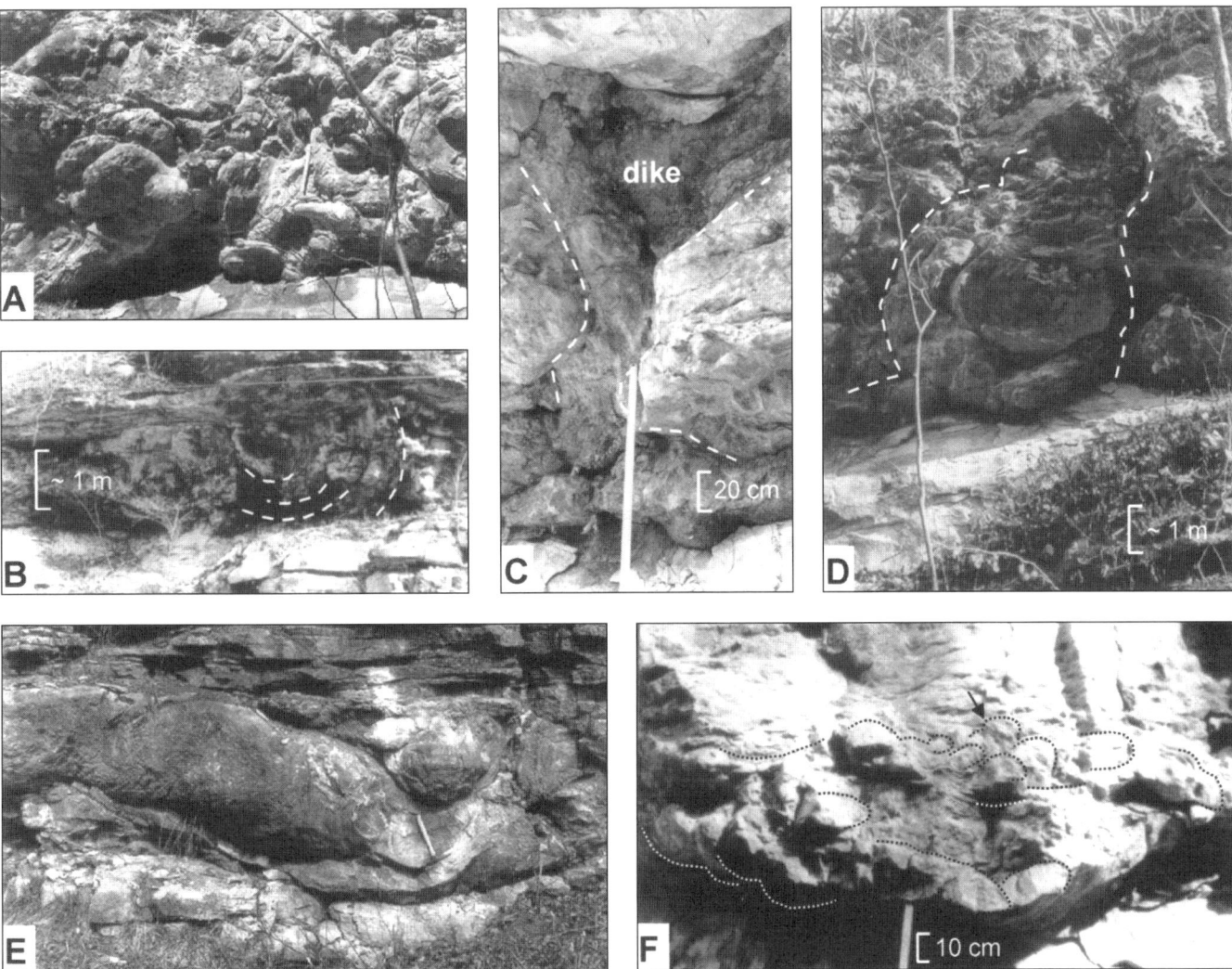

Figure 13. Bedding features of the Big Sinking bed. A: Multiple flow rolls of different sizes. Hammer is 35 cm long. B: Giant flow roll encompassing entire interval. C: Vertical dike with chert clasts (dark spots). D: Stacked flow rolls between vertical pipes. E: Slump-form masses above a scour at the base of the unit, overlain by relatively undeformed strata within the Big Sinking bed. Hammer is 35 cm long. F: Scour overlain by flow rolls and relatively transitional contact with overlying undeformed strata. Arrow indicates location of several small flow rolls in the top of the deformed zone that protrude upward into undeformed strata.

were deposited aseismically by combined flows or storm waves in shallow marine environments. But if the sediment were resuspended by a seism or tsunami, then the valves could settle in essentially quiet water following the shock, in a convex-down position (Seilacher, 1984). This stacking pattern has not been noted above other flow-roll horizons in the basin. It also has not been documented from a modern earthquake or tsunami.

DISCUSSION

Each of the study examples has textural properties and is found in facies that could be liquefied, fluidized, or hydroplastically deformed during an earthquake, and all occur in a basin that was tectonically active during the Carboniferous. Most examples were chosen because of their position along structures that had previously been inferred to have moved near to the time of deposition. For strata older than the Holocene, however, it becomes increasingly unlikely that a deposit can be positively tied to a seism from a specific source, because dating methods commonly show increasing margins of error for increasingly older strata, and because the margins of error are well outside the range required to record a single shock (Wheeler, this volume). But if correlations can be made that link several deposits of possible seismic origin together, the possibility of seismic origin should not be ignored. For many of the beds studied, and similar beds elsewhere, a seismic trigger interpretation must be qualified as "possibly" or "probably" because other mechanisms of formation are also possible.

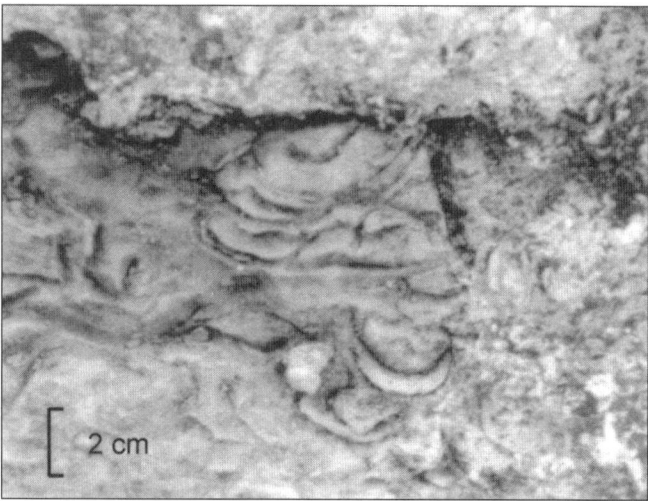

Figure 14. Convex-down brachiopod valves at the top of the Big Sinking bed, Pond School, Jackson County, Kentucky.

Evaluation of the study horizons by criteria

Widespread deformation trends. Numerous investigations have shown that the greatest earthquake effects occur near earthquake epicenters or energy-source regions, and trends of deformation decrease away from the source. In the case of the Big Sinking bed, there appears to be widespread deformation at a correlative horizon south of the Irvine–Paint Creek fault and above the Locust Branch fault trend. In this case, the unusually large and widespread scale of different types of deformation features can be linked to a possible trend of deformation, decreasing away from a fault, which was part of a structural hingeline during the Carboniferous. The much smaller zone of deformation within the Pine Creek sandstone at the Poison Honey bed and down-dip ball-and-pillow horizon could also indicate a weak trend of deformation relative to a fault. Lack of lateral outcrop exposure for some of the features studied precludes the possibility of testing to see whether more widespread trends do or do not exist at these horizons. In the case of the Mt. Vernon paleoslump, a widespread surface of deformation can be identified, but because it is a regional unconformity, deformation above the surface may not be contemporaneous.

These examples show the difficulties in demonstrating definitively a trend of deformation in ancient deposits of possible seismic origin. Even if a trend can be determined, it simply indicates a point source of a high-energy event, and might not preclude slope or storms as possible triggering mechanisms. In the case of the Poison Honey bed paleo–debris flow and downdip ball-and-pillow structures, different types of deformation occur at a single horizon and appear to cut across depositional facies. Deformation that cuts across facies at a correlative horizon might be expected if deformation was earthquake induced; but more widespread deformation might also be expected. In the case of the trend of deformation in the Big Sinking bed, deformation is widespread, and orientations of flow rolls do not parallel paleoslopes.

It is tempting to infer that the epicenters of ancient large-magnitude earthquakes were positioned along the structures bordering the deformation features studied. Even with trends toward structures, however, this cannot always be interpreted definitively. For example, if deformation in the Big Sinking bed was seismogenically initiated, was the paleoearthquake along the Locust Branch fault, the Irvine–Paint Creek fault (Fig. 11, A and B), or possibly a deep-seated structure lacking surface expression? Trends in liquefaction features formed during modern, large-magnitude earthquakes in California do not always point toward the surface rupture of the fault that actually caused the earthquake, but often point to other faults because of fault dips and site characteristics (Wheeler, this volume). For this reason, several investigations of pre-modern earthquakes have made the distinction for trends toward "energy-source regions" or "energy centers," analogous to meizoseismal regions, rather than to earthquake hypocenters or epicenters along specific faults (e.g., Obermeier, 1996).

Seismic likelihood—Alternative mechanisms. Possibly the most important criteria that must be addressed when interpreting a seismic origin for ancient soft-sediment deformation structures or mass movements is the likelihood of seismic origin for that feature in its particular place in time. Unless the preponderance of data collected indicates that the features are analogous to modern proven types of seismites, or that deformation occurs in a facies that can be shown in its modern environment to undergo seismic events at greater frequency than slope-, climate-, tidal-, storm-, or loading-induced events, then seismic triggering interpretations should probably be qualified.

For example, in southern California, hundreds to thousands of landslides may accompany major earthquakes (>M 6.2) (Keefer, 1984), but hundreds to thousands of landslides may also occur annually because of normal channel processes, wave action at the shoreline, and seasonal fluctuation in rainfall. In addition, large landslides may be reactivated through time by seasonal climate changes and periodic earthquakes (Ward and Page, 1991). With this in mind, it is perhaps not surprising, that the paleolandslides investigated herein, could not be shown to be triggered by earthquakes exclusively.

Likewise, for tidal-estuarine facies, rapid sedimentation, rapid water-level changes, and in some cases slope failure are difficult to negate as alternative triggering mechanisms for landslides and soft-sediment deformations. The Poison Honey bed, lateral ball-and-pillow structures, and the paleoslump at Mt. Vernon all occur in tidal-estuarine strata. The Poison Honey bed horizon can be interpreted as being unlike other tidal-estuarine facies in the basin, but tidal processes were active. The paleo–debris flow flowed in the direction of downthrow. This would have been parallel to local paleoslope. For any landslide in tidal facies, slope and artesian pressures may be difficult triggers to negate as at least contributing causal mechanisms.

Great storms are also more common at many low latitudes than earthquakes (Morton, 1988). For much of the Carboniferous, the areas studied in this investigation were within tropical latitudes (Golanka et al., 1994), which may have experienced hurricanes. For marine facies, an effort should be made to rule out the possibility that wave-induced pore-fluid pressure changes and deformation were as likely as seismic activity in the deposit's place and time. In the Big Sinking bed, other types of storm-wave stratification are lacking, but the surrounding carbonates were deposited in environments with obvious wave energy, such that wave action, although perhaps improbable, cannot be completely discounted. More work is needed in modern environments to see how widespread the theoretical effects of both wave-induced and seismogenically induced soft-sediment deformations may be following storms and earthquakes.

What is interesting about the features that lend the most credence to the possibility that the Big Sinking bed and the ball-and-pillow structures in the Pine Creek sandstone are seismites is that both are very subtle and only locally evident. The convex-down shell stacking was only noted at one location in the Big Sinking bed, possibly because of the rarity of brachiopods in the deposit, or if seismic in origin, because of the relative strength of a seism at that location. In the case of the ball-and-pillow structures, the opposing orientation of pillow folding is not present across the entire exposure. The overall orientation is random. In fact, there are several listric glide planes above the uneven topography at the base of the unit; these could be used to infer a slope trigger. The opposing orientations of pillow folding, however, seem more likely to have been produced by a horizontally oscillating force.

Identifying these types of subtle features, in addition to correlating widespread zones of deformation that are atypical of a particular facies and cut across sedimentary facies, may be needed to negate or lessen the relative probability of alternative triggering mechanisms in many ancient deposits. For the Big Sinking bed, widespread deformation can be demonstrated; for the ball-and-pillows and Poison Honey bed, only local deformation is shown. Even when numerous lines of evidence are gathered for a possible seismic origin, alternative mechanisms may still have caused or contributed to deformation. Alternatively, when evidence favoring other causal mechanisms is gathered, seismic triggering may still be possible, just less likely. In the future, a systematic methodology for categorizing the relative possible, probable, or definite origin of a deposit as a seismite, based on multiple lines of evidence, may aid in standardizing criteria or methods used to infer seismogenic origins. Additionally, more comparative studies of soft-sediment deformations and various types of landslides known to have been initiated by different triggering mechanisms in modern environments may aid in differentiating the mechanisms that triggered the various processes in the rock record.

ACKNOWLEDGMENTS

We thank J.D. Kiefer and E. Woolery of the Kentucky Geological Survey (KGS) for discussions about the possible earthquake origins of some of the structures described in this manuscript; M.L. Smath at KGS for grammatical editing; S.F. Obermeier of the U.S. Geological Survey for discussions concerning liquefaction features and for providing the reference to Lignier et al. (1998), which shared similarities to one of the deposits we analyzed; and R.L. Martino, W.J. Nelson, and N. Rast for helpful reviews.

REFERENCES CITED

Allen, J.R.L., 1982, Sedimentary structures: Their character and physical basis: Amsterdam, Elsevier, Developments in Sedimentology, v. 30A, 593 p.; v. 30B, 663 p.

Allen, J.R.L., and Banks, N.L., 1972, An interpretation and analysis of recumbent-folded deformed crossbedding: Sedimentology, v. 19, p. 257–283.

Ammerman, M.L., and Keller, G.R., 1979, Delineation of Rome trough in eastern Kentucky by gravity and deep drilling data: American Association of Petroleum Geologists Bulletin, v. 63, p. 341-353.

Andrews, W.M., Hower, J.C., Ferm, J.C., Evans, S.D., Sirek, N.S., Warrell, M., and Eble, C.F., 1996, A depositional model for the Taylor coal bed, Martin and Johnson Counties, eastern Kentucky: International Journal of Coal Geology, v. 31, p. 151–167.

Archer, A.W., and Greb, S.F., 1995, An Amazon-scale drainage system in the Early Pennsylvanian of central North America: Journal of Geology, v. 103, p. 611–627.

Bjerrum, J., 1973, Geotechnical problems involved in foundations of structures in the North Sea: Geotechnique, v. 23, p. 319–358.

Brenchley, P.J., and Newall, G., 1977, The significance of contorted bedding in Upper Ordovician sediments of the Oslo Region, Norway: Journal of Sedimentary Petrology, v. 47, p. 819–833.

Bristol, H.M., and Howard, R.H., 1971, Paleogeologic map of the sub-Pennsylvanian Chesterian (Upper Mississippian) surface in the Illinois basin: Illinois State Geological Survey Circular 458, 14 p.

Bristol, H.M., and Howard, R.H., 1980, Sub-Pennsylvanian valleys in the Chesterian surface of the Illinois basin, in Luther, M.K., ed., Proceedings of the technical sessions, Kentucky Oil and Gas Association 36th and 37th Annual Meetings, 1972 and 1973: Kentucky Geological Survey, ser. 11, Special Publication 2, p. 55–71.

Butts, C., 1922, The Mississippian Series of eastern Kentucky: Kentucky Geological Survey, ser. 6, v. 7, 188 p.

Chesnut, D.R., Jr., 1992a, Geological highway cross section: Interstate Highway 75, Conway, Kentucky–Jellico, Tennessee: Kentucky Geological Survey, ser. 11, Map and Chart Series 3.

Chesnut, D.R., Jr., 1992b, Stratigraphic and structural framework of the Carboniferous rocks of the Central Appalachian basin in Kentucky: Kentucky Geological Survey, ser. 11, Bulletin 3, 42 p.

Cita, M.B., and Lucchi, F.R., 1984, Preface: Seismicity and sedimentation: Marine Geology, v. 55, 4 p.

Cobb, J.C., Chesnut, D.R., Jr., Hester, N.C., and Hower, J.C., 1981, Coal and coal-bearing rocks of eastern Kentucky Guidebook and roadlog for Coal Division of Geological Society of America Field Trip no. 14: Kentucky Geological Survey, ser. 11, 169 p.

Dalrymple, R.W., 1979, Wave-induced liquefaction: A modern example from the Bay of Fundy: Sedimentology, v. 26, p. 835–844.

Dever, G.R., Jr., 1977, The lower Newman Limestone: Stratigraphic evidence for Late Mississippian tectonic activity, in Dever, G.R., Jr., Hoge, H.P., Hester, N.C., and Ettensohn, F.R., eds., Stratigraphic evidence for late Paleozoic tectonism in northeastern Kentucky, Guidebook and roadlog for the field trip held in conjunction with the 5th annual meeting of the Eastern Section of the American Association of Petroleum Geologists and the 1977 field conference of the Geological Society of Kentucky: Kentucky Geological Survey, ser. 10, p. 8–18.

Dever, G.R., Jr., 1986, Mississippian reactivation along the Irvine–Paint Creek Fault System in the Rome Trough, east-central Kentucky: Southeastern Geology, v. 27, no. 2, p. 95–105.

Dever, G.R., Jr., 1990, Tectonically influenced deposition and erosion in Mississippian limestones of south-central Kentucky, in Dever, G.R., Jr., Greb, S.F., Moody, J.R., Chesnut, D.R., Jr., Kepferle, R.C., and Sergeant, R.E., eds., Tectonic implications of depositional and erosional features in Carboniferous rocks of south-central Kentucky, Guidebook and roadlog for Geological Society of Kentucky 1990 field conference: Kentucky Geological Survey, ser. 11, p. 1–18.

Dever, G.R., Jr., 1999, Tectonic implications of erosional and depositional features in upper Meramecian and lower Chesterian (Mississippian) rocks of south-central and east-central Kentucky: Kentucky Geological Survey, ser. 11, Bulletin 5, 67 p.

Dever, G.R., Jr., Hester, N.C., Ettensohn, F.R., and Moody, J.R., 1979, Stop 3: Newman Limestone (Mississippian) of east-central Kentucky and Lower Pennsylvanian slump structures, in Ettensohn, F.R., and Dever, G.R., Jr., eds., Carboniferous geology from the Appalachian basin to the Illinois basin through eastern Ohio and Kentucky, Guidebook and roadlog of the Ninth International Congress of Carboniferous Stratigraphy and Geology Field Trip no. 4: Lexington, University of Kentucky, p. 175–181.

Drahovzal, J.A., and Noger, M.C., 1995, Preliminary map of the Precambrian surface in eastern Kentucky: Kentucky Geological Survey, ser. 10, Map and Chart Series 8, 9 p., plus 1 sheet.

Dzulynski, S., and Walton, E.K., 1965, Sedimentary features of flysch greywackes: Amsterdam, Elsevier, Developments in Sedimentology, v. 7, 274 p.

Elliott, C.G., and Williams, P.F., 1988, Sedimentary slump structures: A review of diagnostic criteria and application to an example from Newfoundland: Journal of Structural Geology, v. 10, p. 171–182.

Ettensohn, F.R., 1977, Effects of synsedimentary tectonic activity on the upper Newman Limestone and Pennington Formation, in Dever, G.R., Jr., Hoge, H.P., Hester, N.C., and Ettensohn, F.R., eds., Stratigraphic evidence for late Paleozoic tectonism in northeastern Kentucky, Guidebook and roadlog for the field trip held in conjunction with the 5th annual meeting of the Eastern Section of the American Association of Petroleum Geologists and the 1977 field conference of the Geological Society of Kentucky: Kentucky Geological Survey, ser. 10, p. 18–29.

Golanka, J., Ross, M.I., and Scotese, C.R., 1994, Phanerozoic paleogeographic and paleoclimates modeling maps, in Embry, A.F., Beauchamp, B., and Glass, D.J., eds., Pangea: Global environments and resources: Canadian Society of Petroleum Geologists Memoir 17, p. 1-47.

Greb, S.F., and Archer, A.W., 1998, Annual sedimentation cycles in rhythmites of Carboniferous tidal channels, in Tidalites: Processes and products: SEPM (Society for Sedimentary Geology) Special Publication 61, p. 75–83.

Greb, S.F., and Chesnut, D.R., Jr., 1990, Sedimentological evidence for a seismic event in the Lower Pennsylvanian of the Eastern Kentucky Coal Field, in Dever, G.R., Jr., Greb, S.F., Moody, J.R., Chesnut, D.R., Jr., Kepferle, R.C., and Sergeant, R.E., eds., Tectonic implications of depositional and erosional features in Carboniferous rocks of south-central Kentucky, Guidebook and roadlog for Geological Society of Kentucky 1990 field conference: Kentucky Geological Survey, ser. 11, p. 31–37.

Greb, S.F., and Chesnut, D.R., Jr., 1992, Transgressive channel filling in the Breathitt Formation (upper Carboniferous), eastern Kentucky Coal Field, U.S.A.: Sedimentary Geology, v. 75, p. 209-221.

Greb, S.F., and Chesnut, D.R., Jr., 1994, Paleoecology of an estuarine sequence in the Breathitt Formation (Pennsylvanian), Central Appalachian basin: Palaios, v. 9, p. 388–402.

Greb, S.F., and Chesnut, D.R., Jr., 1996, Lower and lower Middle Pennsylvanian fluvial to estuarine deposition, central Appalachian basin: Effects of eustasy, tectonics, and climate: Geological Society of America Bulletin, v. 108, p. 303–317.

Greb, S.F., and Weisenfluh, G.A., 1996, Paleoslumps in coal-bearing strata of the Breathitt Group (Pennsylvanian) in the Eastern Kentucky Coal Field, U.S.A.: International Journal of Coal Geology, v. 31, p. 115–134.

Greb, S.F., Chesnut, D.R., Jr., Davidson, O.B., and Rodriguez, R., 1990, An anomalous mass-flow deposit in the Lee Formation (Pennsylvanian), Eastern Kentucky Coal Field: Southeastern Geology, v. 31, p. 79–92.

Greb, S.F., Eble, C.F., and Hower, J.C., 1999, Depositional history of the Fire Clay coal bed (Late Duckmantian), eastern Kentucky, USA: International Journal of Coal Geology, v. 40, p. 255–280.

Guiraud, M., and Plaziat, J.-C., 1993, Seismites in the fluviatile Bima sandstones: Identification of paleoseisms and discussion of their magnitudes in a Cretaceous synsedimentary strike-slip basin (upper Benue, Nigeria): Tectonophysics, v. 225, p. 493–522.

Heezen, B., and Ewing, M., 1952, Turbidity currents and submarine slumps in the 1929 Grand Banks earthquake: American Journal of Science, v. 250, p. 849–873.

Henkel, D.J., 1970, The role of waves in causing submarine landslides: Geotechnique, v. 20, p. 75–80.

Hester, N.C., and Brant, R.A., 1981, Paleoslumps: A coal mine roof hazard, in Cobb, J.C., Chesnut, D.R., Jr., Hester, N.C., and Hower, J.C., eds., Coal and coal-bearing rocks of Eastern Kentucky, Annual Geological Society of America Coal Division Field Trip Guidebook: Kentucky Geological Survey, ser. 11, p. 120–126.

Hester, N.C., and Taylor, F., 1977, Corbin Sandstone Member of the Lee Formation, in Dever, G.R., Jr., Hoge, H.P., Hester, N.C., and Ettensohn, F.R., eds., Stratigraphic evidence for late Paleozoic tectonism in northeastern Kentucky, Guidebook and roadlog for the field trip held in conjunction with the 5th annual meeting of the Eastern Section of the American Association of Petroleum Geologists and the 1977 field conference of the Geological Society of Kentucky: Kentucky Geological Survey, ser. 10, p. 35–38.

Horne, J.C., 1979, Estuarine deposits in the Carboniferous of the Pocahontas Basin, in Ferm, J.C., and Horne, J.C., eds., Carboniferous depositional environments in the Appalachian basin: Columbia, University of South Carolina, p. 428–435.

Horne, J.C., and Ferm, J.C., 1978, Sedimentary responses to contemporaneous tectonism, in Horne, J.C., and Ferm, J.C., eds., Carboniferous depositional environments, Eastern Kentucky and southern West Virginia: Columbia, South Carolina, University of South Carolina, Department of Geology, Field Trip Guidebook, p. 27–36.

Jibson, R.W., and Keefer, D.K., 1993, Analysis of the seismic origin of landslides: Examples from the New Madrid Seismic Zone: Geological Society of America Bulletin, v. 105, p. 521–536.

Jorgensen, P.J., and Fielding, C.R., 1999, Debris-flow deposits in an alluvial-plain succession: The upper Triassic Callide coal measures of Queensland, Australia: Journal of Sedimentary Research, v. 69, p. 1027–1040.

Kastens, K.A., and Cita, M.B., 1981, Tsunami-induced sediment transport in the abyssal Mediterranean Sea: Geological Society of America Bulletin, v. 92, p. 845–857.

Keefer, D.K., 1984, Landslides caused by earthquakes: Geological Society of America Bulletin, v. 95, p. 406–421.

Kelling, G., and Williams, B.P.J., 1966, Deformation structures of sedimentary origin in the Lower Limestone Shales (basal Carboniferous) of South Pembrokeshire, Wales: Journal of Sedimentary Petrology, v. 36, p. 927–939.

Kuenen, P.H.H., 1958, Experiments in geology: Transactions of the Geological Society of Glasgow, v. 23, p. 1–28.

Kuribayashi, E., and Tatsuoka, F., 1975, Brief review of liquefaction during earthquakes in Japan: Soils and Foundation, v. 15, p. 81–92.

Li, Y., Craven, J., Schweig, E.S., and Obermeier, S.F., 1996, Sand boils induced by the 1993 Mississippi River flood: Could they one day be misinterpreted as earthquake-induced liquefaction?: Geology, v. 24, p. 171–174.

Lignier, V., Beck, C., and Chapron, E., 1998, Caractérisation géométrique et texturale de perturbations synsédimentaires attribuées à des séismes, dans une formation quaternaire glaciolacustre des Alpes (les Argiles du Trièves): Comptes Rendus de l'Académie des Sciences, Serie 2, Sciences de la Terre et des Planètes, Earth and Planetary Sciences, v. 327, p. 645–652.

Lowe, D.R., 1975, Water-escape structures in coarse-grained sediments: Sedimentology, v. 22, p. 157–204.

Martino, R.L., and Sanderson, D.D., 1993, Fourier and autocorrelation analysis of estuarine tidal rhythmites, lower Breathitt Formation (Pennsylvanian) eastern Kentucky, USA: Journal of Sedimentary Petrology, v. 63, p. 105–119.

McKee, E.D., and Goldberg, M., 1969, Experiments on formation of contorted structures in mud: Geological Society of America Bulletin, v. 80, p. 231–244.

Mills, P.C., 1983, Genesis and diagnostic value of soft-sediment deformation structures: A review: Sedimentary Geology, v. 35, p. 83–104.

Morgenstern, N., 1967, Submarine slumping and the initiation of turbidity currents, in Richards, A.F., ed., Marine Geotechnique: Urbana, University of Illinois Press, p. 189–220.

Morton, R.A., 1988, Nearshore responses to great storms, in Clifton, H.E., ed., Sedimentary consequences of convulsive geologic events: Geological Society of America Special Paper 229, p. 7–22.

Nelson, W.J., and Devera, J.A., 1995, Geology of the Cobden Quadrangle: Illinois State Geological Survey, IGQ-16, 1:24 000, 1 sheet.

Nelson, W.J., Eggert, D.L., DiMichele, W.A., and Stecyk, A.C., 1985, Origin of discontinuities in coal-bearing strata at Roaring Creek (basal Pennsylvanian of Indiana): International Journal of Coal Geology, v. 4, p. 355–399.

Nelson, W.J., Devera, J.A., Jacobson, R.J., Lumm, D.K., Peppers, R.A., Trask, B., Weibel, C.P., Follmer, L.R., Riggs, M.H., Esling, S.P., Henderson, E.D., and Lannon, M.S., 1991, Geology of the Eddyville, Stonefort, and Creal Springs Quadrangles, Southern Illinois: Illinois State Geological Survey, Bulletin 96, 85 p.

Nemec, W., and Steele, R.J., 1984, Alluvial and coastal conglomerates: Their significant features and some comments on gravelly mass-flow deposits, in Koster, E.H., and Steel, R.J., eds., Sedimentology of gravels and conglomerates: Canadian Society of Petroleum Geologists, v. 10, p. 1–31.

Obermeier, S.F., 1996, Use of liquefaction-induced features for paleoseismic analysis: An overview of how seismic liquefaction features can be distinguished from other features and how their regional distribution and properties of source sediment can be used to infer the location and strength of Holocene paleo-earthquakes: Engineering Geology, v. 44, p. 1–77.

Obermeier, S.F., 1998, Liquefaction evidence for strong earthquakes of Holocene and latest Pliocene ages in the states of Indiana and Illinois, USA: Engineering Geology, v. 50, p. 227–254.

Palmer, H.D., 1976, Sedimentation and ocean engineering structures, in Stanley, D.J., and Swift, D.J.P., eds., Marine sediment transport and environmental management: New York, Wiley, p. 519–534.

Plint, A.G., 1985, Possible earthquake-induced soft-sediment faulting and remobilization in Pennsylvanian alluvial strata, southern New Brunswick, Canada: Canadian Journal of Earth Sciences, v. 22, p. 907–912.

Pope, M.C., Read, J.F., Bambach, R., and Hoffman, H.J., 1997, Late Middle to Late Ordovician seismites of Kentucky, southwest Ohio, Virginia: Sedimentary recorders of earthquakes in the Appalachian basin: Geological Society of America Bulletin, v. 109, p. 489–503.

Potter, P.E., 1957, Breccia and small-scale Lower Pennsylvanian overthrusting in southern Illinois: American Association of Petroleum Geologists Bulletin, v. 41, no. 12, p. 2695–2709.

Quinlan, G.M., and Beaumont, C., 1984, Appalachian thrusting, lithospheric flexure, and the Paleozoic stratigraphy of the Eastern Interior of North America: Canadian Journal of Earth Science, v. 21, p. 973–996.

Rascoe, B., Jr., 1975, Tectonic origin of preconsolidation deformation in Upper Pennsylvanian rocks near Bartlesville, Oklahoma: American Association of Petroleum Geologists Bulletin, v. 59, p. 1626-1638.

Rast, N., and Ettensohn, F.R., 1995, Effects of seismic disturbance on epicontinental depositional systems in the Ordovician and Devonian rocks of central Kentucky: Geological Society of America Abstracts with Programs, v. 27, p. A381.

Rice, C.L., Sable, E.G., Dever, G.R., Jr., and Kehn, T.M., 1979, The Mississippian and Pennsylvanian (Carboniferous) Systems in the United States–Kentucky: U.S. Geological Survey Professional Paper 1110–F, 32 p.

Seed, H.B., 1967, Slope stability during earthquakes: Journal of Soil Mechanics, Foundation Engineering Division, American Society of Civil Engineers, v. 93, SM4, p. 299–324.

Seed, H.B., 1968, Landslides during earthquakes due to liquefaction: Journal of Soil Mechanics, Foundation Engineering Division, American Society of Civil Engineers, v. 94, SM5, p. 1053–1122.

Seed, H.B., and Idriss, I.M., 1982, Ground motions and soil liquefaction during earthquakes: Earthquake Engineering Research Investigations Monograph, 134 p.

Seed, H.B., and Wilson, S.D., 1967, The Turnagain Heights landslide, Anchorage, Alaska: Journal of Soil Mechanics, Foundation Engineering Division, American Society of Civil Engineers, v. 93, p. 325–353.

Seed, H.B., Idriss, I.M., and Arango, I., 1983, Evaluation of liquefaction potential using field performance data: Journal of Geotechnical Engineering, American Society of Civil Engineers, v. 109, no. 3, p. 458–482.

Seilacher, A., 1969, Fault-graded beds interpreted as seismites: Sedimentology, v. 13, p. 155–159.

Seilacher, A., 1973, Biostratinomy: The sedimentology of biologically standardized particles, in Ginsberg, G.N., ed., Evolving concepts in sedimentology: Baltimore, Maryland, Johns Hopkins University Press, p. 159–177.

Seilacher, A., 1984, Sedimentary structures tentatively attributed to seismic events: Marine Geology, v. 55, p. 1–12.

Sims, J.D., 1975, Determining earthquake recurrence intervals from deformational structures in young lacustrine sediments: Tectonophysics, v. 29, p. 141–152.

Tankard, A.J., 1986, Depositional response to foreland deformation in the Carboniferous of eastern Kentucky: American Association of Petroleum Geologists Bulletin, v. 70, p. 853–868.

Terzaghi, K., 1950, Mechanisms of landslides, in Paige, S., chairman, Application of geology to engineering practice: Geological Society of America, Berkey Volume, p. 83–123.

Vittori, E., Labini, S.S., and Serva, L., 1991, Palaeoseismology: Review of the state-of-the-art: Tectonophysics, v. 193, p. 9–32.

Ward, P.L., and Page, R.A., 1991, The Loma Prieta earthquake of October 17, 1989, in Baldwin, J.E., II, and Sitar, N., eds., Loma Prieta earthquake: Engineering geologic perspectives: Association of Engineering Geologists Special Publication 1, p. 5-17.

Weaver, J.D., 1976, Seismically induced load structures in the Basal Coal Measures, South Wales: Geological Magazine, v. 113, p. 535–543.

Weaver, J.D., and Jeffcoat, R.E., 1978, Carbonate ball-and-pillow structures: Geological Magazine, v. 115, p. 245–253.

Weisenfluh, G.A., and Ferm, J.C., 1991, Roof control in the Fireclay coal group, southeastern Kentucky: Journal of Coal Quality, v. 10, p. 67–74.

Wunderlich, F., 1967, Die Entstehung von 'convolute bedding' an Platenraendern: Senkenbergiana Lethaea, v. 48, p. 345–349.

Youd, T.L., 1973, Liquefaction, flow, and associated ground failure: U.S. Geological Survey Circular 688, 12 p.

Youd, T.L., and Perkins, D.M., 1987, Mapping of liquefaction severity index: Proceedings, American Society of Civil Engineering, Geotechnical Engineering Division, v. 113, p. 1374–1392.

MANUSCRIPT ACCEPTED BY THE SOCIETY MAY 11, 2001

Late Mississippian paleoseismites from southeastern West Virginia and southwestern Virginia

Kevin G. Stewart
John M. Dennison
Department of Geological Sciences, University of North Carolina at Chapel Hill, Chapel Hill, North Carolina, 27599-3315, USA

Mervin J. Bartholomew
Earth Sciences & Resources Institute, School of the Environment, University of South Carolina, Columbia, South Carolina 29208, USA

ABSTRACT

Late Mississippian sedimentary rocks exposed in the Appalachian Plateau of southeastern West Virginia and southwestern Virginia contain structures attributable to paleo-earthquakes. An exposure of the Hinton Formation (Late Mississippian) east of Princeton, West Virginia, contains >30 clastic dikes which cut bedding at a high angle. Features of the dikes indicate that they were rapidly injected, and it can be shown in some dikes that filling occurred by upward sand injection. Dikes are tabular and do not share characteristics of clastic dikes produced by nonseismic processes. In addition to the dikes, these strata also contain convolute beds, pseudonodules, possibly penecontemporaneous faults, beds showing evidence of lateral flow, and slumps. Nearby outcrops in overlying Upper Mississippian Bluestone Formation also contain numerous slumps. Paleoseismites in these rocks formed over several million years. The likely source of the stress responsible for the earthquakes is the incipient collision of Africa with North America during the initial stages of the Alleghanian orogeny.

The paleoseismites are found within an area with a diameter of ~50 km, which, following palinspastic restoration, coincides with the northwestern edge of the modern-day Giles County, Virginia, seismic zone. The Giles County seismic zone has had several historic earthquakes with magnitude greater than 4, the largest being 5.8. The coincidence of the Late Mississippian seismicity with the Giles County seismic zone may indicate that the reactivated Precambrian faults, interpreted as the source for modern-day Giles County seismicity, may have been reactivated earlier in the compressional stress field generated by the Alleghanian orogeny.

INTRODUCTION

Large earthquakes (usually M ≥ 6) can produce a wide variety of syndepostional structures caused by liquefaction of unconsolidated sediments (Obermeier, 1996) and by lateral movements, such as slumps and landslides (Seilacher, 1984). Recognizing paleoseismites in ancient sedimentary rocks can be challenging because of their similarity to syndepostional structures produced by depositional processes, such as loading or nonseismic slumping (Mills, 1983; Obermeier, 1996; Pope et al., 1997). Most workers agree, however, that tabular clastic dikes that show evidence of rapid, upward injection of unconsolidated sediments from a lower, liquefied layer provide some of the strongest evidence for past seismic shaking (Tuttle and Seeber, 1991; Bourgeois and Johnson, 2001; Obermeier, 1996).

Stewart, K.G., Dennison, J.M., and Bartholomew, M.J., 2002, Late Mississippian paleoseismites from southeastern West Virginia and southwestern Virginia, *in* Ettensohn, F.R., Rast, N., and Brett, C.E., eds., Ancient seismites: Boulder, Colorado, Geological Society of America Special Paper 359, p. 127–144.

In this chapter, we describe paleoseismites in exposures of Late Mississippian sedimentary rocks in southeastern West Virginia and southwestern Virginia (Fig. 1) that include upwardly injected clastic dikes and a wide variety of associated penecontemporaneous structures, such as pseudonodules, convolute bedding, and slumps. These structures occur within the Upper Mississippian Mauch Chunk Group in an interval that represents a few million years, at most. The structures occur in a geographically restricted area centered near Princeton, West Virginia. We discuss possible origins of the seismicity and the relationship between the Late Mississippian seismicity and the modern-day Giles County, Virginia, seismic zone.

STRATIGRAPHIC SETTING

The strata containing evidence of paleoseismicity are latest Mississippian in age (Fig. 2), about 328 m.y. old (Palmer, 1983; Harland et al., 1990). The part of the Hinton Formation above the marine Little Stone Gap Limestone (called Avis Limestone in early stratigraphic literature) is dominated by redbeds, mostly nonmarine shale interbedded with sandstone layers and some channel sandstones, which represent small distributary channels and crevasse splays on a flat delta plain (Reger, 1925a, 1925b; Thomas, 1959; Donaldson and Shumaker, 1981). The pervasive Princeton Sandstone represents a marine transgression and is overlain by dark Pride Shale, which was deposited in a prodeltaic setting (Miller and Eriksson, 1997, 2000), but locally (as in the Route 460 roadcut, locality 1, Fig. 1) contains lycopod plant stems. The Bluestone Formation above its basal Pride Shale Member returns to mostly redbed mudstones with some sandstone (such as the rather conspicuous Glady Fork Sandstone). Thin coal beds are present in the upper Bluestone Formation, and some marker beds of greenish and dark shale, indicating humid conditions rather than the dry conditions of the redbeds. Nodular limestones within the Bluestone and Hinton Formations in the Princeton area are mostly caliche zones. A prominent imprint of climatostratigraphic zones or sea-level change(?) cycles is present in these lower delta plain to marginal marine deposits (Beuthin and Neal, 1998).

The top of the Bluestone Formation is marked by an unconformity with the Green Valley Paleosol Complex (Beuthin, 1997; Beuthin and Neal, 1998) and a broad valley fill. These paleosols formed during the global lowstand of sea level, which separates the Kaskaskia and Absaroka sequences (Monday Creek Discontinuity).

Superimposed on this pattern of widely traceable stratigraphy, and within a narrow stratigraphic range in the late Mississippian (perhaps spanning no more than 7 m.y.; Miller and Eriksson, 2000) are penecontemporaneously deformed beds. These features are known only within 45 km of Princeton, West Virginia, and are generally situated near the present-day Giles County seismic zone. It is not known whether these features were originally restricted to this area or if their present-day distribution is also a function of where rocks of this age are exposed. Wheeler (1995) has suggested that the modern seismic activity reflects a reactivation of old rift faults accompanying the opening of the Iapetus Ocean about 565 m.y. ago (see Thomas, 1991). Local Mississippian seismic activity is midway in timing between the Iapetus rifting and the modern activity in the Giles County seismic zone.

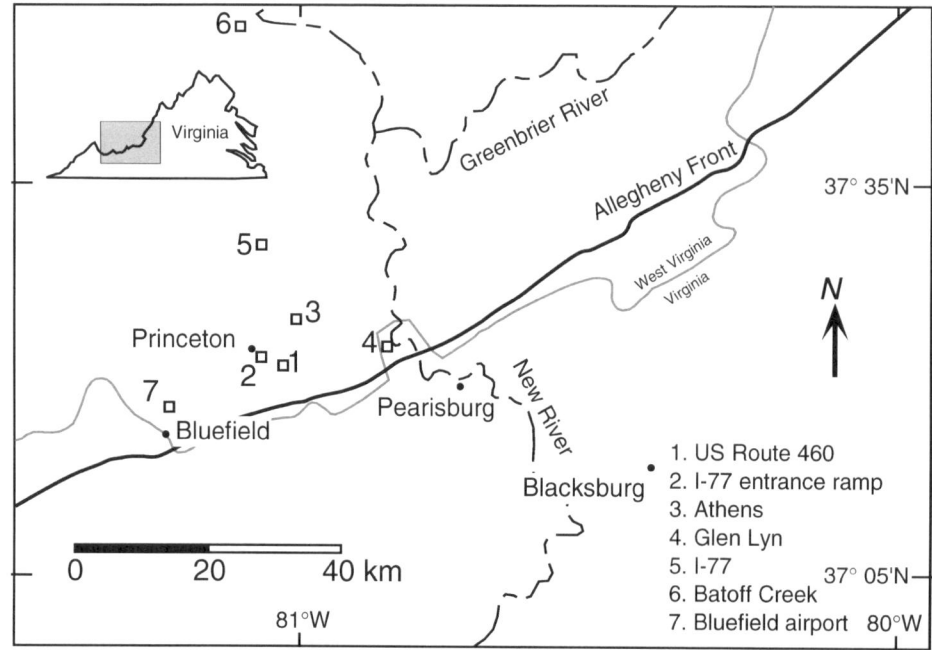

Figure 1. Paleoseismite localities. Allegheny Front corresponds to the transition from steeply dipping beds to shallowly dipping beds across the Glen Lyn syncline.

DEPOSITIONAL TECTONICS

The late Paleozoic site of the present Appalachian Plateau and Allegheny Front boundary with the western Valley and Ridge Province was a foredeep basin (Donaldson and Shumaker, 1981) receiving siliciclastic sediment from the southeast, east, and northeast of the present directional orientations. By the time of the Mississippian-Pennsylvanian boundary, the Alleghanian orogeny had begun in the Piedmont province to the southeast, with faulting and igneous intrusion (e.g., Hatcher et al., 1989). Presumably, continental collision of Africa and North America had initiated orogenic thrusting in the Piedmont, and possibly the Blue Ridge. Farther to the northwest in the Appalachian basin, a deepening foredeep basin was rapidly filled with siliciclastic sediments and rare marine shales and limestones. The center of this foredeep basin was at or southeast of the currently preserved southeastern limit of Upper Mississippian strata.

The specific sites of seismites and possible seismites in our study were developing regionally traceable layers of strata with a marked overprint of sea-level changes and climatostratigraphy, producing time banding within the strata. A dominance of low-energy muds in deposits accumulating near sea level indicates very flat depositional surfaces.

PALEOSEISMITES

Upper Hinton Formation east of Princeton, West Virginia

The strongest evidence we have discovered to date for Late Mississippian paleoseismites is in outcrops of interbedded sandstone and shale from the upper part of the Hinton Formation (Upper Mississippian Mauch Chunk Group). Two exposures, along opposing sides of U.S. Route 460, are 3 km southeast of the intersection with Interstate Highway (I-77) near Princeton, West Virginia (locality 1, Fig. 1). The rocks are gently dipping (<3°) and are located within the Appalachian Plateau about 1.6 km northwest of the Allegheny Front.

The prominent sandstone shown in Figure 3 is the Princeton Sandstone, which separates underlying redbeds of the Hinton Formation from the overlying dark gray Pride Shale Member of the basal Bluestone Formation (Reger, 1925a; John D. Beuthin, 1999, personal commun.). We have not recognized any paleoseismites in the Pride Shale in this outcrop although later in this chapter we describe features in the Pride Shale from outcrops along I-77 north of Princeton that may have been generated by seismic shaking.

Clastic dikes. We have documented 33 separate clastic dikes in these two outcrops. Most are filled with sand (e.g., Fig. 4); two are filled with mud. The fill material is homogeneous in composition, although some of the sand dikes that traverse red shale have been stained red after emplacement, giving the appearance of layering. The dikes range in thickness from about 0.5 to 8 cm, and in length from 0.2 to 6 m. All dikes are generally tabular, and individual dikes commonly exhibit variable thickness, dip, and strike. The dips are generally steep to vertical (60°–90°), and the

strike of a single dike varies as much as 20° along its length. The strike data (Fig. 5) were obtained by measuring a part of the dike that appeared to mimic the average strike estimated by visual observation. Although the dikes have variable strikes, there is a strong preferred orientation at about N45°E. The strata are essentially horizontal, so the steep dip of the dikes shows that most are nearly perpendicular to bedding.

Rapid cyclic shearing of unconsolidated sediments by earthquake waves can lead to compaction and a concomitant increase in pore pressure, causing liquefaction of the sediments (Obermeier and Pond, 1998). This phenomenon is most evident in sand-sized material (Valera et al., 1994) and can result in the formation of clastic dikes in an overlying nonliquefied layer that is fractured by the elevated fluid pressure in the liquefied sand, by

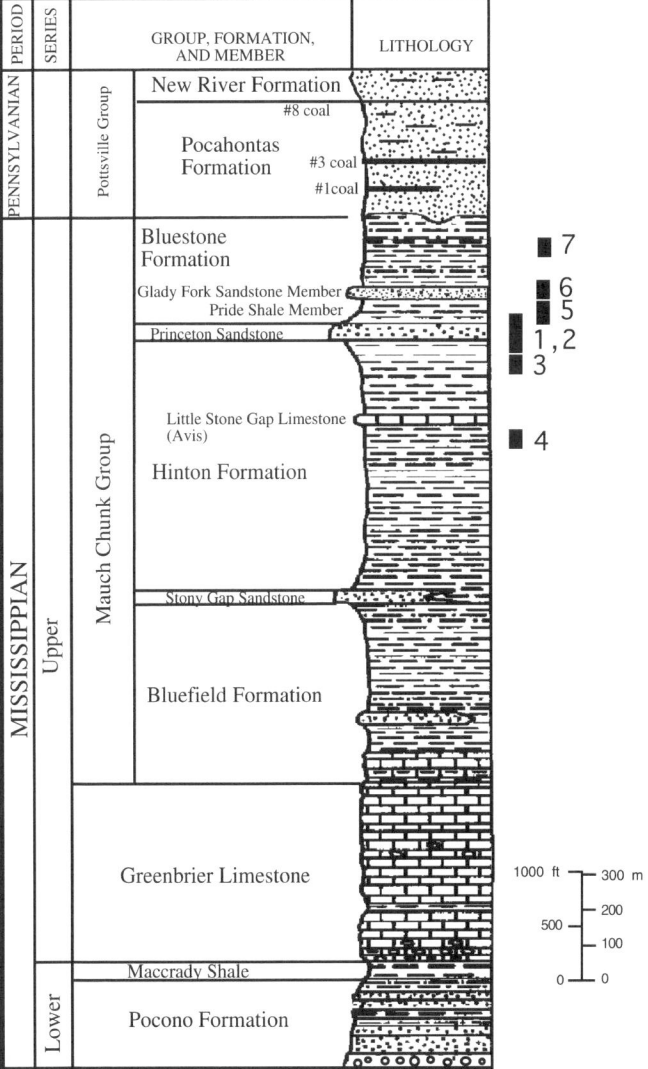

Figure 2. Stratigraphic column of Upper Mississippian and Lower Pennsylvanian units of southeastern West Virginia and southwestern Virginia. Numbers indicate stratigraphic positions of paleoseismite localities shown in Figure 1. Modified from McDowell and Schultz (1990).

Figure 3. Southeast end of outcrop along westbound lane of U.S. Route 460, showing typical interbedded sandstone and shale of the Upper Mississippian Hinton Formation and overlying Princeton Sandstone. Prominent normal fault has about 1 m of displacement and may be penecontemporaneous. Convolute bedding includes a diapiric sandstone plug (light colored) and a rotated block of coherently bedded shale immediately to the left. Clastic dike is sand-filled.

Figure 4. Sand-filled dike cutting shale, U.S. Route 460 outcrop. Pick end of hammer is about 12 cm long.

surface oscillations, or by lateral spreading, or the overlying layer may contain preexisting fractures (Obermeier, 1996; Obermeier and Pond, 1998).

Clastic dikes can also form by nonseismic processes, and distinguishing between the two can be difficult. For example, sand boils that formed adjacent to Mississippi River levees during the 1993 floods shared many features with seismically induced sand blows (Yong et al., 1996). One characteristic of the flood-induced sand boils is that they typically erupt through tubular conduits, usually exploiting preexisting holes made by tree roots and burrowing crayfish. In contrast, seismically induced clastic dikes generally are tabular (Fig. 6) (e.g., Obermeier, 1996). This evidence, in conjunction with the lack of any sedimentologic evidence that would suggest the dikes were injected into flood-plain deposits adjacent to levees, rules out the possibility of the dikes having formed in response to flooding during Late Mississippian time.

Clastic dikes can also form by passive infilling of nonseismically induced fissures. Such dikes fill relatively slowly with sediment from the walls of the fissure and from layers above. Dikes formed by seismically induced liquefaction, however, fill from below, and the fluidized sand is deposited rapidly into the fracture. For most of the dikes in the Route 460 exposure, the filling direction of the dikes cannot be determined unequivocally. In one dike, preserved flow structures indicate that the dike originated in a layer of clean sand and was injected upward into a fracture through both sand and clay layers (Fig. 7). Several of the dikes terminate upward, again indicating that they were filled from below.

The rate at which the dikes were filled cannot be determined absolutely, but an estimate of the velocity, whether rapid due to seismically induced liquefaction or slow by passive fill, can be inferred from the nature of the fill material and the dimensions of the dikes. The dikes have uniform fill material, either sand or mud, and commonly cut both sandstone and shale layers. If the dikes had filled by passive collapse of the dike walls into the open fissure, there would be a mixture of sand and clay in the dike accompanied by layering. No layering was observed within the dikes, and the uniformity of the fill indicates that the dike was filled rapidly before the weak material making up the dike walls collapsed. The dikes tend to be long and narrow, with length-to-thickness ratios from about six to nearly 600, the average being about 80. In cross-sectional view, the dikes with the largest length-to-thickness ratios are fairly irregular and locally show bifurcation and shallow dips (Fig. 8). These long, delicate fissures could not have remained open very long and therefore must have been immediately filled by rapid injection of sand.

Faults. The beds that contain the clastic dikes in the upper Hinton Formation along U.S. Route 460 are also cut by numerous faults (Fig. 3). The faults have variable orientations, but they

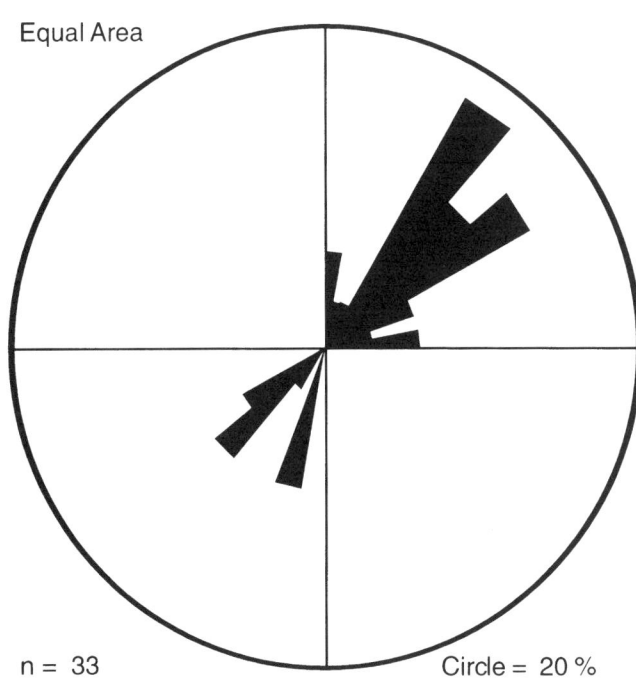

Figure 5. Rose diagram of dike strike orientations from U.S. Route 460 outcrop.

Figure 6. Sand-filled dike from U.S. Route 460 outcrop showing tabular form typical of dikes formed by seismically induced liquefaction. Artesian features and nonseismic sand boils typically erupt through cylindrical conduits. Dike is about 3 cm wide.

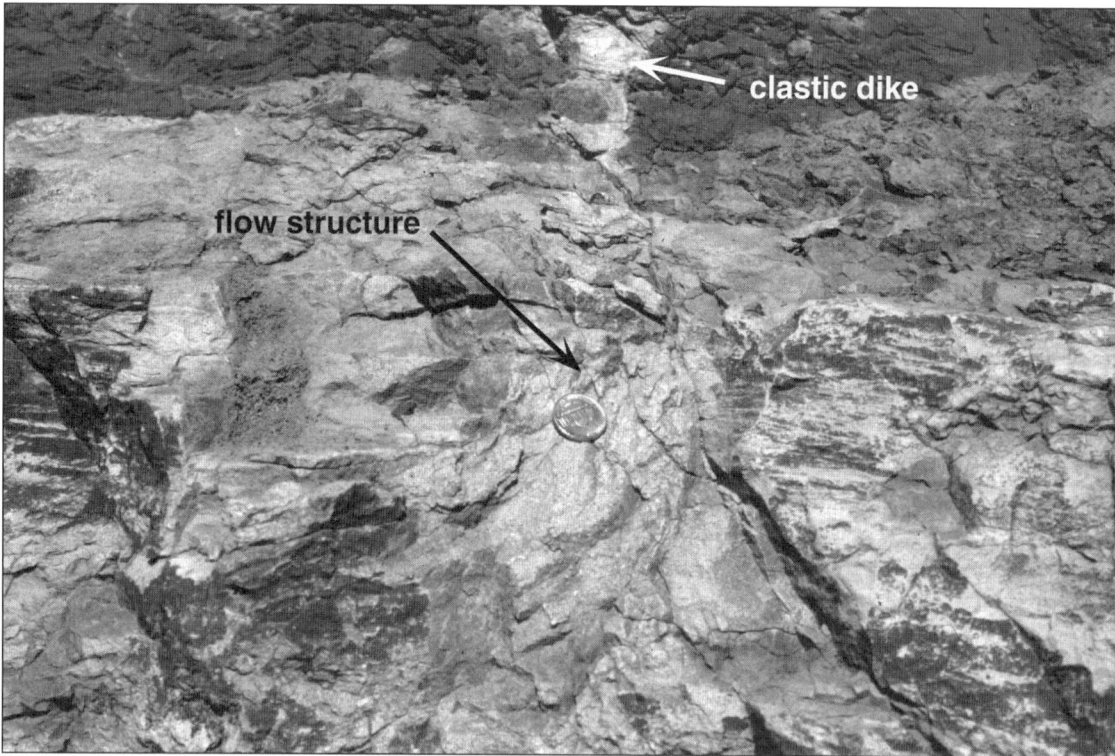

Figure 7. Liquefied sand layer showing flow structures indicating upward injection of sand-filled clastic dike. U.S. Route 460 outcrop. Coin is about 2 cm in diameter.

strike generally northeast and dip northwest and southeast (Fig. 9). All of the observed faults show at least a component of normal dip-slip motion, and a few of the faults preserve faint slickenlines in the shale layers, showing nearly pure dip-slip motion. The dip-slip displacement on the faults ranges from a few centimeters to 1 m. The faults are most commonly isolated structures, such as the one shown in Figure 3, or in conjugate pairs (Dennison and Stewart, 1998, Fig. 38). One part of the outcrop contains abundant fractures and faults (Fig. 10) and rotation of bedding. Immediately above this part of the outcrop, the beds are not exposed; however, continuously exposed beds a few meters higher show no evidence of faulting.

Bedding adjacent to the faults remains largely coherent, although internal laminations in sandstone layers locally are disrupted or obliterated, suggesting that faulting may have occurred prior to lithification. Displacement on most of the observed faults appears to die out abruptly upward, and unfaulted layers are commonly visible above the faults. Nearly all of the faults terminate upward at a stratigraphic position about 3 m below the base of the Princeton Sandstone, in cuts on both sides of the highway. The distinct basal contact of the Princeton Sandstone is not offset by any of the faults. This would appear to indicate that the faulting occurred prior to the deposition of the Princeton Sandstone. We cannot rule out the possibility, however, that the abrupt upward termination of displacement on the faults is only apparent. Some faults splay at their terminations or root into bedding planes and become difficult to trace. Many of the observed faults appear to lose their displacement in shale beds, and our inability to trace the faults may be due to the difficulty in discriminating shale partings and fractures from nonslickensided fault surfaces. If the normal faulting occurred during deposition of the Hinton Formation and produced surface breaks, we would expect to see small colluvial wedges on the hanging-wall block adjacent to the fault surface arising from erosion of the fault scarp. We have not recognized any such deposits, so if the faulting was syndepositional, the faults did not produce significant surface scarps.

Although we have been unable to find definitive proof of syndepositional faulting, the orientation of the normal faults indicates roughly northwest-southeast extension of these layers, which is the same as the extension direction indicated by the orientation of the clastic dikes. Regional depositional paleoslope during Hinton Formation sedimentation was toward the northwest (e.g., Dennison and Wheeler, 1975; Donaldson and Shumaker, 1981), perhaps due to syndepositional tectonic tilting (Thomas, 1966), and earthquake shaking would probably have caused lateral movement toward the northwest, down the paleoslope, producing northeast-striking normal faults and clastic dikes.

Convolute bedding. Several beds within the Hinton Formation contain internally disrupted laminations and other kinds of convolute bedding. Figure 11 shows laminations within 1-m-thick sandstone bed that have remained coherent but are strongly folded into a pair of recumbent, isoclinal folds. The upper contact of this bed with an overlying shale layer is relatively planar, indicating that the convolutions were the result of internal flow within the sand layer. This bed is the source for several clastic dikes observed in this outcrop, and we interpret the convoluted laminations as further evidence that these clastic dikes were a result of liquefaction of the underlying sand bed.

In the example of convolute bedding in this outcrop shown in Figure 3, part of a 0.5-m-thick sandstone layer has flowed upward, producing a diapir-shaped intrusion of sand into the overlying shale beds. Bedding is preserved in the shale immediately to the left of the light-colored sand intrusion, indicating that only the sand layer was liquefied. Sandstone and shale layers immediately above the liquefied sand layer are continuous and only slightly warped, indicating that this sand layer was less than 1 m below the ground surface when the structure formed. This also provides a tight constraint on timing of this liquefaction event within the depositional history of the Hinton Formation exposed in this outcrop. Other structures, such as clastic dikes and faults, clearly cut the beds above this particular feature, which means that these sediments were affected by more than one liquefaction event.

Pseudonodules. The irregular, rounded clasts of sand in a matrix of sand and mud shown in Figure 12 resemble "pseudonodules" that have been described from other localities and interpreted as having both seismic (e.g. Sims, 1975, Khullar et al., 1997) and nonseismic (e.g. Obermeier, 1996) origins. The pseudonodules in the Route 460 outcrop are present within a 1.5-m-thick zone that extends for a length of about 10 m at the southeast end of the southwest road cut. At its southeast end, the gently southeast-dipping pseudonodule zone disappears below the level of the outcrop; at its west end, the pseudonodule zone thins and seems to originate along a shallow scarp in the top of the sandstone layer shown in Figure 11. The pseudonodule zone is bound above by a continuous sandstone layer marked by the hammer head in Figure 12. This pseudonodule zone consists of hundreds of individual nodules ranging in size from a few centimeters to about 20 cm across (Fig. 13). The nodules consist primarily of fine- to very-fine sand with millimeter-scale laminations of mud and are surrounded by a mud and sand matrix. The nodules have been internally deformed, commonly exhibiting tight and isoclinal folding of the original sedimentary layering (Fig. 14). Above the continuous sandstone layer, the upper part of the pseudonodule zone contains fewer, larger (as much as 30 cm across) nodules, which appear to be boudins of a once-continuous sandstone layer. Mud and sand have been injected between the boudins. As in the smaller nodules shown in Figure 13, the sedimentary layering in the larger nodules has been folded.

Pseudonodules can form by rapid deposition of sand on a muddy substrate (Obermeier et al., 1990). The sandy sediment detaches and sinks into the underlying mud, forming isolated kidney-shaped bodies surrounded by the mud. Ball-and-pillow structures have similar forms as pseudonodules but are typically kidney-shaped bodies of fine or silty sand surrounded by silty sand. Although these structures can form by nonseismic mechanisms,

Figure 8. Sand-filled clastic dike showing bifurcation-rejoining at about knee-level of geologist. Geologist's feet are on liquefied sand layer that was the source for the dike. Below the sand layer is a layer with disturbed bedding (shown in detail in Fig. 15). From U.S. Route 460 outcrop. Geologist is 1.85 m tall.

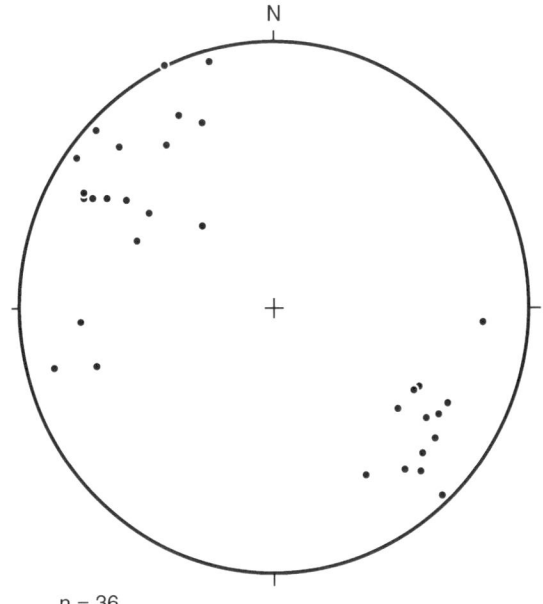

Figure 9. Equal-area plot of poles to normal faults from U.S. Route 460 outcrop.

Figure 10. Apparently penecontemporaneous normal faults from U.S. Route 460 outcrop. Hammer handle is 20 cm long.

Figure 11. Convolute bedding within a liquefied sand bed that was the source for several of the clastic dikes in the U.S. Route 460 outcrop. The top of this bed is horizontal, roughly parallel to the top edge of the photograph. Lens cap is 5 cm in diameter.

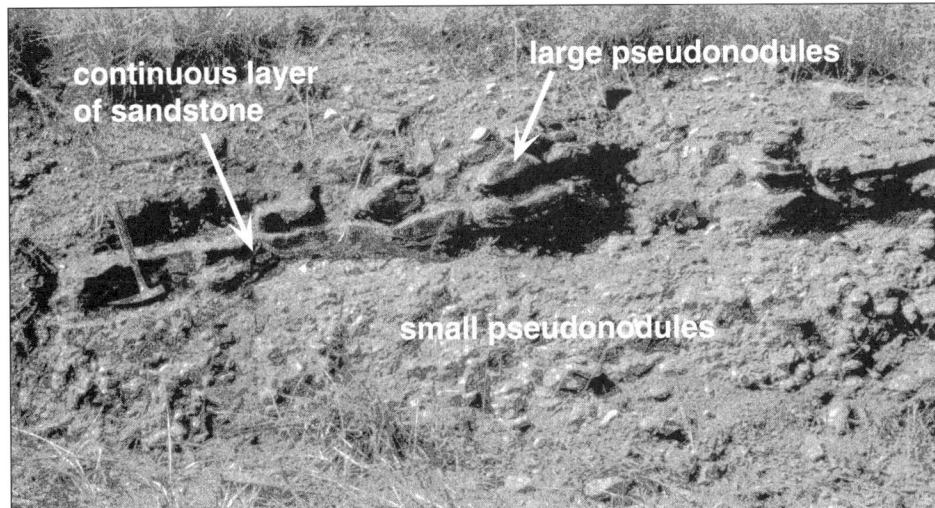

Figure 12. Layer with abundant pseudonodules, U.S. Route 460 exposure. Area below the continuous sand layer contains pseudonodules that are typically 10 cm in their longest direction. We estimate that there are more than 700 small nodules in an exposed area of about 30 m^2 in this layer. Above the continuous sand layer is another layer of pseudonodules that appear to be boudins of a once-continuous sand layer. Hammer handle is 20 cm long.

Figure 13. Close-up of small pseudonodules shown in Figure 12. Pencil is 15 cm long.

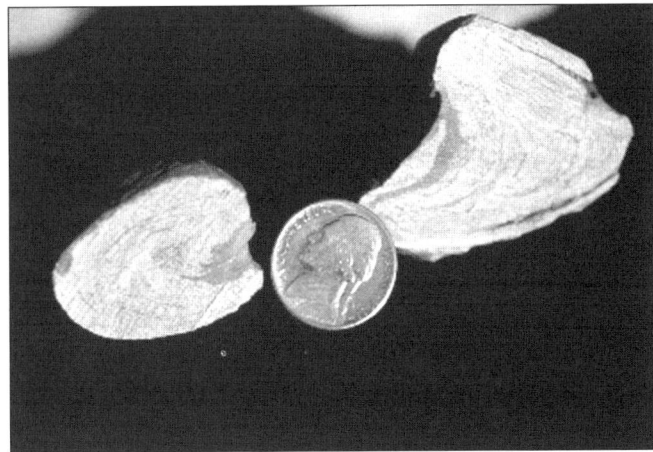

Figure 14. Small pseudonodules from layer shown in Figure 12 cut in half to reveal internal deformation. Light-colored areas are sand, dark areas are mud. Coin is about 2 cm in diameter.

examples of seismically generated pseudonodules and ball-and-pillow structures have been reported (Kuenen, 1958; Sims, 1975; Ringrose, 1989; Pope et al., 1997). Obermeier (1996) and Obermeier et al. (1990) described criteria they used to conclude that pseudonodules in sediments near New Madrid, Missouri, were the result of synsedimentary loading and not seismic shaking. These include the presence of pseudonodules along with load-cast ripples both grading into sand lenses; pseudonodule layers that are laterally equivalent to undeformed sand lenses; and the presence of several pseudonodule layers, indicating a repeated sequence of rapid sand deposition on a mud substrate. The pseudonodule layer in the Route 460 outcrop contains none of the features cited by Obermeier (1996) and Obermeier et al. (1990) as evidence of sedimentary processes. The pseudonodule zone is an isolated bed, and we have not observed load structures in the other sand bodies in this outcrop. These characteristics lead us to conclude that these pseudonodules are the result of seismically induced liquefaction. Ball-and-pillow load-cast structures in Appalachian basin sandstones are usually much larger than the pseudonodules in this outcrop.

Mesoscopic flow structures. The final example of penecontemporaneous disturbance that we have observed in the U.S. Route 460 outcrops is a 0.5-m-thick layer (Fig. 15) that contains clasts of mud and rotated blocks of coherently bedded silt and shale in a predominantly sand matrix. This layer is at the base of the thick sand layer shown in Figure 8. It appears to show evidence of flow of the sand and entrainment of the large and small clasts of surrounding sediments. This could be a sill of liquefied sand or a debris flow. Its origin is unclear because most of the layer, as well as the underlying sediments, is concealed by a thick apron of modern talus in the roadcut.

As noted by Obermeier (1996), clastic dikes showing rapid upward injection of liquefied sand provide the strongest evidence for past earthquakes. Although the other structures described above, such as pseudonodules and convolute bedding, can have sedimentary or other nonseismic origins, their association with

Figure 15. Disturbed layer from U.S. Route 460 outcrop. Within a sand matrix are both clasts of mud and rotated blocks of coherently bedded material. This layer is also shown in Figure 8. Lens cap is 5 cm in diameter.

the clastic dikes in this outcrop indicates that they, too, probably originated from seismic shaking.

In the next sections we describe other features in nearby outcrops within the upper part of the Hinton Formation and the immediately overlying Bluestone Formation that define a region where past seismicity may have been focused. Although the structures described below can form by nonseismic processes, their close association in time and space to the paleoseismites described above argues strongly for a seismic origin.

Slumped Hinton Formation at Princeton, West Virginia

West (3.2 km) of the U.S. Route 460 outcrops is an outcrop (Fig. 16) of the uppermost Hinton Formation and overlying Princeton Sandstone located along the east side of the entrance ramp from Route 460 onto northbound Interstate Highway 77 (locality 2, Fig. 1). The Hinton beds strike N30°E and dip 36°SE and maintain this orientation for at least 100 m along the face of the outcrop. It appears that the beds were rotated, probably by large-scale slumping, shortly before deposition of the Princeton Sandstone. The timing of this deformation overlaps the timing of the formation of the structures observed at the Route 460 outcrops.

Slumped Hinton Formation near Athens, West Virginia

Thomas (1959) described an outcrop about 3.2 km north of Athens, West Virginia, along West Virginia Highway 20 (locality 3, Fig. 1; outcrop sketch in Fig. 17) that contained steeply dipping beds of the Hinton Formation overlain by a gently dipping, although irregularly bedded, sandstone. The sandstone is about 25 m below the level of the Princeton Sandstone (section 26 in Thomas, 1959) and is probably the Falls Mills Sandstone (Reger, 1925a). The disturbed beds crop out over a distance of about 100 m; the current condition of the outcrop is poor. The sketches in Thomas (1959) show details of a much fresher road cut than is now visible, although the tilted beds and remnants of limestone are still visible. Within the steeply dipping beds is a 3-m-thick limestone bed, which, according to Thomas (1959), thins abruptly across the hinge of a syncline. It is not clear which limestone this is. If the original thickness of the limestone layer was 3 m and the reduced thickness on the opposing limb of the syncline is the result of slumping, then it is the Little Stone Gap Limestone (Avis Limestone of earlier workers) because no other limestone in the Hinton Formation is more than 2 m thick (Reger, 1925a; Thomas, 1959; Dennison and Wheeler, 1975). On the other hand, if the original thickness is represented by the thin limb and the thick limb has been thickened by slumping, then there are several candidates for the limestone layer (Falls Mills Limestone, Tallery Limestone, and Low Gap Limestone). On the basis of known stratigraphic thicknesses in this area (Reger, 1925a; Thomas, 1959; Dennison and Wheeler, 1975), the Little Stone Gap Limestone is expected to be about 50–80 m below the Falls Mills Sandstone. Incorporation of Little Stone Gap Limestone beds into a slump would require a detachment surface that extended from sediments immediately below the Falls Mills Sandstone to below the level of the Little Stone Gap Limestone. In addition, the slumping would have had to produce folds with amplitudes on the order of 25–40 m. Although this is possible, we have no direct evidence for such large-scale slump folding at this locality. If the limestone in this exposure is a slump-thickened bed from higher in the section (e.g. Falls Mills, Tallery, or Low Gap Limestone), the amplitude of the slump fold could be significantly less than 25 m.

Convolute bedding near Glen Lyn, Virginia

In a sequence of interbedded sandstones and shales of the Hinton Formation near Glen Lyn, Virginia (locality 4, Fig. 1), we have observed several examples of convolute bedding and other penecontemporaneous deformation features within sandstone layers (Fig. 18). Parts of this outcrop show distorted layering within sandstone beds.

SLUMPS IN THE BLUESTONE FORMATION

The Late Mississippian Bluestone Formation contains a lower shale member, the Pride Shale, which is overlain by the Glady Fork Sandstone member (Fig. 2). The Bluestone Formation immediately overlies the Princeton Sandstone, and penecontemporaneous structures in the Bluestone Formation are no more than about 3 m.y. younger than the structures in the underlying Hinton Formation (Miller and Eriksson, 2000).

Slumps in the Pride Shale Member

Along Interstate Highway 77 about 14.5 km north of Princeton, West Virginia, is a long series of outcrops in the Late Mississippian Pride Shale Member of the Bluestone Formation, first described by Cooper et al. (1961) and later by Englund (1989) and Miller and Eriksson (1997). The Pride Shale is a transgressive deposit over the Princeton Sandstone. Miller and Eriksson (1997) interpreted the fine laminations in the shales as semidiurnal tidal cycles formed in a distal subtidal setting of a prodelta. They concluded that the approximately 50 m of Pride Shale exposed here thus represents only a few centuries of Mississippian sedimentation.

Miller and Eriksson (1997) interpreted surfaces such as those shown in Figure 19 as infilled slump scars created by subaqueous gravity sliding of large coherent blocks of sediment. The detachment surfaces for these slumps extend for hundreds of meters and represent the removal of blocks as much as 15 m thick (Miller and Eriksson, 1997). The detachment surfaces commonly contain small rotated blocks of coherently bedded sediment (Fig. 20), blocks that are either remnants of the original slide block or pieces of the escarpment that have broken off and slid down the detachment surface. Cooper et al. (1961) and Cooper (1971) also interpreted these features as slumps formed penecontemporaneously by down-slope movement of consolidated sediments toward the axis of a syndepositional syncline to the northwest.

Figure 16. Outcrop along entrance ramp of I-77 (locality 2, Fig. 1). Slumped, dipping beds of uppermost Hinton Formation are overlain by nearly horizontal Princeton Sandstone. Field book is 20 cm long.

Slumps in Glady Fork Sandstone

A large penecontemporaneous slump in the Glady Fork Sandstone, described by McColloch (1986), is along Batoff Creek 22 km northeast of Beckley, West Virginia (locality 6, Fig. 1). In this outcrop the top of the Pride Shale Member is at the base of a large slump block involving at least 10 m of sandstone, siltstone, and shale of the Glady Fork Sandstone Member. This block is overlain by undeformed sandstone of the Glady Fork Member.

Slumps in upper Bluestone Formation

Thomas (1959, 1966) recognized a slumped interval (Fig. 21) in the upper part of the Bluestone Formation, about 140 m above the Glady Fork Sandstone of the Bluestone Formation near Bluefield, West Virginia (locality 7, Fig. 1). Similar to the Cooper et al. (1961) interpretation for the slumps in the Pride Shale, Thomas (1959) thought this and other slumps in the area represented sliding from both fold limbs toward the trough of a syncline. In Thomas's sketch and in our observations, gently dipping sandstone beds cap the steeply dipping slumped layers shown in Figure 21.

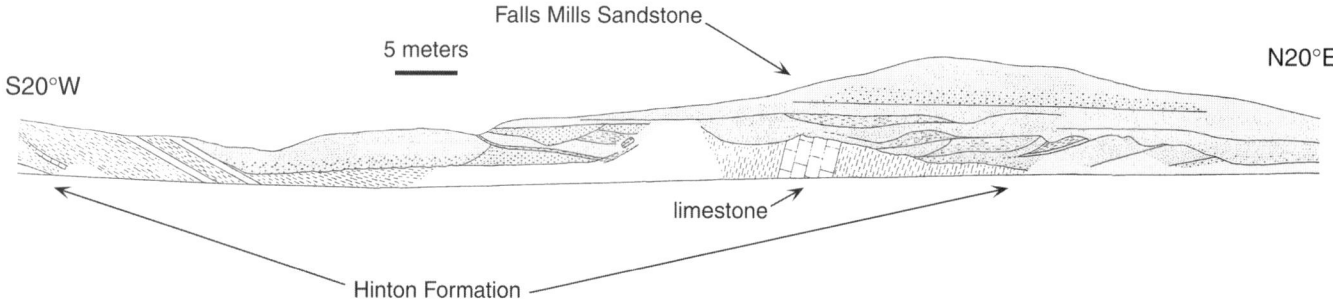

Figure 17. Sketch of outcrop near Athens, West Virginia (locality 3, Fig. 1). Steeply dipping beds of Hinton Formation are overlain by shallowly dipping Falls Mills Sandstone. From Thomas (1959).

TIMING OF FORMATION OF PALEOSEISMITES

Two of the outcrops described above (U.S. Route 460 and I-77 on-ramp; localities 1 and 2, respectively, in Fig. 1) are within the uppermost Hinton Formation and are capped by undeformed or only slightly deformed Princeton Sandstone. The slump structures at the I-77 on-ramp must have formed during latest Hinton deposition, because the beds involved in the deformation are in sharp contact with the overlying undeformed Princeton Sandstone. The structures exposed at the Route 460 outcrop are at several levels within 10–15 m of Hinton Formation that is exposed below the Princeton, including beds immediately beneath the Princeton, although none of the structures appear to cut the Princeton or affect the Pride Shale exposed above. Reger (1925a) showed the eastern edge of the outcrop belt of the Princeton Sandstone in the vicinity of this outcrop and, in fact, the Princeton Sandstone thins at the southeastern end of the southwestern roadcut (Fig. 22). One possible explanation for this thinning is that fault movement during deposition of the uppermost Hinton resulted in gentle warping and approximately 2 m of uplift of Hinton sedimentary strata, creating a slight topographic high that caused the thinning of the Princeton Sandstone. If this scenario is correct, it constrains penecontemporaneous deformation in the Hinton at the U.S. Route 460 outcrop to include the time of deposition of the uppermost Hinton, similar to the I-77 on-ramp exposure.

The disturbed beds of sandstone in the Hinton Formation cropping out near Glen Lyn are lower (older) in the section than the outcrops described at the U.S. 460 locality. By examination of the geologic map of Monroe County, West Virginia (Reger, 1925b), we place this outcrop at a stratigraphic level ~50 m below the Little Stone Gap Limestone (called Avis Limestone by Reger, 1925b). Although we have no absolute age controls on these strata, the 100–200 m of Hinton Formation between the strata at Glen Lyn and the strata at the outcrops directly below the Princeton Sandstone probably represent no more than a few million years (Miller and Eriksson, 2000).

The outcrop of Pride Shale along I-77 (locality 5, Fig. 1) is in strata immediately above the Princeton Sandstone. If Miller and Eriksson's (1997) interpretation of the sedimentation rate of the Pride Shale is correct, then the sediments affected by the slumps may have been deposited within a few hundred years of deposition of the Princeton Sandstone. The slump at Batoff Creek is within the lower part of the Glady Fork Sandstone Member of the Bluestone Formation (McColloch, 1986), which is about 60–80 m stratigraphically above the Princeton Sandstone and immediately overlies the Pride Shale Member. If Miller and Eriksson's (1997) sedimentation rates are accurate, this slump could have occurred several hundred years after the slumps in the Pride Shale.

With the exception of the outcrops at Glen Lyn and the Bluefield airport, all the paleoseismites in this study formed within a stratigraphic interval that begins about 10 m below the Princeton Sandstone and ends within the Glady Fork Sandstone, about 60–80 m above the Princeton Sandstone (Fig. 2). If we use the deposition rates of the Pride Shale from Miller and Eriksson (1997), this stratigraphic interval represents anywhere from perhaps several hundred thousand to a few million years. Including the strata involved at Glen Lyn and the Bluefield airport extends this range to about 5–7 m.y. (Miller and Eriksson, 2000). We have not observed paleoseismites in the strata above or below this interval, which indicates either a restricted time interval of paleoseismicity of sufficient energy to generate paleoseismites or a change in the nature of the sediments during this time that permitted the formation of seismically induced structures.

ORIGIN OF LATE MISSISSIPPIAN SEISMICITY

The most likely source of the seismicity that generated the paleoseismites of this study was the onset of collisional tectonics associated with the Alleghanian orogeny, which began in late Mississippian time. Hatcher et al. (1989) placed the onset of Alleghanian tectonism in the southern Appalachians during the Mississippian, possibly as early as 345 Ma. Geochronology of an Alleghanian fault zone in the Inner Piedmont of North Carolina indicates that thrusting began by mid-Mississippian time (~330 Ma; Hibbard et al., 1998, Wortman et al., 1998). Goldberg and Dallmeyer (1997) reported an Ar-Ar cooling age of 327 Ma

Figure 18. Disturbed layering in Hinton Formation sandstone at Glen Lyn, Virginia (locality 4, Fig. 1). Rotated block with preserved laminations is underlain by sandstone with disrupted lamination and overlain by sandstone with preserved laminations. Lens cap is 5 cm in diameter.

Figure 19. Outcrop of Pride Shale along I-77 (locality 5, Fig. 1). Horizontally bedded rocks at the base of the exposure are cut by paleoslump scar, which was then infilled.

from a shear zone in metamorphic rocks of the North Carolina Blue Ridge, which also supports Mississippian onset of thrusting associated with the Alleghanian orogeny.

There is a significant coincidence between the restricted area in which we have found evidence for Late Mississippian paleoseismicity and the modern-day Giles County, Virginia, seismic zone (Fig. 23). The Giles County seismic zone lies within a seismically active region known as the southern Appalachian seismic zone (Bollinger, 1973) and is defined by an area with a distinct swarm of earthquakes (Bollinger and Wheeler, 1983).

Figure 20. Close-up of rotated block along slump scar from I-77 outcrop of Pride Shale (locality 5, Fig. 1). Lens cap is 5 cm in diameter.

Figure 21. Steeply dipping layers in penecontemporaneously deformed Bluestone Formation near Bluefield, West Virginia, airport (locality 7, Fig. 1). This outcrop was first described by Thomas (1959). The steeply dipping rocks are overlain by gently dipping sandstone layers of the upper Bluestone Formation.

Figure 22. Photograph showing thinning of the Princeton Sandstone over a gentle fold in the underlying Hinton Formation (U.S. Route 460 outcrop, locality 1, Fig. 1). Most of the clastic dikes and other penecontemporaneous deformation structures in this outcrop are in the Hinton Formation that is part of the fold.

This seismic zone is the site of numerous well-documented microearthquakes as well as the second largest historic earthquake in the southeastern United States, which occurred on May 31, 1897, at Pearisburg, Virginia (Modified Mercalli Intensity VIII; body wave magnitude (m_b) 5.8; Bollinger and Hopper, 1971). A study of microearthquakes in the Giles County seismic zone revealed a tabular zone of seismic activity with a northeastward strike and a near-vertical dip (Bollinger, 1981). The earthquake hypocenters show that this tabular zone of microseismic activity is approximately 40 km long, 10 km wide, and 5–26 km deep, which places the activity within the Precambrian basement rocks below the sedimentary cover and below the regional decollement within the Cambrian Rome Formation (Bollinger, 1981; Gresko, 1985; Bollinger and Wheeler, 1988; Wheeler, 1995).

In order to evaluate how closely the paleoseismite localities shown in Figure 23 coincide with the location of the active Giles County seismicity, they must be palinspastically restored to their Late Mississippian position. It is not certain how much Alleghanian northwestward transport these rocks have undergone; estimates range as high as 25 km (Couzens and Dunne, 1994). Figure 23 shows a more conservative estimate of 10 km of northwestward transport by the end of the Alleghanian collision. This restoration brings these localities to within a 40 km radius of the modern-day Giles County seismic zone. Bollinger and Wheeler (1983, 1988) and Wheeler (1995) have proposed that the Giles County seismic zone earthquakes are the result of reactivation of a fault within the Precambrian basement that was formed by late Precambrian–Early Cambrian Iapetan rifting. A possible origin of the Late Mississippian seismicity is reactivation of similar Precambrian structures induced by a northwest-southeast compressive stress caused by the early stages of the collision between North America and Africa at the beginning of the Alleghanian orogeny. Such a stress orientation is consistent with the structures observed in the outcrop along U.S. Route 460. Clastic dikes and normal faults in that outcrop strike northeast (Figs. 5 and 9), which may indicate that these features formed in response to extension along the outer arc of a gentle fold, trending northeast, that was rising in this area in the Late Mississippian. A study by Whitaker and Bartholomew (1999) in this area indicates that the maximum compressive stress direction during the earliest phase of the Alleghanian orogeny for this part of the southern Appalachians was oriented at 311°, which supports this model.

PALEO-EARTHQUAKE MAGNITUDE

Magnitudes of paleo-earthquakes have been estimated in other areas by noting the regional extent and size of liquefaction features and comparing these data to known earthquakes in similar settings. Obermeier et al. (1993) used this kind of information to arrive at estimates of earthquake strengths in the Wabash Valley of southern Indiana and Illinois. The paleoseismic features in our study define an area with a diameter of 50 km; however, we cannot use this size to estimate magnitudes, for two reasons. First, relevant liquefaction structures (clastic dikes) are present at only one outcrop. Most of our paleoseismic features are the result of submarine and near-shore slumping, and to our knowledge, there are no studies that relate the distribution of these kinds of features to earthquake magnitude. The second reason is that the features we have described most likely record separate earthquakes spread over hundreds of thousands or a few millions of years. Until we can identify features originating from a single event, it will not be possible to estimate magnitude on the basis of regional extent of liquefaction features.

The method developed by Ishihara (1985) uses the thickness of a liquefied sand layer and the length of an associated clastic dike to arrive at an estimate of the ground acceleration associated with the earthquake. His method is based on the assumption that the dike was injected during hydraulic fracturing. According to Obermeier (1996), dikes resulting from hydraulic fracturing are typically a few millimeters to 10 cm wide, which is consistent with the width of dikes we measured at the Route 460 outcrop. Dike width, however, is not a definitive characteristic of this mode of failure. Fracturing caused by surface oscillations or injection of liquefied sand into pre-existing fractures can also produce dikes that are less than 10 cm thick. Although it remains a possibility that the dikes in the Route 460 outcrop are hydraulic fractures, there is not enough evidence at this time to warrant using Ishihara's method.

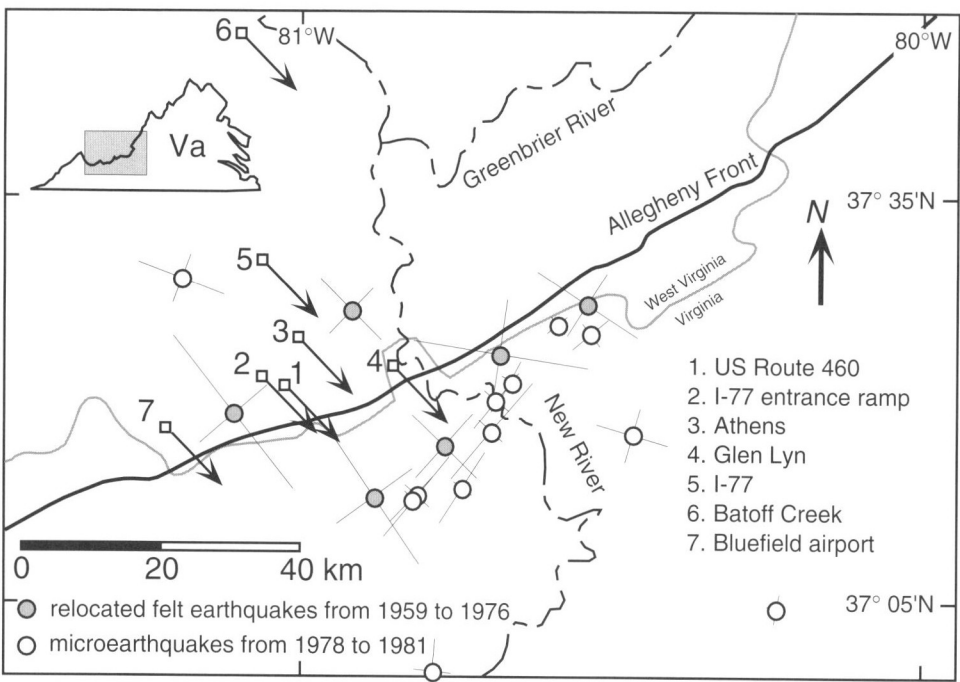

Figure 23. Locations of recent earthquakes (1959–1981), with error bars, in the Giles County seismic zone. Arrows indicate estimated pre–Alleghanian orogeny location of paleoseismite localities from this study. The paleoseismite localities appear to have been located very near the modern-day Giles County seismic zone during Late Mississippian time. Earthquake locations are from Bollinger and Wheeler (1988).

Historic earthquakes in the Giles County seismic zone can have body-wave magnitudes (m_b) as great as 5.8 (Bollinger and Hopper, 1971) and earthquakes of this magnitude can easily generate liquefaction features over a radius of about 8–10 km from the epicenter (see Fig. 42 in Obermeier, 1996). At this time, we can only speculate that if the Late Mississippian earthquakes were generated near the modern-day Giles County seismic zone, the magnitudes were probably on the order of 6, or larger.

CONCLUSIONS

Clastic dikes within Late Mississippian Hinton Formation sandstone and shale near Princeton, West Virginia, have features characteristic of seismically induced dikes from other areas. The dikes are tabular and locally show evidence of rapid, upward injection of material from a lower, liquefied layer of sand. There is no evidence favoring a nonseismic origin for these dikes, such as flooding, landsliding, or artesian flow of groundwater.

The association of these dikes, both in time and space, with a variety of other penecontemporaneous structures permits us to define a region affected by Late Mississippian paleoseismicity. At the same stratigraphic level of the Hinton Formation as the clastic dikes are examples of convolute bedding, pseudonodules, possibly penecontemporaneous faulting, beds showing evidence of lateral flow, and slumps. In addition to the features in the uppermost Hinton Formation, there are numerous slumps in the overlying Bluestone Formation, which possibly represents a time only a few hundred or thousand years after the deposition of the uppermost Hinton. The likely source of the stress responsible for the earthquakes is the incipient collision of Africa with North America during the initial stages of the Alleghanian orogeny.

All of these features define a restricted area with a diameter of about 50 km, which, following palinspastic restoration, coincides with the northwestern edge of the modern-day Giles County seismic zone. That seismic zone is the site of several historic earthquakes with magnitudes greater than 4, the largest being 5.8. The coincidence of the Late Mississippian seismicity with the Giles County seismic zone may indicate that the reactivated Precambrian faults interpreted as the source for modern-day Giles County seismicity also may be part of a family of faults that had been reactivated earlier in the northwest-southeast compressional stress field generated by the Alleghanian collision.

ACKNOWLEDGMENTS

We thank W.A. Thomas, M. Chapman, and N. Rast for thorough reviews, which greatly improved this paper. This study was partially supported by the University Research Council of the University of North Carolina at Chapel Hill.

REFERENCES CITED

Beuthin, J.D., 1997, Paleopedological evidence for a eustatic Mississippian-Pennsylvanian (Mid-Carboniferous) unconformity in southern West Virginia: Southeastern Geology, v. 37, no. 1, p. 25–37.

Beuthin, J.D., and Neal, D.W., 1998, Upper Mississippian paleosols as indicators of allocyclic and autocyclic events in southern West Virginia: Guidebook for field trip in conjunction with annual meeting of the Southeastern Section of Geological Society of America, Charleston, West Virginia, 16 p.

Bollinger, G.A., 1973, Seismicity of the southeastern United States: Bulletin of the Seismological Society of America, v. 63, p. 1785–1808.

Bollinger, G.A., 1981, The Giles County, Virginia seismic zone: Configuration and hazard assessment, *in* Beavers, J.E., ed., Earthquakes and earthquake engineering, Volume 1, The Eastern United States: Ann Arbor, Michigan, Science Publishers, p. 277–308.

Bollinger, G.A., and Hopper, M.G., 1971, Virginia's two largest earthquakes: December 22, 1875 and May 31, 1897: Bulletin of the Seismological Society of America, v. 61, p. 1033–1039.

Bollinger, G.A., and Wheeler, R.L., 1983, The Giles County, Virginia seismic zone: Science, v. 219, p. 1063–1065.

Bollinger, G.A., and Wheeler, R.L., 1988, The Giles County, Virginia seismic zone: Seismological results and geological interpretations: U.S. Geological Survey Professional Paper 1355, 85 p.

Bourgeois, J., and Johnson, S.Y., 2001, Geologic evidence of earthquakes at the Snohomish Delta, Washington, in the past 1200 yr.: Geological Society of America Bulletin, v. 113, p. 482–494.

Cooper, B.N., 1971, The Appalachian structural and topographic front between Narrows and Beckley, Virginia and West Virginia, in Lowry, W.D., ed., Guidebook to Appalachian tectonics and sulfide mineralization of southwestern Virginia: Blacksburg, Virginia Polytechnic Institute and State University, Guidebook No. 5, Field Trip No. 3, p. 87–142.

Cooper, B.N., Arkle, T., Jr., and Latimer, I.S., 1961, Grand Appalachian Field Excursion: Blacksburg, The Virginia Polytechnic Institute, Engineering Extension Series, Geologic Guidebook No. 1, 187 p.

Couzens, B.A., and Dunne, W.M., 1994, Displacement transfer at thrust terminations: The Saltville thrust and Sinking Creek anticline, Virginia, U.S.A.: Journal of Structural Geology, v. 16, p. 781–793.

Dennison, J.M., and Stewart, K.G., editors, 1998, Geologic field guide to extensional structures along the Allegheny Front in Virginia and West Virginia near the Giles County seismic zone: West Virginia Geological Survey Open File Report OF9805, 101 p.

Dennison, J.M., and Wheeler, W.H., 1975, Stratigraphy of Precambrian through Cretaceous strata of probable fluvial origin in southeastern United States and their potential as uranium host rocks: Southeastern Geology Special Publication 5, 210 p.

Donaldson, A.C., and Shumaker, R.C., 1981, Late Paleozoic molasse of central Appalachians, in Miall, A.D., ed., Sedimentation and tectonics in alluvial basins: Geological Association of Canada Special Paper 23, p. 99–124.

Englund, K.J., 1989, Camp Creek interchange section, in Cecil, C.B., and Eble, C., eds., Carboniferous geology of the Eastern United States: 28th International Geological Congress, Field Trip Guidebook T143, p. 91–93.

Goldberg, S.A., and Dallmeyer, R.D., 1997, Chronology of Paleozoic metamorphism and deformation in the Blue Ridge thrust complex, North Carolina and Tennessee: American Journal of Science, v. 297, p. 488–526.

Gresko, M.J., 1985, Analysis and interpretation of compressional (P-wave) and shear (SH-wave) reflection seismic and geologic data over the Bane dome, Giles County, Virginia [Ph.D. thesis]: Blacksburg, Virginia Polytechnic Institute and State University, 74 p.

Harland, W.B., Armstrong, R.L., Cox, A.V., Craig, L.E., Smith, A.G., and Smith, D.G., 1990, A geologic time scale 1989: Cambridge, UK, Cambridge University Press, 261 p.

Hatcher, R.D., Jr., Thomas, W.A., Geiser, P.A., Snoke, A.W., Mosher, S., and Wiltschko, D.V., 1989, Alleghanian orogen, in Hatcher, R.D., Jr., Thomas, W.A., and Viele, G.W., eds., The Appalachian-Ouachita orogen in the United States: Boulder, Colorado, Geological Society of America, Geology of North America, v. F-2, p. 233–318.

Hibbard, J.P., Shell, G.S., Bradley, P.J., Samson, S.D., and Wortman, G.L., 1998, The Hyco shear zone in North Carolina and southern Virginia: Implications for the Piedmont Zone-Carolina Zone boundary in the Southern Appalachians: American Journal of Science, v. 298, p. 85–107.

Ishihara, K., 1985, Stability of natural soils during earthquakes: Proceedings of the Eleventh International Conference on Soil Mechanics and Foundation Engineering, San Francisco, v. 1, p. 321–376.

Khullar, V.K., Gadhoke, S.K., and Kumar, G., 1997, Earthquake induced structures in Quaternary sediments, Indogangetic Plain, Uttar Pradesh, India: Geoscience Journal, v. 18, no. 1, p. 73–78.

Kuenen, P.H., 1958, Experiments in geology: Transactions of the Geological Society of Glasgow, v. 23, p. 1–28.

McColloch, G.H., Jr., 1986, Batoff Creek section of Pennsylvanian–Mississippian strata, Raleigh County, West Virginia: Geological Society of America Centennial Field Guide, Southeastern Section, p. 109–112.

McDowell, R.C., and Schultz, A.P., 1990, Structural and stratigraphic framework of the Giles County area, a part of the Appalachian basin of Virginia and West Virginia, U.S. Geological Survey Bulletin 1839-E, p. E1-E24 (with geologic map, scale 1:125 000, 4 sheets).

Miller, D.J., and Eriksson, K.A., 1997, Late Mississippian prodeltaic rhythmites in the Appalachian basin: A hierarchical record of climate and tidal periodicities: Sedimentary Research, v. 67, p. 653–660.

Miller, D.J., and Eriksson, K.A., 2000, Sequence stratigraphy of Upper Mississippian strata in the Central Appalachians: A record of glacioeustasy and tectonoeustasy in a foreland basin setting: American Association of Petroleum Geologists Bulletin, v. 84, no. 2, p. 210–233.

Mills, P.C., 1983, Genesis and diagnostic value of soft-sediment deformation structures: A review: Sedimentary Geology, v. 35, p. 83–104.

Obermeier, S.F., 1996, Use of liquefaction-induced features for paleoseismic analysis: An overview of how seismic liquefaction features can be distinguished from other features and how their regional distribution and properties of source sediment can be used to infer the location and strength of Holocene paleo-earthquakes: Engineering Geology, v. 44, p. 1–76.

Obermeier, S.F., and Pond, E.C., 1998, Issues in using liquefaction features for paleoseismic analysis: U.S. Geological Survey Open-File Report 98-28, 38 p.

Obermeier, S.F., Jacobson, R.B., Smoot, J.P., Weems, R.E., Gohn, G.S., Monroe, J.E., and Powars, D.S., 1990, Earthquake-induced liquefaction features in the coastal setting of South Carolina and in the fluvial setting of the New Madrid seismic zone: U.S. Geological Survey Professional Paper 1504, 44 p.

Obermeier, S.F., Martin, J.R., Frankel, A.D., Youd, T.L., Munson, P.J., Munson, C.A., and Pond, E.C., 1993, Liquefaction evidence for one or more strong earthquakes in the Wabash Valley of southern Indiana and Illinois, with a preliminary estimate of magnitude: U.S. Geological Survey Professional Paper 1536, 27 p.

Palmer, A.R., 1983, The Decade of North American Geology geologic time scale: Geology, v. 11, p. 503–504.

Pope, M.C., Read, J.F., Bambach, R.K., and Hofmann, H.J., 1997, Late Middle to Late Ordovician seismites of Kentucky, Southwest Ohio and Virginia: Sedimentary recorders of earthquakes in the Appalachian Basin: Geological Society of America Bulletin, v. 109, p. 489–503.

Reger, D.B., 1925a, Map of Mercer County showing general and economic geology: West Virginia Geological Survey, scale 1:62 500, 1 sheet.

Reger, D.B., 1925b, Map of Monroe County showing general and economic geology: West Virginia Geological Survey, scale 1:62 500, 1 sheet.

Ringrose, P.S., 1989, Paleoseismic(?) liquefaction event in late Quaternary lake sediments at Glen Roy, Scotland: Terra Nova, v. 1, p. 57–62.

Seilacher, A., 1984, Sedimentary structures tentatively attributed to seismic events: Marine Geology, v. 55, p. 1–12.

Sims, J.D., 1975, Determining earthquake recurrences intervals from deformational structures in young lacustrine sediments: Tectonophysics, v. 29, p. 141–152.

Thomas, W.A., 1959, Upper Mississippian stratigraphy of southwestern Virginia [Ph.D. thesis]: Blacksburg, Virginia Polytechnic Institute, 322 p.

Thomas, W.A., 1966, Late Mississippian folding of a syncline in the western Appalachians, West Virginia and Virginia: Geological Society of America Bulletin, v. 77, p. 473–494.

Thomas, W.A., 1991, The Appalachian-Ouachita rifted margin of southeastern North America: Geological Society of America Bulletin, v. 103, p. 415–431.

Tuttle, M., and Seeber, L., 1991, Historic and prehistoric earthquake-induced liquefaction in Newbury, Massachusetts: Geology, v. 19, p. 594–597.

Valera, J.E., Traubenik, M.L., Egan, J.A., and Kaneshiro, J.Y., 1994, A practical perspective on liquefaction of gravels, in Prakash, S., and Dakoulas, P., eds., Ground failures under seismic conditions: American Society of Civil Engineers Geotechnical Special Publication 44, p. 241–257.

Wheeler, R.L., 1995, Earthquakes and the cratonward limit of Iapetus faulting in eastern North America: Geology, v. 23, p. 105–108.

Whitaker, A.E., and Bartholomew, M.J., 1999, Layer parallel shortening: A mechanism for determining deformation timing at the junction of the central and southern Appalachians: American Journal of Science, v. 299, p. 238–254.

Wortman, G.L., Samson, S.D., and Hibbard, J.P., 1998, Precise timing constraints on the kinematic development of the Hyco shear zone: Implications for the central Piedmont shear zone, Southern Appalachian orogen: American Journal of Science, v. 298, p. 108–130.

Yong, L., Craven, J., Schweig, E.S., and Obermeier, S.F., 1996, Sand boils induced by the 1993 Mississippi River flood: Could they one day be misinterpreted as earthquake-induced liquefaction?: Geology, v. 24, p. 171–174.

MANUSCRIPT ACCEPTED BY THE SOCIETY MAY 11, 2001

Printed in the U.S.A.

Geological Society of America
Special Paper 359
2002

Anomalous paleoflow orientations: A potential methodology for determining recurrence rates and magnitudes in paleoseismic studies

Gerald J. Smith
Robert D. Jacobi
Department of Geology, 876 Natural Science Complex, University at Buffalo, State University of New York, Amherst, New York 14260-3050, USA

ABSTRACT

In southwestern New York State paleoflow data collected from Upper Devonian sandstones in areas of known north-striking, syndepositionally active faults of the Clarendon-Linden fault system contain anomalous north-south–trending paleoflow orientations. If the faults were active during deposition, then topographic relief generated as a result of faulting may have influenced the paleoflow direction in the immediate area of the active fault.

On the basis of a simplistic model, we assume that the normally west-directed flows were redirected north-south along a north-striking fault scarp. Using anomalous-trending paleoflows, the height of the scarp (fault displacement) can be estimated from the uncompacted thickness of stratigraphic section that contains the sequence of redirected flows. We can estimate the maximum magnitude of the seismic event from the fault scarp height (assuming the fault scarp was developed from a single seismic event). Recurrrence rates can be estimated from the uncompacted thickness of stratigraphic section with west-directed flows between successive sequences of north-south flows.

Using uncompacted stratigraphic thickness as an estimate of the surface offset, magnitudes of potential earthquakes on the Clarendon-Linden fault system are estimated to be between 6.0 and 6.9. Using a range of accumulation rates of 10 to 30 cm/k.y., the average recurrence rate for an earthquake of magnitude 6 or greater during the Late Devonian was estimated to be between 8100 and 24 300 yr. This rate is comparable to the average recurrence rate of 4800 to 14 400 yr obtained from seismites in the stratigraphic section.

INTRODUCTION

On land, faults, fault scarps, and fracture systems are known to influence the direction of current flow from rivulets to rivers (e.g., Keller et al., 1982). Can a similar effect of faults be found in the subaqueous setting of the Devonian Catskill sea (where the seafloor is thought to have been generally featureless (e.g., Woodrow and Isley, 1983)? We collected and analyzed paleoflow data gathered from more than 1500 sites in the Upper Devonian Catskill Delta complex, southwestern New York State. Although paleoflow was generally to the west (Table 1), analysis of paleoflow data revealed anomalous northward paleoflow trends in regions of known and suspected north-striking faults. Activity of the faults prior to, or contemporaneous with, the generation of the flow feature may account for the anomalous paleoflow trends. Examination of anomalous-paleoflow indicators within the stratigraphic section may provide a methodology for the estimation of paleoseismicity during the depositional period. Past studies have shown that other sedimentologic features can be of value in seismicity investigations (Sims, 1975; Doig, 1990). Recurrence

Smith, G.J., and Jacobi, R.D., 2002, Anomalous paleoflow orientations: A potential methodology for determining recurrence rates and magnitudes in paleoseismic studies, *in* Ettensohn, F.R., Rast, N., and Brett, C.E., eds., Ancient seismites: Boulder, Colorado, Geological Society of America Special Paper 359, p. 145–164.

TABLE 1. PREVIOUS PALEOFLOW STUDIES

Who	When	Where	What	Paleoflow mean (°)	No. of measurements
Burtner	1963	Pennsylvania	Catskill Fm.	270	
Leeper	1963	Pennsylvania	Chemung and Catskill Fms.	270	
Manspeizer	1963	Allegany County, New York	Canaseraga Fm.	270	73
Manspeizer	1963	Allegany County, New York	Caneadea Fm.	275	94
Manspeizer	1963	Allegany County, New York	Rushford Fm.	25	21
Manspeizer	1963	Allegany County, New York	Machias Fm.	20	54
Colton	1967	Western New York	Upper Devonian	273	1400
Meckel	1967	Pennsylvania	Pottsville Fm. and Olean Congl.	270	
Walker and Sutton	1967	Central New York	Sonyea Grp.	281–290	
Woodrow	1968	Bradford County, Pennsylvania	Wiscoy and Towanda Fms.	298	
Sutton, Bowen and McAlester	1970	Central New York	Sonyea Grp.	277	
Lundegard et al.	1985	Appalachians	Brallier and Scherr	263	700
Gutmann and Jacobi	1988–89	Erie County, New York	South Wales Fm.	291.6	525
Smith and Jacobi (this report)	1999	Allegany County, New York	U. Dev. Allegany County	292.4	1727
Smith and Jacobi (this report)	1999	Allegany County, New York	Canadaway Grp.	293.1	1692
Smith and Jacobi (this report)	1999	Allegany County, New York	West Falls Grp.	272.2	35
Smith and Jacobi (this report)	1999	Allegany County, New York	Machias Fm.	344.0	583
Smith and Jacobi (this report)	1999	Allegany County, New York	Rushford Fm.	305.0	129
Smith and Jacobi (this report)	1999	Allegany County, New York	Caneadea Fm.	285.0	553
Smith and Jacobi (this report)	1999	Allegany County, New York	Hume Fm.	271.4	20
Smith and Jacobi (this report)	1999	Allegany County, New York	Mills-Mills Fm.	283.3	110
Smith and Jacobi (this report)	1999	Allegany County, New York	South Wales Fm.	265.8	166
Smith and Jacobi (this report)	1999	Allegany County, New York	Dunkirk Fm.	272.6	131
Smith and Jacobi (this report)	1999	Allegany County, New York	Wiscoy Fm.	275.9	24
Smith and Jacobi (this report)	1999	Allegany County, New York	Hanover Fm.	265.8	11

rates for seismic events are obtainable from examination of seismically generated features such as seismites (Sims, 1975). However, the use of anomalous-paleoflow indicators may also provide a method for estimating fault displacement and thus allow the magnitude of the seismic event to be approximated.

Our field area covers ten 7½-minute U.S. Geological Survey topographic quadrangles along the northern and western borders of Allegany County, New York (Fig. 1). The stratigraphic section contains the upper part of the Frasnian West Falls Group and the Famennian Canadaway Group (Fig. 2). The location of the study area is coincident with the Clarendon-Linden fault system, a prominent north-trending fault zone that has been active since the Precambrian (Rickard, 1973; Fakundiny et al., 1978a, 1978b; Jacobi and Fountain, 1996, 1998). Fakundiny et al. (1978b) referred to the Clarenon-Linden as the longest and oldest fault system in New York. In the region of Allegany County, it comprises at least 17 parallel, high-angle, segmented faults that separate fault blocks that exhibited semi-independent motion through geologic time (Jacobi and Fountain, 1996, 1998). This fault system has been shown to have been syndepositionally active in Late Cambrian–Middle Ordovician time (Van Tyne, 1975; Jacobi and Fountain, 1993, 1996, 1998) and in Early and Late Devonian time (Van Tyne et al., 1980a, 1980b, 1980c, 1980d, 1980e, 1980f, 1980g, 1980h, 1980i, 1980j, 1980k; Beinkafner, 1983; Jacobi and Fountain, 1993, 1996, 1998; Smith and Jacobi, 1998, 1999, and 2001). The motion on individual faults reversed periodically over the long history of the fault system, on the basis of surface stratigraphy, seismic data, and well-log analyses. Growth fault geometries have been inferred in several different formations at the surface, the uncompacted stratigraphic thickness differential ranging from a few centimeters to a few meters (Smith and Jacobi, 2001).

OBSERVATIONAL DATA FROM PALEOFLOW INDICATORS

Methodology

We measured 1727 paleoflow orientations from approximately 1500 sites in the Upper Devonian West Falls and Canadaway Groups. The data were sorted by stratigraphic formation and subdivided by the type of paleoflow indicator. The first subgroup consists of unidirectional paleoflow indicators that exhibit a distinct paleoflow direction, such as ripples, cross-bedding, trough cross-sets, and flute casts. The second subgroup contains nonunique unidirectional data from unidirectional paleoflow indicators that do not show which one of the two potential directions is the correct flow direction—e.g., grooves, striations, furrows (gutter casts), parting lineations, and some plant fragments. A subset of the nonunique unidirectional data is bidirectional oscillatory flow features such as paleoflow data from symmetrical ripples. For our study, both the nonunique and bidirectional paleoflow data are combined in one group, labeled "bidirectional." The last subgroup consists of all data from observed sole marks;

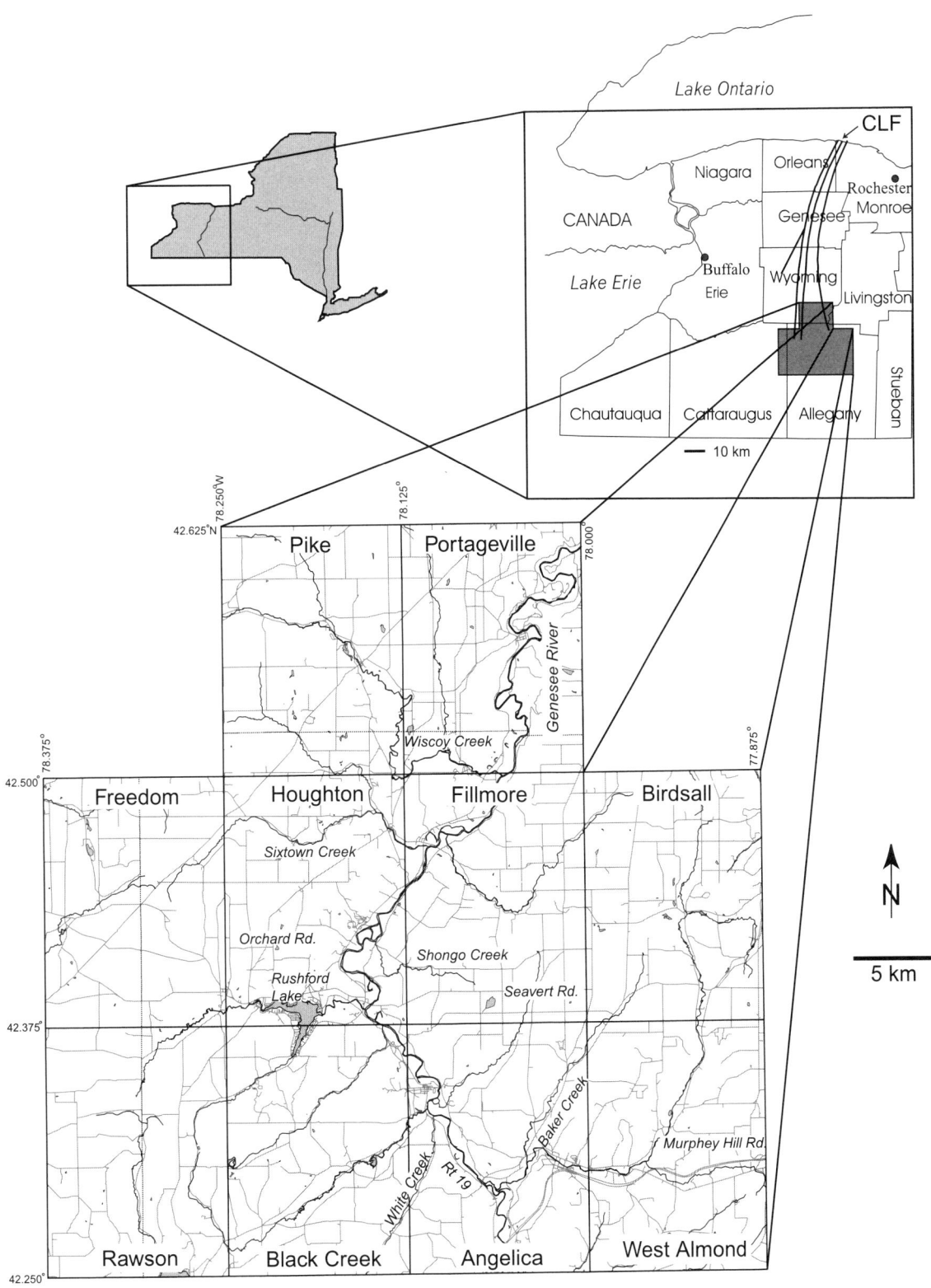

Figure 1. Location of the field area. Enlarged map of western New York shows the location of the field area relative to the Clarendon-Linden fault system (CLF) (from Van Tyne, 1975); detailed map shows the road network and major streams in the area as well as locations mentioned in the text.

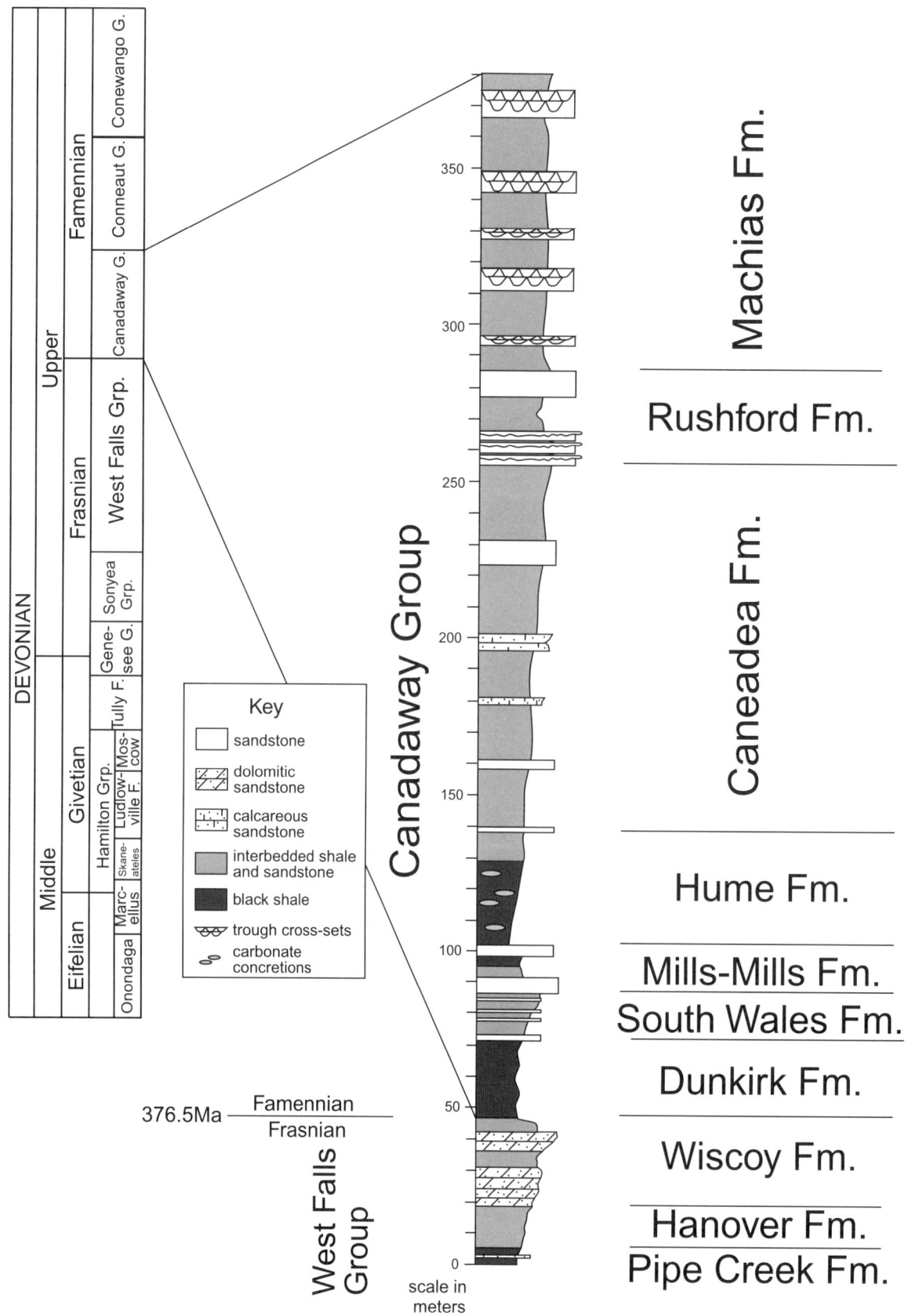

Figure 2. Stratigraphic column for the field area in northern Allegany County, New York. The column is constructed from data of more than 1500 sites measured in northern Allegany County (Smith et al., 1998). F., Fm.—Formation; G., Grp.—Group.

paleoflow indicators such as flute casts, grooves, and striations; and furrows (gutter casts). The collected data were then used to generate rose diagrams and calculate the vector mean and magnitude for each formation and its subgroups. A final calculation is made to determine the circular standard deviation based on the vector magnitude (Krause and Geijer, 1987).

Results

The average observed paleoflow orientation for all paleoflow indicators measured in the study area is typically toward the west to west-northwest (vector mean is 292.4°) (Fig. 3 and Table 2). This value is slightly misleading, because the greater variation observed in the Rushford and Machias formations is due to storm and tidal effects on the shallow-water–deposited sediments. Examination of sole-mark measurements displays a consistent westward flow for all of the examined formations in the study area (vector mean of sole marks is 269.9°). This is consistent with previous paleoflow studies in the Upper Devonian Catskill Delta complex that yielded paleoflow trends generally orthogonal (west to west-northwest) to the assumed Acadian orogenic front (Table 1).

Paleoflow determined from nonunique unidirectional data and bidirectional oscillatory paleoflow data also trends east-west (vector mean is 271.0°) (Fig. 3 and Table 2). In contrast, the unidirectional paleoflow indicators indicate a relatively high variability, although they are generally directed northwestward (vector mean is 318.6°) (Fig. 3 and Table 2). Comparison of the paleoflow orientations among individual formations shows that the vector mean for unidirectional paleoflow rotates from west to northwest up section (Fig. 3 and Table 2). However, the vector mean for nonunique unidirectional data and bidirectional data remains consistently west trending up section. The rotation of the unidirectional paleoflow reflects, in part, a general up-section shallowing of the depositional environment (particularly in the Rushford and Machias formations). This shallowing resulted in (1) a stronger influence of northwest combined flow (i.e., storm waves impinged upon the sea floor in the upper units as a result of shallower depths), and (2) increasing tidal influence observed in the Machias Formation (i.e., both east- and west-trending trough cross-sets are in sand packets). Because of these depositional factors, it is difficult to recognize "anomalous" paleoflow in the Rushford and Machias that might be due to fault-scarp redirection of paleoflow. We therefore regard the flow data from the Machias as less diagnostic for determination of seismic activity. However, we have included the data from the Rushford and Machias formations to better cover the western extent of the Clarendon-Linden fault system.

In contrast to the normal western paleoflow observed throughout western New York, we have noted 348 instances of north-south paleoflow at 167 sites in Allegany County (Fig. 4) that cannot be ascribed to aseismic deposition processes such as combined flow. For the study area, we defined anomalous paleoflows as north-south–trending paleoflow (within 30° of 0° or 180°) typically occurring outside the range of the standard deviation. One feature common among the anomalous north-south trending paleoflows is that the north-south–trending paleoflow features are in north-south zones that coincide with both suspected and confirmed north-striking faults associated with the Clarendon-Linden fault system (Jacobi and Fountain, 1993, 1996; Jacobi and Zhao, 1996; Smith and Jacobi, 1998) (Fig. 4). Figure 4 shows that nearly all of the anomalous-paleoflow measurements are located within, or on strike with, the gray bands that represent north-south–trending fracture intensification domains (FIDs) (Jacobi and Zhao, 1996). These FIDs generally indicate faults observed at depth rather than at the surface. We propose that this variation of paleoflow orientation could represent the interaction of fault-influenced topography and paleocurrents. The spatial coincidence of anomalous-paleoflow markers with the location of observed and suspected faults and fault zones suggests that syndepositional fault activity influenced and/or controlled localized paleoflow. Examination of the anomalous data reveals that evidence of north-south–trending paleoflow is not restricted to any one formation, or even groups of formations. North-south–trending paleoflow evidence was observed in unidirectional, nonunique unidirectional, and bidirectional features.

NORTH-SOUTH–TRENDING PALEOFLOWS

Comparison of paleoflow measurements for the South Wales Formation between Allegany and Erie counties provides the clearest example that the north-south paleoflow orientations are anomalous and are coincident with the Clarendon-Linden fault system, suggesting fault control on the paleoflow. In Erie County, the South Wales Formation paleoflow indicators described part of a regularly varying fan with fairly constant orientations ranging between 258° and 318°. The vector mean is 291.6°, and a standard deviation is 11.11° (Fig. 5 and Table 3) (Gutmann, 1989; Jacobi et al., 1994). The typically low standard deviation at individual sites ranged from 0.81° to 8.76°, indicating that for each site the variation in paleoflow was extremely low (Table 3).

In contrast, for seven outcrops of the South Wales Formation in Allegany County (Fig. 5), the vector mean varies from 219.3° to 314.7°, and standard deviations vary from 26.7° to 42.4° (Table 3). On a more detailed scale, in a 2 km transect along Wiscoy Creek in Allegany County, the South Wales Formation paleoflow measurements cover a wide range of orientations from 180° to 350°, and have a vector mean of 242.4° with a standard deviation of 35.5°. In general, all formations that occur along Wiscoy Creek have a vector mean of 257.0° and a standard deviation of 45.3° (Figs. 6 and 7). The vector mean for the South Wales Formation throughout the entire Allegany study area is 265.8°, with a standard deviation of 42.5°.

The anomalous north-south–trending paleoflows are observed in outcrops that are close to faults of the Clarendon-Linden system, as determined by seismic reflection lines and structure mapping (Jacobi and Fountain, 1996). For example, near the northern border of the Houghton quadrangle, a series of anomalous

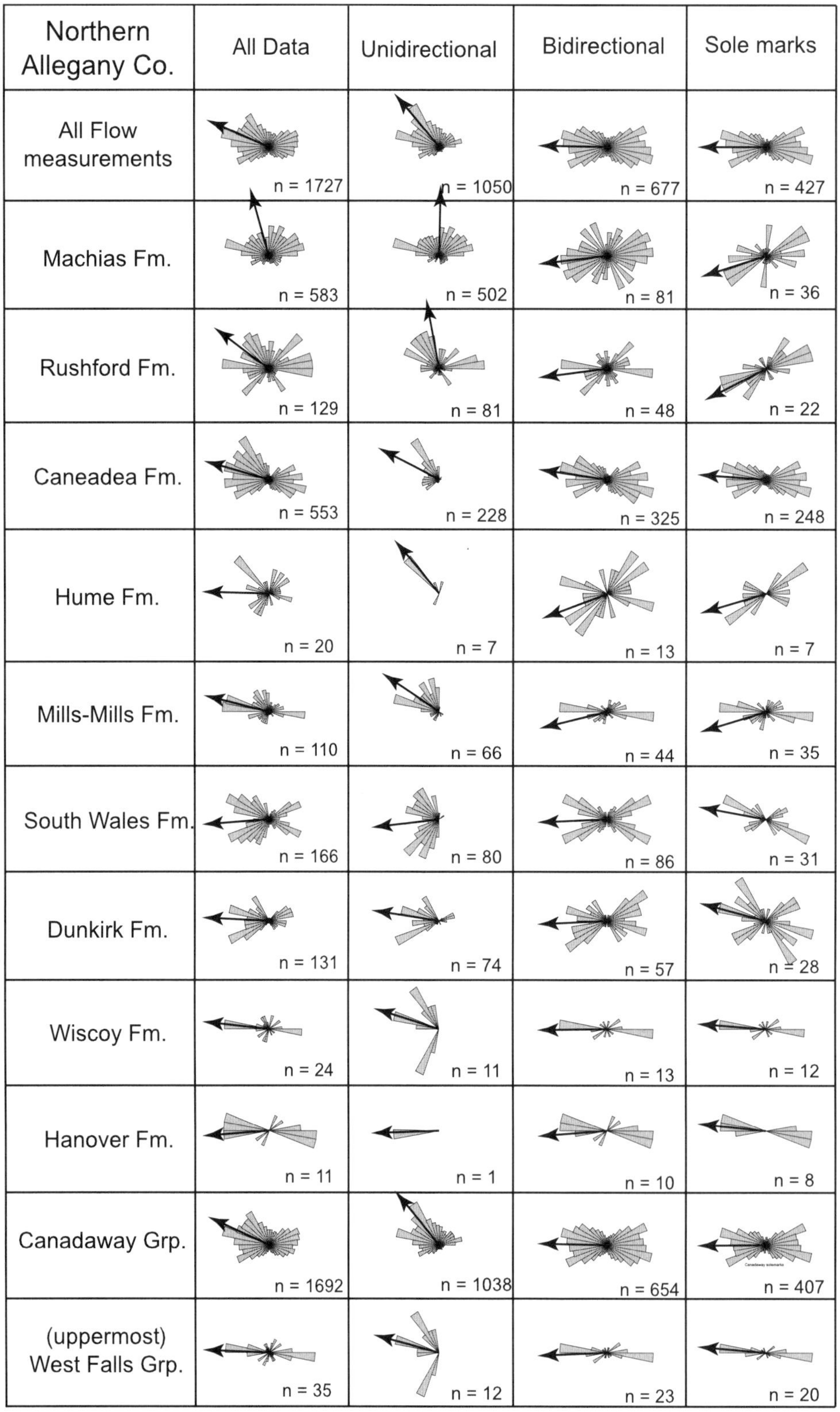

Figure 3. Rose diagrams compiled for each formation, each group, and the entire study area. Arrows represent the vector mean for each rose diagram. Sole-mark paleoflow data include grooves, striations, flute casts, and furrows (gutter casts). Bidirectional paleoflow data (nonunique unidirectional and oscillatory paleoflow) includes grooves, striations, furrows (gutter casts), parting lineations, plant fragments, and symmetrical ripples. Unidirectional paleoflow data includes ripples and flute casts. All paleoflow data combine unidirectional, bidirectional, and oscillatory paleoflow data. "n" indicates the number of measurements in each rose diagram.

TABLE 2. PALEOFLOW MEASUREMENTS FOR NORTHERN ALLEGANY COUNTY

Formation	All				Unidirectional				Bidirectional				Solemarks			
	No.	Vector mean (°)	Vector length (m)	Std. dev.	No.	Vector mean (°)	Vector length (m)	Std. dev.	No.	Vector mean (°)	Vector length (m)	Std. dev.	No.	Vector mean (°)	Vector length (m)	Std. dev.
Machias	583	344	0.3416	65.7	502	1	0.3952	63.0	81	263	0.731	42.0	36	253	0.7789	38.1
Rushford	129	305	0.3762	64.0	81	349	0.379	63.9	48	263	0.7102	43.6	22	244	0.8372	32.7
Caneadea	553	285	0.7752	38.4	228	296	0.7174	43.1	325	278	0.8315	33.3	248	274	0.8338	33.0
Hume	20	271	0.6469	48.1	7	319	0.7273	42.3	13	250	0.7841	37.6	7	254	0.8883	27.1
Mills-Mills	110	283	0.71	43.6	66	303	0.7369	41.6	44	257	0.8209	34.3	35	254	0.8002	36.2
S. Wales	166	266	0.7249	42.5	80	263	0.6373	48.8	86	268	0.8073	35.6	31	280	0.8772	28.4
Dunkirk	131	273	0.6894	45.2	74	279	0.5848	52.2	57	267	0.8327	33.1	28	285	0.7828	37.8
Wiscoy	24	276	0.7365	41.6	11	287	0.6165	50.2	13	269	0.8534	31.0	12	274	0.9054	24.9
Hanover	11	266	0.9219	22.6	1	N/A	N/A	N/A	10	266	0.9142	23.7	8	276	0.9896	8.3
All	1727	292	0.5238	55.9	1050	319	0.4239	61.5	677	271	0.8004	36.2	427	270	0.8182	34.5
West Falls	35	272	0.7919	37.0	12	285	0.6445	48.3	23	268	0.8794	28.1	20	275	0.939	20.0
Canadaway	1692	293	0.5193	56.2	1038	319	0.4226	61.6	654	271	0.7976	36.5	407	270	0.8125	35.1

north-south paleoflows are present along Sixtown Creek; these coincide with north-south–trending faults observed in the cross section for Sixtown Creek (Fig. 8). Soft-sediment deformation offers further evidence that these faults were syndepositionally active. Large (>1 m diameter) ball-and-pillow structures are present in the same area of Sixtown Creek. Ball-and-pillow structures may result from ground shaking caused by seismic events (Sorauf, 1965).

In the model we propose (Fig. 9), the syndepositional faulting creates a minor topographic impediment to the normal paleoflow, thus diverting the flow along the fault scarp. As deposition continues and fault activity ceases, the relief of the fault scarp decreases and eventually will not be large enough to cause a change in the flow direction.

Criteria we used to distinguish fault-controlled anomalous paleoflow include: (1) the anomalous-paleoflow trend observed in sole marks in the site (reducing the possibility of a combined-flow causal mechanism); (2) clusters of anomalous-paleoflow trends between units with east-west paleoflow at a single site (eliminating as a causal mechanism a channel avulsion event); (3) faults observed in the same outcrop as the anomalous paleoflow; and (4) ball-and-pillow structures adjacent or within sites that contain the north-south trending data, suggesting ground-shaking events (seismites). The coincidence of the anomalous data with north-trend fracture-intensification domains (Fig. 4) suggests that faulting may control or influence paleoflow.

RECURRENCE RATES AND MAGNITUDES ESTIMATED FROM ANOMALOUS PALEOFLOW

If the anomalous north-south trending paleoflows formed in response to local fault activity, then it is possible to use the detailed stratigraphic data from the sections containing evidence of the anomalous paleoflow to estimate the magnitude of the earthquake as well as the recurrence rates of seismic events. For stratigraphic sections with detailed paleoflow information, the surface displacement of the fault may be approximated from the uncompacted stratigraphic thickness between beds containing typical (west-trending) paleoflow indicators and beds containing anomalous-trending paleoflow indicators, as shown in the simplistic model (Fig. 9). Surface displacement or scarp height of a fault can be related to the magnitude of the earthquake (Slemmons, 1982; Bonilla et al., 1984; dePolo and Slemmons, 1990; Wells and Coppersmith, 1994). Thus, by assuming that the flow diversion resulted from fault-block uplift during a single event, we can estimate a maximum magnitude. A minimum fault-scarp height can be estimated from the uncompacted total stratigraphic thickness of units with anomalous-paleoflow indicators. A maximum surface displacement can be estimated from the uncompacted total stratigraphic thickness between typical "normal" west-trending paleoflow indicators that occur above and below the anomalous paleoflow feature(s) (Fig. 10). The data yield a maximum event, if we assume that the fault scarp was produced from a single seismic event.

The paleoflow and stratigraphic data for Allegany County revealed nine sections that allowed the estimation of earthquake magnitude. Decompaction of the stratigraphic sections was based on compaction estimates of the Lower Cretaceous Colorado Shales determined by Leckie and Potocki (1988) and Upper Devonian Chattanooga Shale by Lobza and Schieber (1999) (Table 4). Of the nine sections, four sections, such as the example from Wiscoy Creek shown in Figure 10, contain paleoflow data that allow both maximum and minimum estimates of surface displacement. The remaining five sections provide minimum estimates of surface displacement.

In order to convert surface displacement to earthquake magnitudes, we used the surface displacement versus magnitudes curves of Bonilla et al. (1984), dePolo and Slemmons (1990), and Wells and Coppersmith (1994). The data used in the construction of the Bonilla et al. (1984), dePolo and Slemmons (1990), and Wells and Coppersmith (1994) curves were compiled from earthquake data collected globally from different types of faults. Note

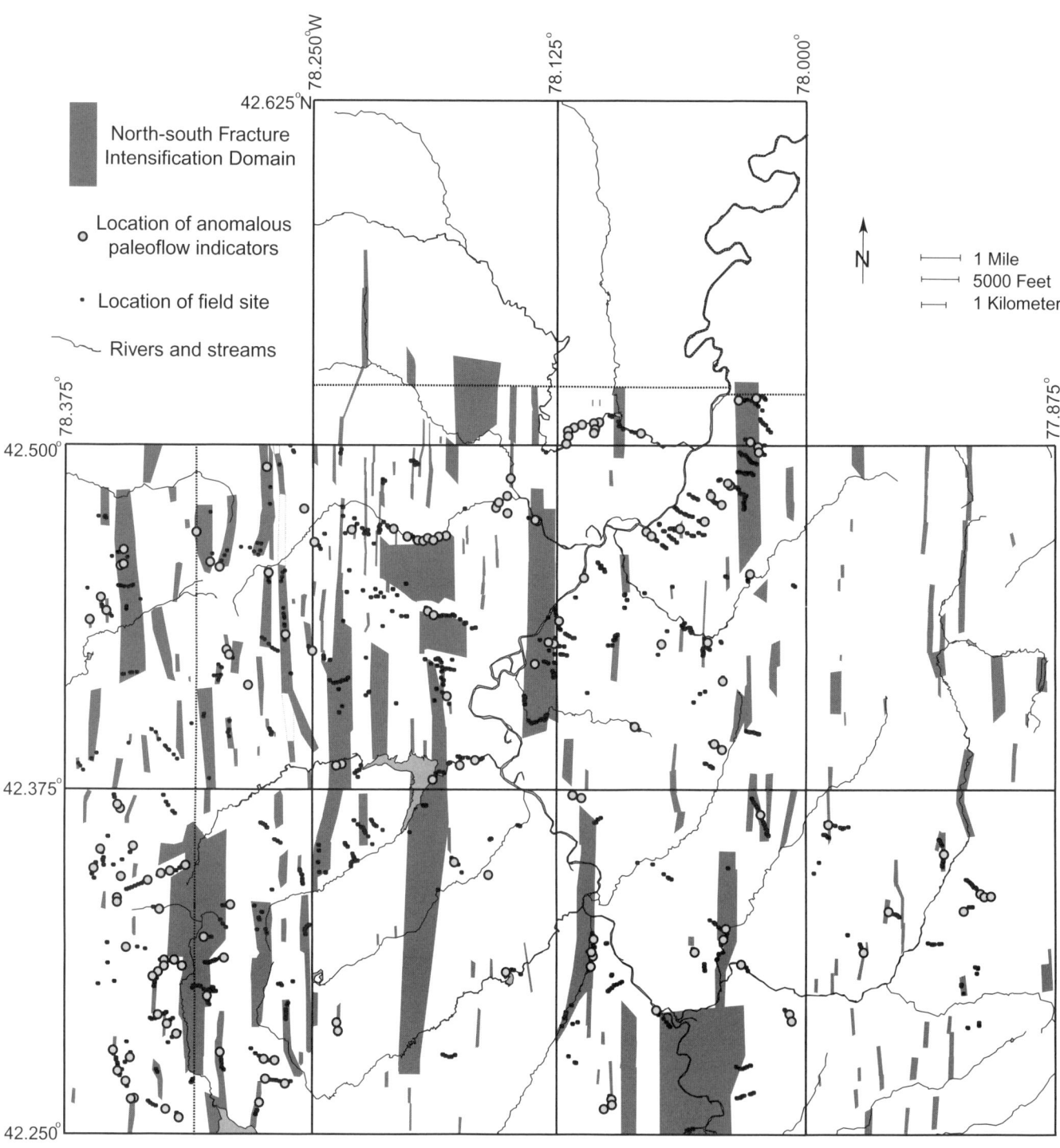

Figure 4. Locations of all anomalous north-south trending paleoflow and the relationship to the north-south fracture intensification as determined by Jacobi and Zhao (1996), Peters (1998), and Zack (1998).

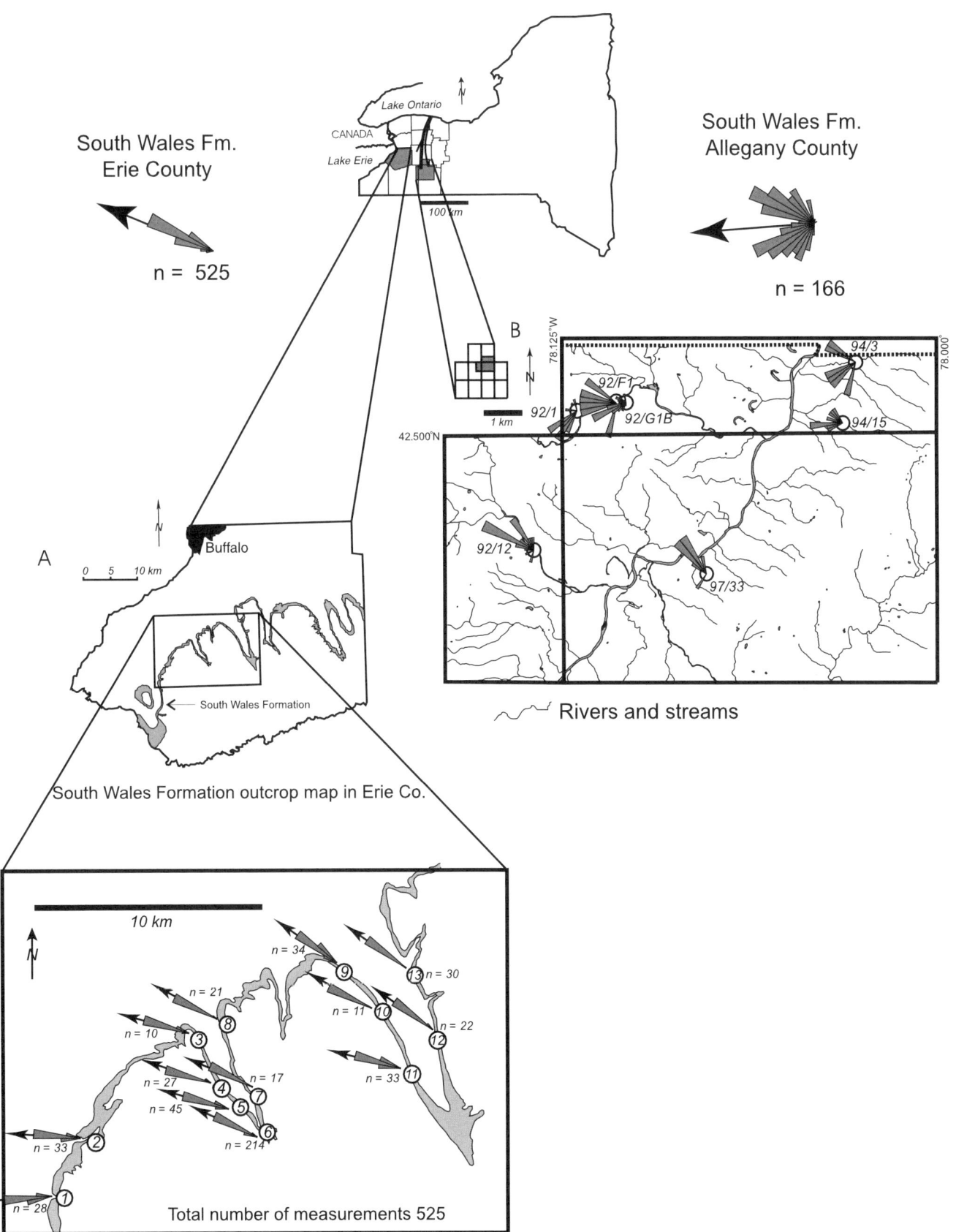

Figure 5. A: Location of Gutmann (1989) and Jacobi et al. (1994) study area of the South Wales Formation. Rose diagrams at locations represent measured paleoflow indicators; arrows are the vector mean for each site; numbers are sites listed in Table 3. Two rose diagrams above compare the paleoflow of the South Wales Formation in Erie County (Gutmann, 1989) and Allegany County (this chapter), New York. Note that the Erie County data are distinctly unimodal, whereas Allegany County can be considered bimodal, possibly polymodal. B: South Wales data from Allegany County. The large circles and rose diagrams with adjacent site numbers denote locations for sites listed in Table 3.

TABLE 3. COMPARISON OF PALEOFLOW INDICATORS IN SOUTH WALES FORMATION, ERIE COUNTY AND ALLEGANY COUNTY, AND ALL FORMATIONS IN WISCOY CREEK

	No. of indicators	Vector mean (°)	Vector length (m)	Std. dev.
Erie County				
S.Wales	525	291.6	0.9812	11.1101
Site 1	28	265.6	0.9949	5.7866
Site 2	33	273.9	0.9920	7.2474
Site 3	10	285.0	0.9944	6.0636
Site 4	27	287.8	0.9943	6.1175
Site 5	45	285.4	0.9966	4.7247
Site 6	214	294.4	0.9974	4.1317
Site 7	17	293.1	0.9999	0.8103
Site 8	21	297.6	0.9992	2.2918
Site 9	34	304.5	0.9892	8.4207
Site 10	11	298.8	0.9997	1.4035
Site 11	33	284.0	0.9883	8.7646
Site 12	22	304.7	0.9977	3.8860
Site 13	30	304.6	0.9992	2.2918
Allegany County				
92/1	28	219.3	0.8900	26.8741
92/12	30	302.0	0.8778	28.3252
92/F1	7	264.2	0.8633	29.9586
92/G1B	53	254.6	0.8056	35.7261
94/15	9	286.5	0.8772	28.3947
94/3	8	254.1	0.7257	42.4376
97/33	10	314.7	0.8615	30.1552
All S.Wales	166	265.8	0.7249	42.4994
Wiscoy Creek Formation				
Hume	2	309.0	0.4695	59.0174
Mills-Mills	43	283.7	0.6962	44.6613
S. Wales	94	242.4	0.8083	35.4771
Dunkirk	76	256.9	0.6426	48.4412
Wiscoy	20	276.8	0.6845	45.5132
All	235	257.0	0.6871	45.3253

Note: South Wales data from Gutmann (1989).

that the curves are fairly insensitive to fault-scarp height; any measurable offset will yield M ≥ 6, reflecting the amount of seismic energy required to generate a disruption in surface topography. Calculated magnitudes ranged from 6.2 to 6.9, the average estimated magnitude being 6.7, based on the Bonilla et al. (1984) and dePolo and Slemmons (1990) curves, or 6.0 to 6.6, the average estimated magnitude being 6.4, based on the Wells and Coppersmith (1994) curves (Fig. 11 and Table 5).

Recurrence rates of seismic events may be estimated from stratigraphic sections that contain multiple sets of anomalous-trending paleoflows (Fig. 12). Assuming that the stratigraphic thickness of sediment with westward flow between two anomalous-trending paleoflow sections indicates a time of fault quiescence, we can estimate a seismicity recurrence rate, provided the sedimentary accumulation rates or a detailed geochronology of the section are known. Previous accumulation rates for the Late Devonian were based on sequence cyclicity (van Tassell, 1987). No accumulation rates were available for the Upper Devonian Catskill Delta complex in our study area at the time of our research. Additionally, radiometrically dated horizons in the Catskill Delta complex are few and are generally stratigraphically below the studied units (Tucker et al., 1998). The absence of a detailed ash geochronology and the absence of detailed fossil geochronology do not allow the calculation of definitive sedimentation rates for our area.

We therefore used a range of depositional rates based upon sequence stratigraphic studies in the area (van Tassell, 1987), and accumulation rates measured in similar geologic environments (Gautier, 1982; Bridge and Willis, 1994) (Table 4). Sandstones and similar coarser sediments in our study area were deposited geologically instantaneously as turbidites or storm beds. The stratigraphic thickness of remaining shales and interbedded shales and siltstones were used to estimate time duration between anomalous flow events. To reflect the different possible sediment accumulation rates, we calculated the recurrence rates using a range of deposition rates between 10 and 30 cm/k.y. Sixteen sections with multiple sets of paleoflow anomalies and high density of paleoflow data were used to calculate an average recurrence rate (Table 6). Average recurrence rates were found to range from 24 300 to 8100 yr.

We can use the same general methodology to estimate seismicity recurrence rates from the thickness of sediment between ball-and-pillow horizons, assuming they are caused by ground shaking related to seismicity (seismites). We examined 27 seismites in which the stratigraphic section separating them was fully exposed and did not contain erosional unconformities (Table 7). The average calculated recurrence rate ranged from 14 400 to 4800 yr, which is on the same order as that calculated from the anomalous paleoflow. The higher recurrence rates from the seismites may reflect a sampling bias: the anomalous-paleoflow data are primarily from the lower units of the Canadaway Group, whereas the seismites data are primarily from the upper units of the Canadaway. Our previous investigations suggest that the time of upper Canadaway deposition was a more seismically active period (Smith and Jacobi, 1998, 2001).

DISCUSSION

The Clarendon-Linden fault system has been active since the Precambrian (Fletcher and Sykes, 1977; Fakundiny, 1978a, 1978b; Jacobi and Fountain, 1993, 1996, 1998). Growth-fault geometries observed in seismic profiles, in outcrop, and in stratigraphic cross sections have shown that the fault system was active during the Devonian (Jacobi and Fountain, 1996, 1998; Smith and Jacobi, 1998, 2001).

In the Caneadea Formation, the paleoflow indicators do not all trend north-south; those with north-south flow are observed along Sixtown Creek and Rushford Lake and in the Rushford Formation along White Creek, Baker Creek, and the Genesee River in the Angelica topographic quadrangle. In the same outcrops are large ball-and-pillow structures (>1.5 m along the long axis) with north-trending long axes. Thus, the locally deformed sediments are consistent with local fault activity in the area of the

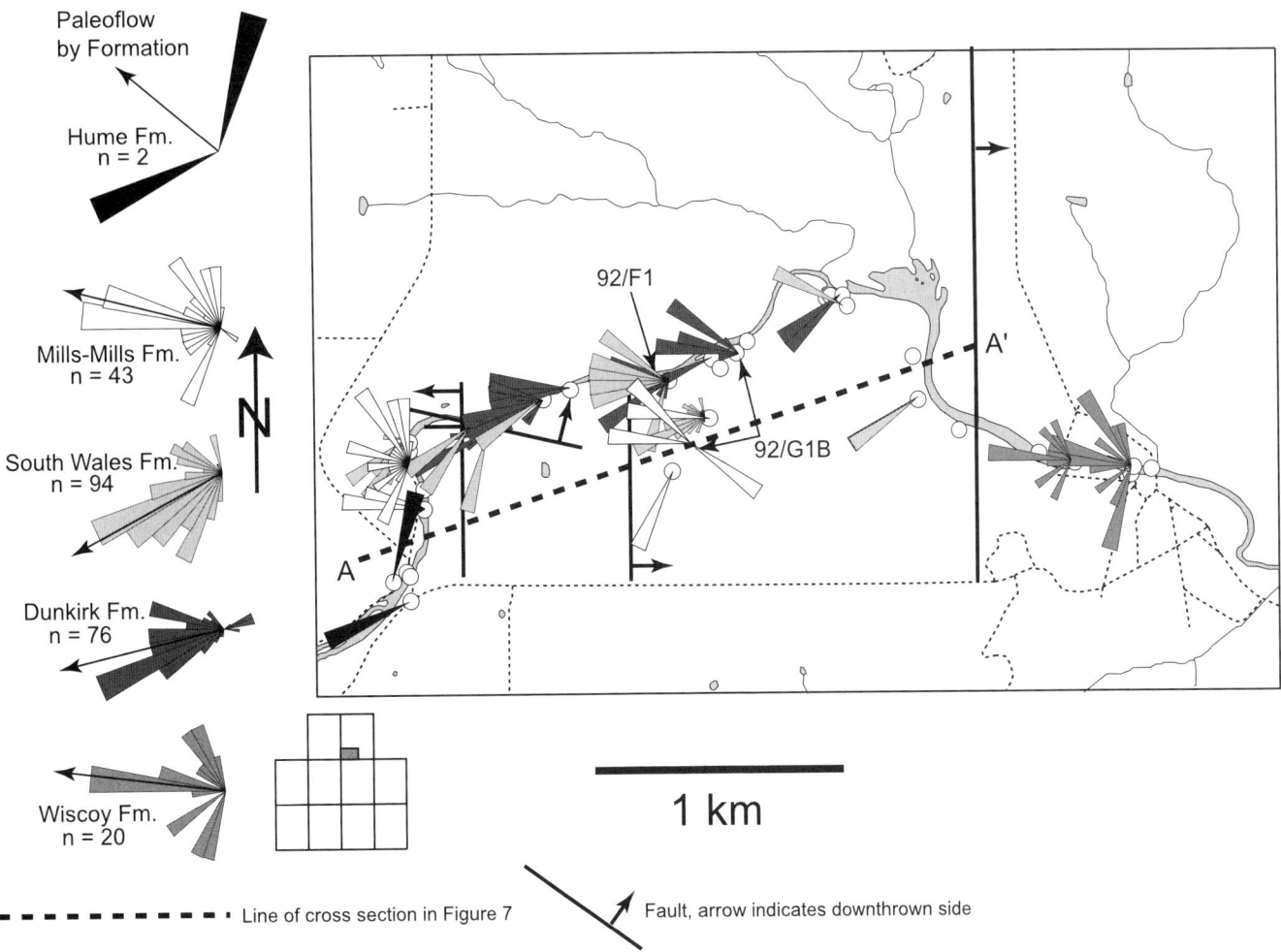

Figure 6. Map of Wiscoy Creek, in southern Portageville topographic quadrangle, displaying rose diagrams of measured paleoflow indicators, by formation and location. Note that prominent north and south paleoflow petals occur adjacent to north-south faults. Labels 92/F1 and 92/G1B are sites in Table 3.

anomalous north-south trending paleoflow. In the Sixtown Creek area, evidence of soft-sediment deformation in the form of large ball-and-pillow structures is present in the same stratigraphic level as the anomalous north-south trending paleoflow. If the ball-and-pillow structures represent ground-shaking events (seismites), then the fault activity was contemporaneous with the anomalous north-south–trending paleoflow.

Our estimated earthquake magnitudes and rates of recurrence for the Late Devonian are consistent with other intracontinental fault systems on extended continental crust (Johnston et al., 1994) and are somewhat comparable to present activity along the Clarendon-Linden fault. The compilation by Johnston et al. (1994) showed that intracontinental faults can undergo M6–M7 seismic events. The largest historical seismic event on the Clarendon-Linden fault was a M 5.2 event in 1929 near Attica, New York (Fletcher and Sykes, 1977). A paleoseismic study of the Clarendon-Linden fault system (Tuttle et al., 1996) found no unequivocal evidence for an earthquake of M ≥ 6 in glacial sediments. If the observations of Tuttle et al. (1996) are correct, then a seismic event of M ≥ 6 probably has not occurred in the past 10 000 years on the Clarendon-Linden fault. The relatively low number of seismic events at present along the fault makes a plot of magnitude versus recurrence difficult to evaluate (e.g., Jacobi and Fountain, 1996). The truncated magnitude-recurrence curve for the Clarendon-Linden fault (Jacobi and Fountain, 1996) could be interpreted to be consistent with the Late Devonian recurrence rates.

SUMMARY AND CONCLUSIONS

Paleoflow data collected from Upper Devonian units in northern Allegany County, western New York State, revealed that although most paleoflows were westward directed, 348 paleoflow indicators (out of a total 1727 measurements) trend within 30° of north or south. These anomalous north-south–trending paleoflows occur in areas where north-south trending faults of the Clarendon-Linden system are thought to be. Activity on the

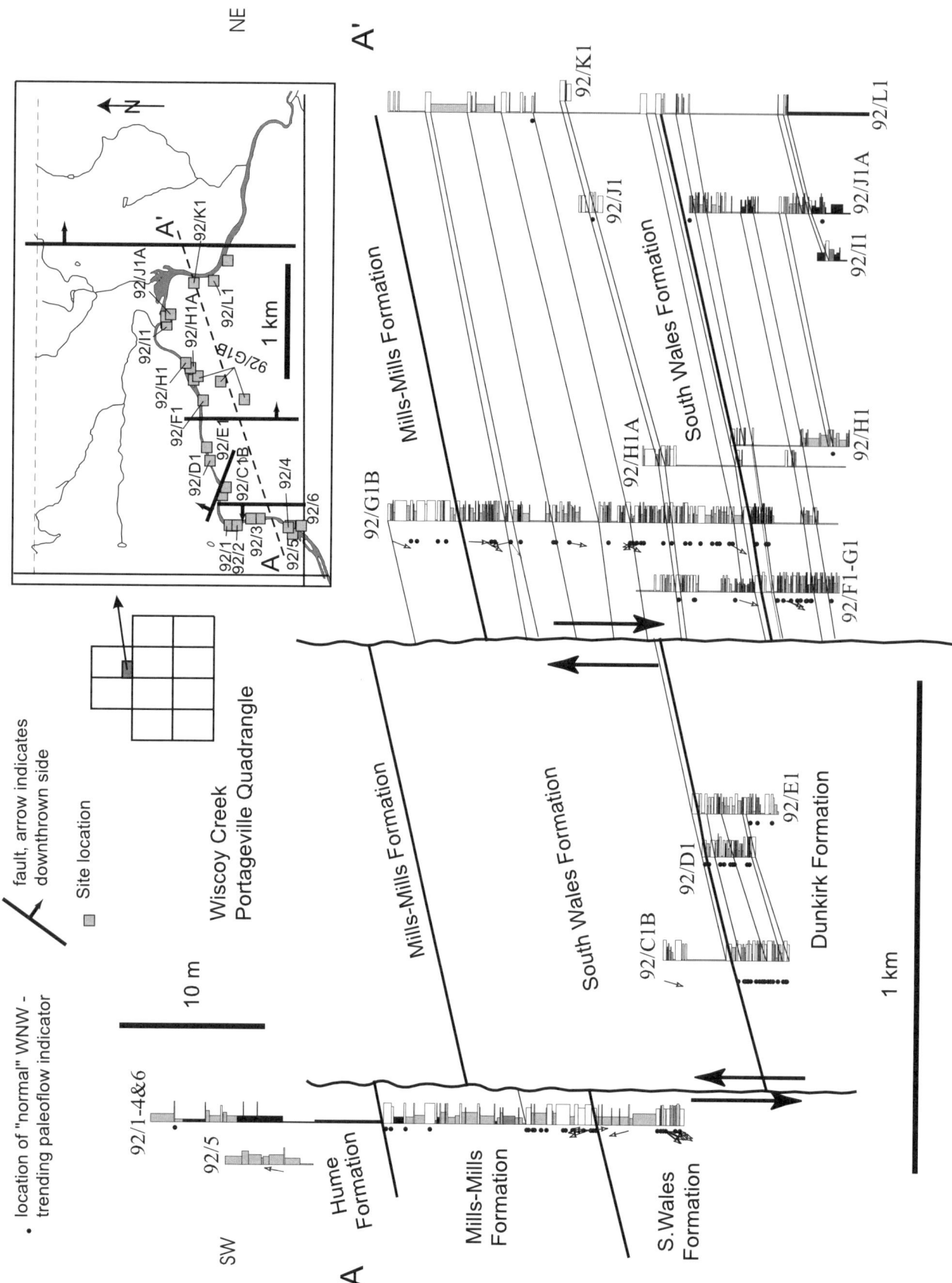

Figure 7. Cross section of Wiscoy Creek. Dots indicate beds that contained paleoflow indicators; arrows represent anomalous north-south–trending paleoflow. Large numbers (e.g., 92/G1B) are site designations.

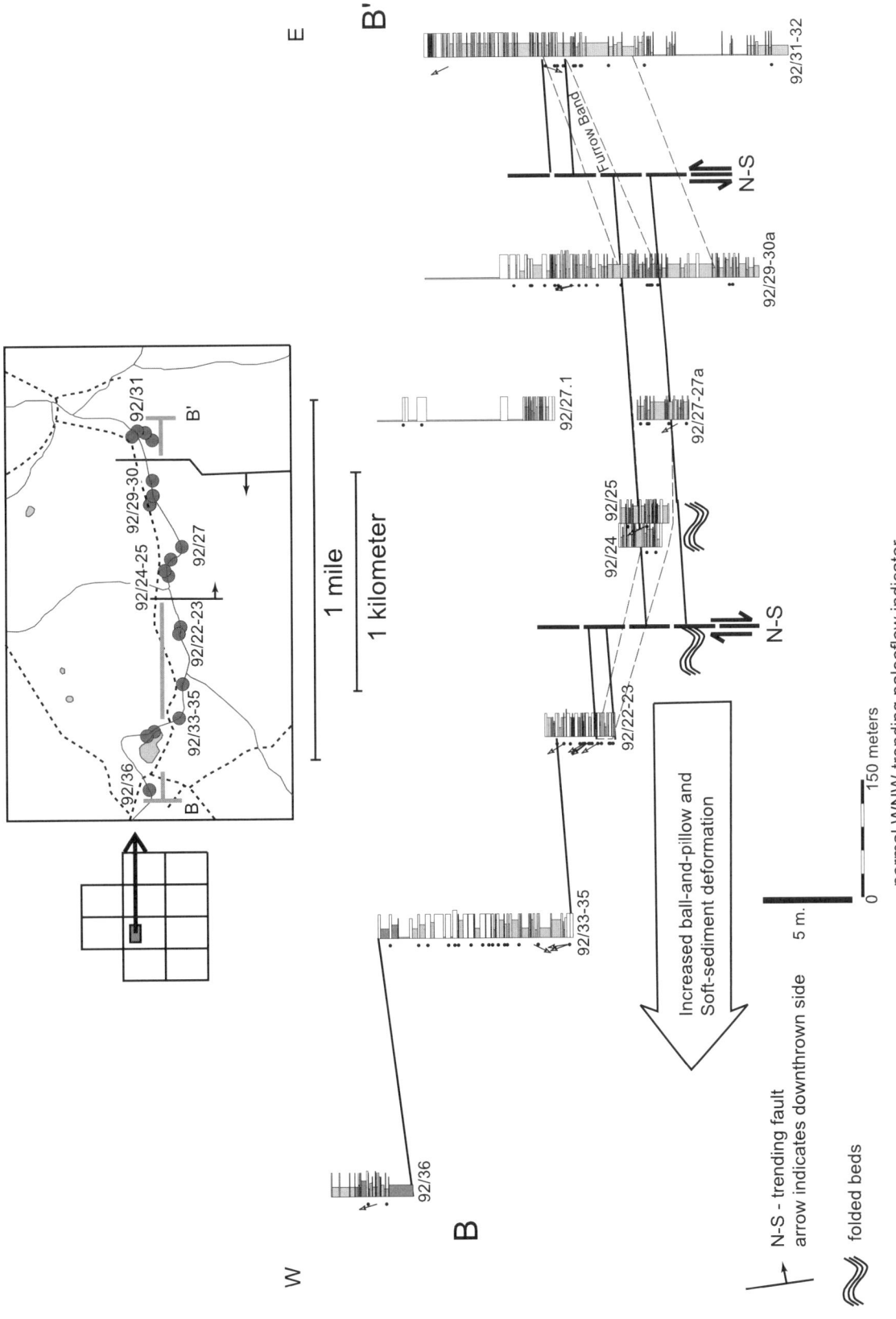

Figure 8. Cross section of the western extent of Sixtown Creek reconstructed for late Acadian time (prior to northeast-trending Alleghanian folding). Arrows represent anomalous north-south–trending paleoflow. Numbers are site locations.

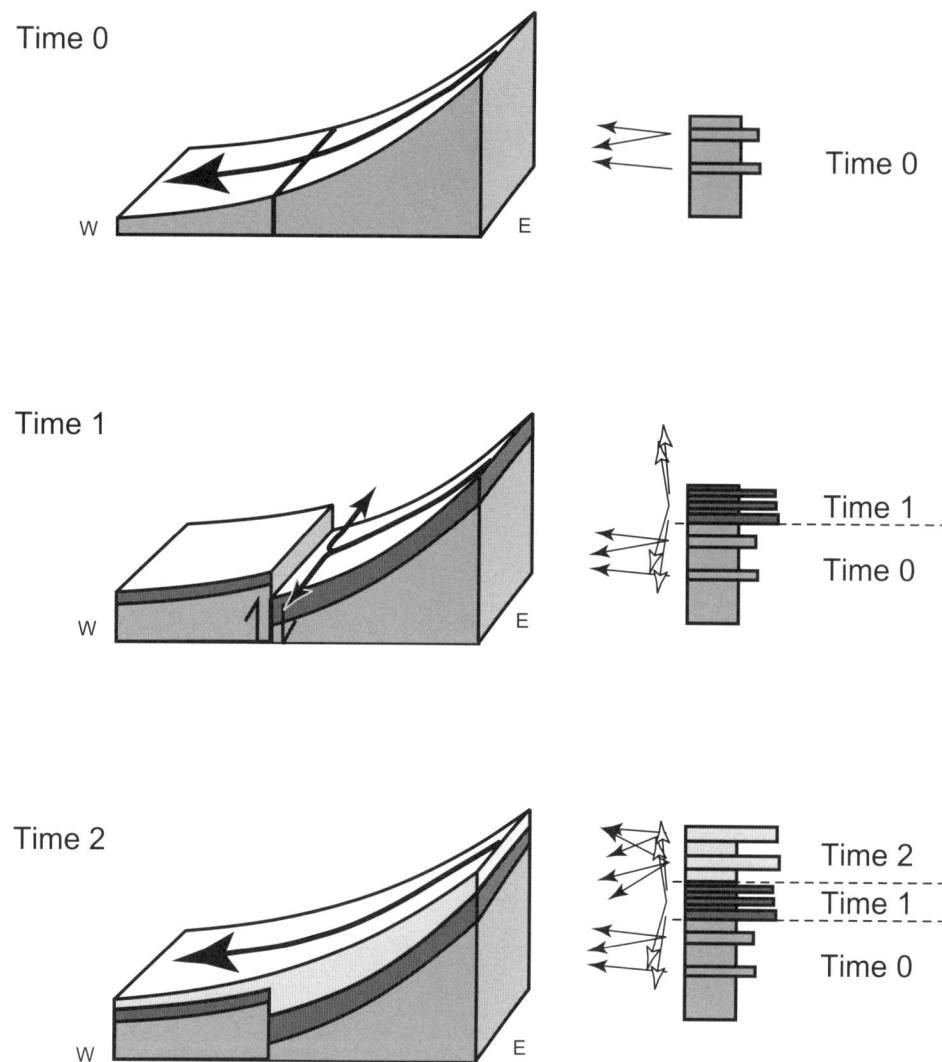

Figure 9. Simplistic model illustrating how fault activity may influence paleoflow directions. Time 0 is a period of unbroken topographic slope and normal west-trending paleoflow; the stratigraphic column displays west-trending indicators. In time 1, faulting produces a small fault-scarp offset, which blocks the normal west-trending paleoflow, redirecting it; the stratigraphic column displays anomalous-trending indicators overlying west-trending indicators. In time 2, continual deposition diminishes the effects of the scarp, allowing the normal western trend of the paleoflow; the stratigraphic column shows the west-trending indicators overlying anomalous-trending paleoflow indicators.

north-south Clarendon-Linden during the Late Devonian has been documented on the basis of (1) growth fault geometries and (2) soft-sediment deformation. We suggest that the anomalous-trending paleoflows resulted from short-lived fault scarps that redirected the paleoflows.

Assuming that fault activity did result in the anomalous-trending paleoflows, we have used the stratigraphic thicknesses of the units for which we have anomalous-paleoflow data to estimate the maximum and minimum fault displacements at the surface. From the maximum and minimum surface displacement, we were able to calculate the earthquake magnitude, using the curves of Bonilla et al. (1984), dePolo and Slemmons (1990), and Wells and Coppersmith (1994). We calculated seismic events of magnitudes between 6.0 and 6.9. The uncompacted stratigraphic thicknesses between anomalous paleoflows suggest that the average recurrence rate for the large earthquakes during the deposition of the Upper Devonian lower Canadaway Group ranged from 24 300 yr (at an accumulation rate of 10 cm/k.y.) to 8100 yr (at an accumulation rate of 30 cm/k.y.). Using the same methodology, we have also used the presence of seismites in the stratigraphic section to obtain an average recurrence rate for the upper Canadaway Group of 4800 to 14 400 yr.

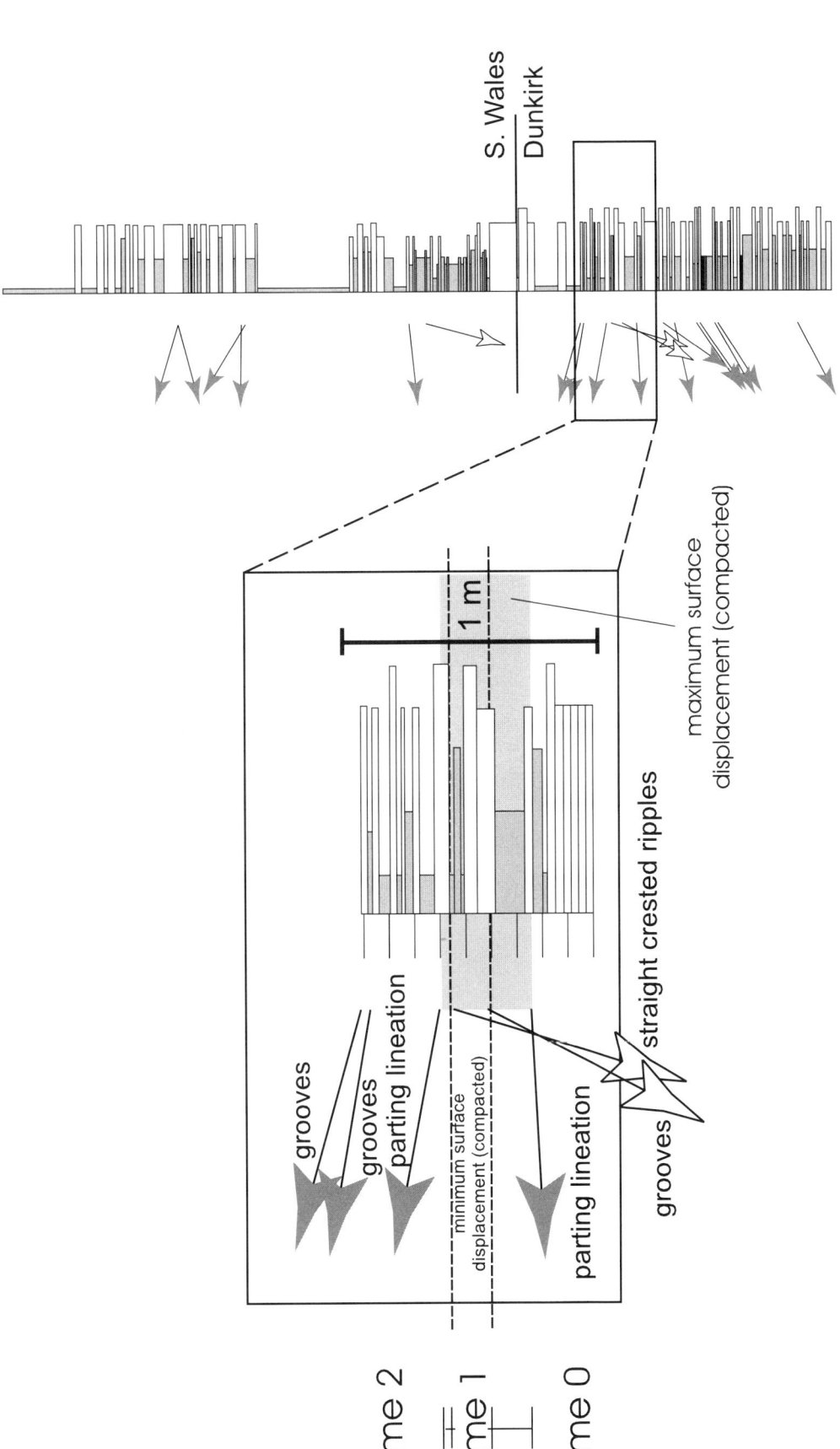

Figure 10. Example of the minimum and maximum surface displacements (both still compacted). The minimum surface displacement is the distance between the first anomalous paleoflow, in this case the straight crested ripples, and the last contiguous anomalous-paleoflow indicator, in this example the groove. Maximum surface displacement is measured from the last occurrence of "normal" paleoflow (lower parting lineation) to the occurrence of the first return to "normal" paleoflow (upper parting lineation). Values for the uncompacted section, minimum and maximum surface displacements, and estimated magnitudes are shown in Table 5.

TABLE 4. VALUES FOR ACCUMULATION AND DECOMPACTION RATES IN DEPOSIT AREAS ANALOGOUS TO STUDY AREA

Source	Year	Accumulation rate (cm/k.y.)	Unit	Geographical location	Time	Lithology
van Tassell	1987	30–140	Foreknobs Fm.	Catskill Delta	Late Devonian	Interbedded
van Tassell	1987	50	Minnehaha Springs Mbr.	Catskill Delta	Late Devonian	Interbedded
van Tassell	1987	65	Back Creek Siltstone Mbr.	Catskill Delta	Late Devonian	Interbedded
Bridge and Willis	1994	10	Moscow Fm.	Catskill Delta	Middle Devonian	Shale
Gautier	1982	9–16	Gammon Shale	Cretaceous Western Interior Seaway	Late Cretaceous	Shale
Source	Year	Decompaction factor	Unit	Geographical location	Time	Lithology
Gautier	1982	1.427	Gammon Shale	Cretaceous Western Interior Seaway	Late Cretaceous	Shale
Leckie and Potocki	1988	1.55–1.71	Colorado Shales	Cretaceous Western Interior Seaway	Late Cretaceous	Shale
Leckie and Potocki	1988	1.25	Colorado Shales	Cretaceous Western Interior Seaway	Late Cretaceous	Sandstone
Lobza and Schieber	1999	1.54	Chattanooga Shale	Appalachian Basin	Late Devonian	Shale

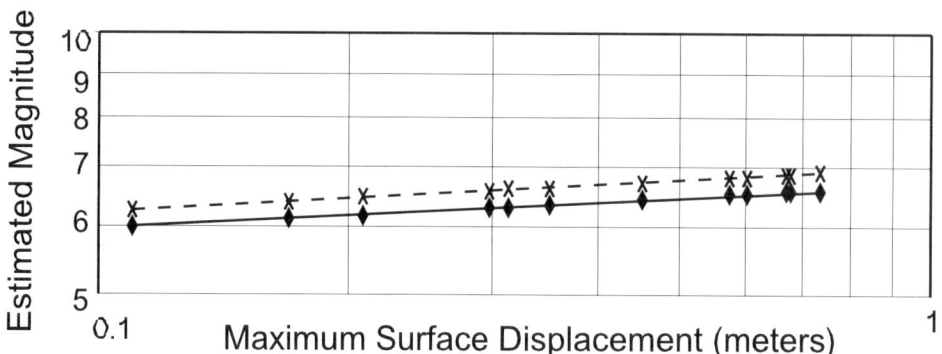

Figure 11. Maximum surface displacement plotted against estimated magnitude of the seismic event. Dashed line represents the curve of Bonilla et al. (1984) and dePolo and Slemmons (1990); solid line represents curve of Wells and Coppersmith (1994). Diamonds and Xs show where our data (Table 5) plot onto each curve.

TABLE 5. ESTIMATION OF EARTHQUAKE MAGNITUDES

Site	Formation	Stratigraphic thickness (cm)				Earthquake magnitude			
		Uncompacted shale	Uncompacted sandstone	Surface displacement (cm)		Minimum displacement		Maximum displacement	
				Minimum	Maximum	D&S	W&C	D&S	W&C
92/7	Wiscoy	7.21	3.75	10.96	N.A.	6.2	6.0	N.A.	N.A.
92/8	Wiscoy	23.32	8.04	31.36	N.A.	6.6	6.3	N.A.	N.A.
92/F1	Dunkirk	35.57	24.72	34.98	60.28	6.6	6.4	6.8	6.5
92/28	Dunkirk	6.21	14.65	20.86	N.A.	6.5	6.2	N.A.	N.A.
92/2	M.Mills	0.00	73.56	67.94	73.56	6.9	6.6	6.9	6.6
92/13	Hume	32.61	2.48	35.09	N.A.	6.6	6.4	N.A.	N.A.
93/267	Caneadea	62.56	4.41	29.68	66.97	6.6	6.3	6.9	6.6
92/29	Caneadea	34.71	10.56	16.94	45.27	6.4	6.1	6.7	6.4
93/280a	Caneadea	44.30	13.09	57.39	N.A.	6.8	6.5	N.A.	N.A.

Note: D&S = magnitudes determined from dePolo and Slemmons (1990); Ms (surface magnitude) = 7.00 + 0.782 (log D_{max}; D_{max} = maximum surface displacement); W&C = magnitudes determined from Wells and Coppersmith (1994); M_S = 6.69 + 0.74 (log D_{max}). N.A.—not applicable.

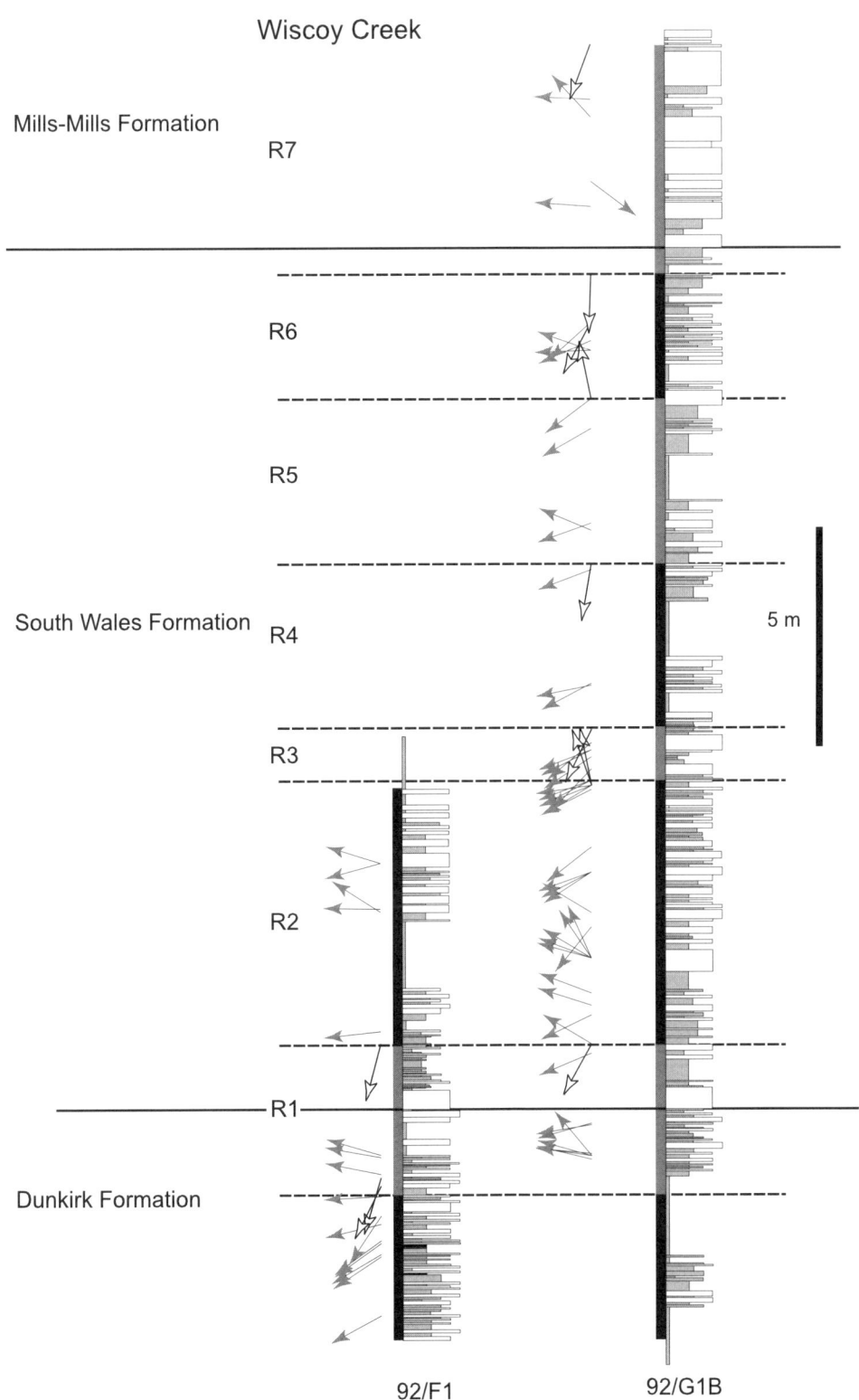

Figure 12. Recurrence rates determined from the stratigraphic sections at Wiscoy Creek. Seven intervals (R1–R7). Each interval is from the first occurrence of anomalous paleoflow (open-head arrows), to the next occurrence of noncontiguous anomalous paleoflow. Values for the uncompacted intervals and recurrence rates are shown in Table 6. Gray arrows represent "normal" west-northwest–trending paleoflow.

TABLE 6. ESTIMATION OF RECURRENCE RATES FOR SEISMIC EVENTS

Location	Site	Shale thickness uncompacted (cm)	Recurrence rates in years for different sedimentary accumulation rates				
			10 (cm/k.y.)	15 (cm/k.y.)	20 (cm/k.y.)	25 (cm/k.y.)	30 (cm/k.y.)
Mills-Mills	92/1	283.76	28 400	18 900	14 200	11 400	9 500
Mills-Mills	92/2	330.41	33 000	22 000	16 500	13 200	11 000
Mills-Mills	92/2	51.01	5 100	3 400	2 600	2 000	1 700
Wiscoy Falls	92/7	80.68	8 100	5 400	4 000	3 200	2 700
Wiscoy Falls	92/7	213.65	21 400	14 200	10 700	8 500	7 100
W. Sixtown	92/33	141.36	14 100	9 400	7 100	5 700	4 700
W. Sixtown	92/31	534.61	53 500	35 600	26 700	21 400	17 800
Wiscoy Creek	92/F1(R1)	287.88	28 800	19 200	14 400	11 500	9 600
Wiscoy Creek	92/G1B (R2)	419.94	42 000	28 000	21 000	16 800	14 000
Wiscoy Creek	92/G1B (R3)	69.70	7 000	4 600	3 500	2 800	2 300
Wiscoy Creek	92/G1B (R4)	385.79	38 600	25 700	19 300	15 400	12 900
Wiscoy Creek	92/G1B (R5)	480.71	48 000	32 000	24 000	19 200	16 000
Wiscoy Creek	92/G1BA (R6)	171.02	17 100	11 400	8 600	6 800	5 700
Wiscoy Creek	92/G1BA (R7)	231.85	23 200	15 500	11 600	9 300	7 700
Baker Creek	93/205a	137.61	13 800	9 200	6 900	5 500	4 600
White Creek	93/182	63.26	6 300	4 200	3 200	2 500	2 100
Average			24 300	16 200	12 100	9 700	8 100

Note: Rates are based on accumulation of shale between anomalous north-south–trending paleoflow indicators. Rate values have been rounded off to the nearest 100 yr.

TABLE 7. ESTIMATION OF RECURRENCE RATES BASED ON SHALE ACCUMULATION BETWEEN SEISMITES

Location	Site	Shale thickness uncompacted (cm)	Recurrence rates in years for different sedimentary accumulation rates				
			10 (cm/k.y.)	15 (cm/k.y.)	20 (cm/k.y.)	25 (cm/k.y.)	30 (cm/k.y.)
W. Sixtown	92/23	9.04	900	600	500	400	300
W. Sixtown	92/23	22.67	2 300	1 500	1 100	900	800
W. Sixtown	92/23	17.39	1 700	1 200	900	700	600
W. Sixtown	92/33	413.92	41 400	27 600	20 700	16 600	13 800
W. Sixtown	92/36	26.29	2 600	1 800	1 300	1 100	900
W. Sixtown	92/36	39.03	3 900	2 600	2 000	1 600	1 300
W. Sixtown	93/4	213.25	21 300	14 200	10 700	8 500	7 100
Shongo Creek	97/14	353.50	35 300	23 600	17 700	14 100	11 800
Shongo Creek	97/15	239.77	24 000	16 000	12 000	9 600	8 000
Shongo Creek	97/16	78.00	7 800	5 200	3 900	3 100	2 600
White Creek	93/182	62.79	6 300	4 200	3 100	2 500	2 100
Genesee River trib., Rt 19	93/191	140.24	14 000	9 300	7 000	5 600	4 700
Genesee River trib., Rt 19	93/191c	136.11	13 600	9 100	6 800	5 400	4 500
Genesee River trib., Rt 19	93/191c	157.40	15 700	10 500	7 900	6 300	5 200
Genesee River trib., Rt 19	93/191c	340.58	34 100	22 700	17 000	13 600	11 400
Baker Creek	93/205	108.11	10 800	7 200	5 400	4 300	3 600
Baker Creek	93/205	73.94	7 400	4 900	3 700	3 000	2 500
Baker Creek	93/205	117.78	11 800	7 900	5 900	4 700	3 900
Baker Creek	93/207	68.17	6 800	4 500	3 400	2 700	2 300
Baker Creek	93/207	27.46	2 700	1 800	1 400	1 100	900
Baker Creek	93/209	196.76	19 700	13 100	9 800	7 900	6 600
Baker Creek	93/217	65.68	6 600	4 400	3 300	2 600	2 200
Baker Creek	93/217	92.98	9 300	6 200	4 600	3 700	3 100
Rushford Lake	93/304b	4.68	500	300	200	200	200
Orchard Road	93/47	287.04	28 700	19 100	14 400	11 500	9 600
Murphey Hill Road	94/79b	61.62	6 200	4 100	3 100	2 500	2 100
Murphey Hill Road	94/81	195.78	19 600	13 100	9 800	7 800	6 500
Seavert Road	97/48	69.11	6 900	4 600	3 500	2 800	2 300
Seavert Road	97/49	545.38	54 500	36 400	27 300	21 800	18 200
Average (yrs)			14 400	9 600	7 200	5 700	4 800

Note: Rates have been rounded off to nearest 100 yr.

REFERENCES CITED

Beinkafner, K.J., 1983, Deformation of the subsurface Silurian and Devonian rocks of the Southern Tier of New York state [unpublished Ph.D. thesis]: Syracuse, New York, Syracuse University, 332 p.

Bonilla, M.G., Mark, R.K., and Lienkaemper, J.J., 1984, Statistical relations among earthquake magnitude, surface rupture length, and surface fault displacement: Bulletin of the Seismological Society of America, v. 74, p. 2379–2411.

Bridge, J.S., and Willis, B.J., 1994, Marine transgressions and regressions recorded in Middle Devonian shore-zone deposits of the Catskill clastic wedge: Geological Society of America Bulletin, v. 106, p. 1440–1458.

Burtner, R.L., 1963, Sediment dispersal patterns within the Catskill facies of southeastern New York and northeastern Pennsylvania, in Shepps, V.C., ed., Symposium on Middle and Upper Devonian Stratigraphy of Pennsylvania and Adjacent States: Pennsylvania Geological Survey Fourth Series Bulletin G39: p. 7–23.

Colton, G.W., 1967a, Late Devonian current directions in western New York, with special references to *Fucoides graphica*: Journal of Geology, v. 75, p. 11–22.

Colton, G.W., 1967b, Orientation of carbonate concretions in the Upper Devonian of New York: U.S. Geological Survey Professional Paper 575-B, p. B57–B59.

dePolo, C.M., and Slemmons, D.B., 1990, Estimation of earthquake size for seismic hazards, in Krinitzsky, E.L., and Slemmons, D.B., Neotectonics in earthquake evaluation: Boulder, Colorado, Geological Society of America Reviews in Engineering Geology, v. 8, p. 1–28.

Doig, R., 1990, 2300 yr history of seismicity from silting events in Lake Tadoussac Charlevoix, Quebec: Geology, v. 18, p. 820–823.

Fakundiny, R.H., Pomeroy, P.W., Pferd, J.W., and Nowack, T.A., 1978a, Structural instability features in the vicinity of the Clarendon-Linden Fault System, Western New York and Lake Ontario: Waterloo, University of Waterloo Press, SM Study No. 13, Paper 4, p. 121–176.

Fakundiny, R.H., Pferd, J.W., and Pomeroy, P.W., 1978b, Clarendon-Linden Fault System of Western New York: Longest (?) and oldest (?) active fault in Eastern United States: Geological Society of America Abstracts with Programs, v. 10, no. 2, p. 42.

Fletcher, J.B., and Sykes, L.R., 1977, Earthquakes related to hydraulic mining and natural seismicity in western New York State: Journal of Geophysical Research, v. 82, p. 3767–3780.

Gautier, D.L., 1982, Siderite concretions: Indications of early diagenesis in the Gammon Shale (Cretaceous): Journal of Sedimentary Petrology, v. 52, p. 859–871.

Gutmann, M.P., 1989, Upper Devonian turbidites in the South Wales Shale Member of the Perrysberg Formation [M.Sc. thesis]: Buffalo, State University of New York at Buffalo, 185 p.

Gutmann, M.P., and Jacobi, R.D., 1988, The Devonian South Wales Shale Member: A suprafan lobe?: Geological Society of America Abstracts with Programs, v. 20, no. 1, p. 24.

Jacobi, R.D., and Fountain, J.C., 1993, The southern extension and reactivations of the Clarendon-Linden Fault System: Géographie Physique et Quaternaire, v. 47, p. 285–302.

Jacobi, R.D., and Fountain, J.C., 1996, Determination of the seismic potential of the Clarendon-Linden Fault System in Allegany County, Final Report: Albany, New York State Energy Research and Development Authority, 2106 p., and 31 oversized maps.

Jacobi, R.D., and Fountain, J.C., 1998, Multiple reactivations of the Clarendon-Linden Fault System, Western New York: Geological Society of America Abstracts with Programs, v. 30, no. 7, p. A249.

Jacobi, R.D., and Zhao, M., 1996, Digital imaging of fractures: Evidence for Appalachian style tectonics in the Appalachian Plateau of Western New York: Geological Society of America Abstracts with Programs, v. 28, no. 3, p. 67.

Jacobi, R.D., Gutmann, M., Piechocki, A., Singer, J., O'Connell, S., and Mitchell, C.E., 1994, Upper Devonian turbidites in western New York: Preliminary observations and implications: New York State Geological Survey Bulletin, no. 481, p. 101–115.

Johnston, A.C., Coppersmith, K.J., Kanter, L.R., and Cornell, C.A., 1994, The earthquakes of stable continental regions, Volume 1, Assessment of large earthquake potential: Final report submitted to Electric Power Research Institute (EPRI), TR-102261, 173 p.

Keller, E.A., Bonkowski, M.S., Korsch, R.J., and Shlemon, R.J., 1982, Tectonic geomorphology of the San Andreas fault zone in the southern Indio Hills, Coachella Valley, California: Geological Society of America Bulletin, v. 93, p. 46–56.

Krause, R.G.F., and Geijer, T.A.M., 1987, An improved method for calculating the standard deviation and variance of paleocurrent data: Journal of Sedimentary Petrology, v. 57, p. 779–780.

Leckie, D.A., and Potocki, D., 1988, Sandstone dikes: An estimate of their depth of injection in the Colorado Shales (Cretaceous), Alberta, in James, D.P., and Leckie, D.A., eds., Sequences, stratigraphy, sedimentology: Surface and subsurface: Canadian Society of Petroleum Geologists, Memoir 15, p. 325–330.

Leeper, W.S., 1963, Interpretation of primary bedding structures in Mississippian and Upper Devonian rocks of southeastern Somerset County, Pennsylvania, in Shepps, V.C., ed., Symposium on Middle and Upper Devonian Stratigraphy of Pennsylvania and Adjacent States: Pennsylvania Geological Survey Fourth Series Bulletin G39, p. 165–181.

Lobza, V., and Schieber, J., 1999, Biogenic sedimentary structures produced by worms in soupy, soft muds: Observations from the Chattanooga Shale (Upper Devonian) and experiments: Journal of Sedimentary Research, v. 69, p. 1041–1049.

Lundegard, P.D., Samuels, N.D., and Pryor, W.A., 1985, Upper Devonian turbidite sequence, central and southern Appalachian basin: Contrasts with submarine fan deposits, in Woodrow, D.L., and Sevon, W.D., eds., The Catskill Delta: Geological Society of America Special Paper 201, p. 107–122.

Manspeizer, W., 1963, A study of the stratigraphy, paleontology, petrology and geologic history of the Canadaway and Conneaut Groups in Allegany County, New York [Ph.D. thesis]: New Brunswick, New Jersey, Rutgers University, 354 p.

Meckel, L.D., 1967, Origin of Pottsville Conglomerates (Pennsylvanian) in the Central Appalachians: Geological Society of America Bulletin, v. 78, p. 223–258.

Peters, T.W., 1998, Geologic mapping of the Rawson 7.5′ quadrangle in New York State: Characterization of multiple fault systems [M.A. thesis]: Buffalo, State University of New York, 161 p.

Rickard, L.V., 1973, Stratigraphy and structure of the subsurface Cambrian and Ordovician carbonates of New York: Albany, New York Museum and Science Service Map and Chart Serial 18, 26 p.

Sims, J.D., 1975, Determining earthquake recurrence intervals from deformational structures in young lacustrine sediments: Tectonophysics, v. 29, p. 141–152.

Slemmons, D.B., 1982, Determinations of design earthquake magnitudes for microzonation, in Proceedings of the 3rd International Earthquake Microzonation Conference: Seattle, Washington, v. 1, p. 119–130.

Smith, G.J., and Jacobi, R.D., 1998, Fault-influenced transgressive incised shoreface model for the Canadaway Group, Catskill Delta Complex: Journal of Sedimentary Research B, v. 68, p. 668–683.

Smith, G.J., and Jacobi, R.D., 1999, Syndepositional fault-controlled reservoir formation: Lowstand shoreface deposits in southwestern New York State: 1999 American Association of Petroleum Geologists Annual Meeting Official Program, v. 8, p. A131.

Smith, G.J., and Jacobi, R.D., 2001, Tectonic and eustatic signals in the sequence stratigraphy of the Upper Devonian Canadaway Group, New York State: American Association of Petroleum Geologists Bulletin, v. 85, p. 325–327.

Smith, G.J., Jacobi, R.D., Peters, T.W., Reay, M.L., Zack, D.L., and Zhao, M., 1998, Stratigraphic and structural analyses of five 7½′ topographic quadrangles in western New York State: Geological Society of America 33rd Annual Northeastern Section Abstracts with Programs, v. 30, p. 75.

Sorauf, J.E., 1965, Flow rolls of Upper Devonian rocks of south-central New York State: Journal of Sedimentary Petrology, v. 35, p. 553–563.

Sutton, R.G., Bowen, Z.P., and McAlester, A.L., 1970, Marine shelf environment of the Upper Devonian Sonyea Group of New York: Geological Society of America Bulletin, v. 81, p. 2975–2992.

Tucker, R.D., Bradley, D.C., Ver Straeten, C.A., Harris, A.G., Ebert, J.R., and McCutcheon, S.R., 1998, New U-Pb zircon ages and the duration and division of Devonian time: Earth and Planetary Science Letters, v. 158, p. 175–186.

Tuttle, M.P., Dyer-Williams, K., and Barstow, N., 1996, Seismic hazard implications of a paleoliquefaction study along the Clarendon-Linden Fault System in western New York State: Geological Society of America Abstracts with Programs, v. 28, n. 3, p. 106.

Van Tassell, J., 1987, Upper Devonian Catskill Delta margin cyclic sedimentation: Brallier, Scherr, and Foreknobs Formations of Virginia and West Virginia: Geological Society of America Bulletin, v. 99, p. 414–426.

Van Tyne, A.M., 1975, Clarendon-Linden structure, western New York: Albany, New York State Geological Survey Open-File Report 10, 12 p.

Van Tyne, A.M., Kamakaris, D.G., and Corbo, S., 1980a, Structure contours on the base of the Dunkirk: Albany, New York State Museum and Science Service, Geological Survey, Alfred Oil and Gas Office, Morgantown Energy Technology Center/Eastern Gas Shale Project series 111, 1 map, scale 1:250 000, 1 sheet.

Van Tyne, A.M., Kamakaris, D.G., and Corbo, S., 1980b, Structure contours on the base of the Java Formation: Albany, New York State Museum and Science Service, Geological Survey, Alfred Oil and Gas Office, Morgantown Energy Technology Center/Eastern Gas Shale Project series 112, 1 map, scale 1:250 000, 1 sheet.

Van Tyne, A.M., Kamakaris, D.G., and Corbo, S., 1980c, Structure contours on the base of the West Falls Formation: Albany, New York State Museum and Science Service, Geological Survey, Alfred Oil and Gas Office, Morgantown Energy Technology Center/Eastern Gas Shale Project series 113, 2 maps, scale 1:250 000, 2 sheets.

Van Tyne, A.M., Kamakaris, D.G., and Corbo, S., 1980d, Structure contours on the base of the Sonyea Group: Albany, New York State Museum and Science Service, Geological Survey, Alfred Oil and Gas Office, Morgantown Energy Technology Center/Eastern Gas Shale Project series 114, 2 maps, scale 1:250 000, 2 sheets.

Van Tyne, A.M., Kamakaris, D.G., and Corbo, S., 1980e, Structure contours on the base of the Genesee Group: Albany, New York State Museum and Science Service, Geological Survey, Alfred Oil and Gas Office, Morgantown Energy Technology Center/Eastern Gas Shale Project series 115, 2 maps, scale 1:250 000, 2 sheets.

Van Tyne, A.M., Kamakaris, D.G., and Corbo, S., 1980f, Isopach of radioactive shale in the Hamilton Group: Albany, New York State Museum and Science Service, Geological Survey, Alfred Oil and Gas Office, Morgantown Energy Technology Center/Eastern Gas Shale Project series 116, 2 maps, scale 1:250 000, 2 sheets.

Van Tyne, A.M., Kamakaris, D.G., and Corbo, S., 1980g, Isopach map of Java Formation: Albany, New York State Museum and Science Service, Geological Survey, Alfred Oil and Gas Office, Morgantown Energy Technology Center/Eastern Gas Shale Project series 117, 1 map, scale 1:250 000, 1 sheet.

Van Tyne, A.M., Kamakaris, D.G., and Corbo, S., 1980h, Isopach map of West Falls Formation: Albany, New York State Museum and Science Service, Geological Survey, Alfred Oil and Gas Office, Morgantown Energy Technology Center/Eastern Gas Shale Project series 118, 1 map, scale 1:250 000, 1 sheet.

Van Tyne, A.M., Kamakaris, D.G., and Corbo, S., 1980i, Isopach map of Sonyea Group: Albany, New York State Museum and Science Service, Geological Survey, Alfred Oil and Gas Office, Morgantown Energy Technology Center/Eastern Gas Shale Project series 119, 2 maps, scale 1:250 000, 2 sheets.

Van Tyne, A.M., Kamakaris, D.G., and Corbo, S., 1980j, Isopach map of Genesee Group: Albany, New York State Museum and Science Service, Geological Survey, Alfred Oil and Gas Office, Morgantown Energy Technology Center/Eastern Gas Shale Project series 120, 2 maps, scale 1:250 000, 2 sheets.

Van Tyne, A.M., Kamakaris, D.G., and Corbo, S., 1980k, Isopach map of radioactive shale in the Perrysburg Formation: Albany, New York State Museum and Science Service, Geological Survey, Alfred Oil and Gas Office, Morgantown Energy Technology Center/Eastern Gas Shale Project series 125, 1 map, scale 1:250 000, 1 sheet.

Walker, R.G., and Sutton, R.G., 1967, Quantitative analysis of turbidites in the Upper Devonian Sonyea Group, New York: Journal of Sedimentary Petrology, v. 37, p. 1012–1022.

Wells, D.L., and Coppersmith, K.J., 1994, New empirical relationships among magnitude, rupture length, rupture width, rupture area, and surface displacement: Bulletin of the Seismological Society of America, v. 74, p. 621–653.

Woodrow, D.L., 1968, Stratigraphy, structure and sedimentary patterns in the Upper Devonian of Bradford County, Pennsylvania: Pennsylvania Geological Survey, 4th Series, General Geology Report, G54, 78 p.

Woodrow, D.L., and Isley, A.M., 1983, Facies, topography, and sedimentary processes in the Catskill Sea (Devonian), New York and Pennsylvania: Geological Society of America Bulletin, v. 94, p. 459–470.

Zack, D.L., 1998, Geologic mapping of the Freedom 7½′ topographical quadrangle in southwestern New York state: Evidence for multiple fault systems in the Appalachian Basin [M.A. thesis]: Buffalo, State University of New York at Buffalo, 163 p.

MANUSCRIPT ACCEPTED BY THE SOCIETY MAY 11, 2001

Seismically induced soft-sediment deformation in some Silurian carbonates, eastern U.S. Midcontinent

Charles M. Onasch
Charles F. Kahle
Department of Geology, Bowling Green State University, Bowling Green, Ohio 43403, USA

ABSTRACT

Middle and Upper Silurian carbonates in western Ohio and southeastern Michigan contain an extensive array of soft sediment deformation structures: concordant and discordant breccias, folds and faults, homogenized zones, sediment intrusions, and fault-graded beds formed by hydroplastic deformation, liquefaction, and fluidization. The variety of behaviors indicated by the range of deformational styles is a consequence of the varying degrees of cementation at the time of deformation. The unlithified nature of the sediments at the time of deformation, the type and large areal distribution of structures, and similarity to structures related to modern earthquakes indicate that the deformation is seismically induced. The source of the seismicity is believed to be nearby faults associated with the Grenville front which are known to have been active during deposition of Silurian strata.

INTRODUCTION

Soft-sediment deformation is a common feature in sedimentary rocks deposited in a wide range of environments. Although most abundant in siliciclastic rocks, it is a common feature in carbonate rocks as well (e.g., Weaver and Jeffcoat, 1978; Plaziat et al., 1990; Dugue, 1995; Pope et al., 1997). It can result from a variety of causes, including: (1) density inversions (Lowe, 1975; Owen, 1987); (2) rapid sediment loading (Lowe, 1975, 1976); (3) groundwater movement (Jeyapalan et al., 1983); (4) wave action (Dalrymple, 1980; Nataraja and Gill, 1983; Seed, 1968); (5) drag by currents (McKee et al., 1962); and (6) seismic activity (Youd, 1978).

Deformation of unlithified or incompletely lithified, water-saturated sediments can occur by one of three mechanisms: hydroplastic deformation, liquefaction, and fluidization (Lowe, 1975). Hydroplastic deformation involves plastic deformation of cohesive sediments. The cohesion can be a result of partial cementation or normal cohesive forces characteristic of clay-rich, fine-grained sediments. Liquefaction and fluidization occur in sediments that behave as a liquid (Allen, 1982). The loss in shear strength is most often a result of an increase in pore-fluid pressure. Although an increase in pore pressure can be externally controlled by fluctuations in groundwater level, more often it is internally controlled through the collapse of framework grains. As framework grains reach a more efficient packing arrangement during compaction, the pore volume is reduced, causing an increase in pore-fluid pressure and a decrease in strength (Terzaghi, 1947). When this breakdown occurs more rapidly than the pore fluids can escape, the sediment undergoes liquefaction as the grains are suspended in the pore fluid until they reach a new packing arrangement (Lowe, 1975; Owen, 1987). Fluidization occurs if the pore fluid is expelled at a velocity sufficient to overcome the weight of the grains (Lowe, 1975; Owen, 1987). The breakdown in framework grain arrangement requires an input of energy, or trigger (Owen, 1987). Triggers include rapid sediment loading, artesian groundwater flow, wave action, slope failure, flood surges, and seismic activity. Once liquefied or fluidized, external forces such as gravity, differential loading, glacial loading, shear from currents, or shear from seismic waves will deform the sediment into a variety of structures.

Seismic activity has been documented or implied as an important cause of soft-sediment deformation (Lowe, 1975;

Owen, 1987; Moss and Howells, 1996; Pope et al., 1997). The term "seismite" was proposed by Seilacher (1969) to refer to beds with seismically induced soft-sediment deformation structures. In Holocene sediments, the recognition of seismites is important for assessing the seismic risk associated with faults. In the rock record, seismites can be used to develop a better understanding of the tectonic history of an area. To date, seismites have been most often recognized in siliciclastic rocks (e.g., Laird, 1968; Seilacher, 1969, 1984; Hempton and Dewey, 1983; Mohindra and Bagati, 1996; Moss and Howells, 1996; Plaziat and Ahmamou, 1998; Vanneste et al., 1999). Relatively few have been reported in carbonate rocks (e.g., Weaver and Jeffcoat, 1978; Alvarez et al., 1985; Plaziat et al., 1990; Pratt, 1994; Dugue, 1995; Pope et al., 1997). Whether this is a function of the relative abundance of the two lithologies or some fundamental difference in mechanical behavior is unclear. In any case, more attention must be directed to the study of soft-sediment deformation structures in carbonates and their possible association with seismic activity.

Certain carbonate rocks in the eastern U.S. Midcontinent contain abundant soft-sediment deformation structures. Many of these structures, such as breccias, have been previously ascribed to a variety of nonseismic processes such as karsting and solution collapse (e.g., Carman, 1946; Sparling, 1970; Johnson, 1974). However, we believe that many characteristics of these rocks support a seismic origin for the soft-sediment deformation. In this chapter, we evaluate the hypothesis that many soft-sediment deformation structures in some Silurian carbonate rocks found in western Ohio and southeastern Michigan are the products of syndepositional seismic activity on regional faults.

GENERAL GEOLOGY AND STRUCTURAL SETTING

Soft-sediment deformation structures are common in middle and upper Silurian rock units exposed along the Cincinnati and Findlay arches in western Ohio and southeastern Michigan (Fig. 1). The units examined include, from oldest to youngest, the Lilley Formation, Tymochtee Dolomite, and Bass Islands Dolomite (Fig. 2 and Table 1). The units examined are exposed in the following stone quarries: (1) Bass Islands Dolomite at Monroe, Michigan; (2) Bass Islands Dolomite at Ottawa Lake, Michigan; (3) Tymochtee and Bass Islands dolomites at Waterville, Ohio; (4) Bass Islands Dolomite at Salisbury, Ohio (also known as the Holland quarry; Carman, 1927); and (5) Lilley Formation at Peebles, Ohio (Fig. 1). Locations 1–4 are on the west flank of the Findlay Arch, and location 5 is on the east flank of the Cincinnati Arch. All of the locations are within 30 km of the Grenville Front (Fig. 1), and the quarries at Ottawa Lake, Waterville, and Salisbury are within 8 km of the Bowling Green fault (Fig. 1). Several faults are exposed in and around the Peebles quarry with displacements on the order of several meters (Schmidt et al., 1961; Swineford, 1985). Of the quarries studied, all are active except Monroe and Salisbury.

SOFT-SEDIMENT DEFORMATION FEATURES

A wide variety of soft-sediment deformation features are present in the rocks examined. Where outcrop permits, it can be shown that these features are restricted to certain horizons that extend for considerable distances. The structures, in order of decreasing abundance, include breccias, folds and faults, homogenized zones, sediment intrusions, and fault-graded beds.

Breccias

Breccias are the most abundant soft-sediment deformation feature in the rocks examined. They are common in all units, but particularly so in the Bass Islands Dolomite, where they locally compose up to 50% of the unit by volume. In the Lilley Formation, soft-sediment deformation is represented almost entirely by breccias. In all units, breccias range from clast to matrix supported (Fig. 3A), and monomict (Fig. 3A) to polymict (Fig. 3B) breccias. Clast size ranges from a few millimeters to 40 cm. Individual clasts vary from subrounded (Fig. 3, A and B) to highly angular (Fig. 3C); many are bent or have tails that taper off into the enclosing matrix (Fig. 3, D and E). Where breccias are clast supported, the clasts tend to have stylolitic contacts. In outcrop, breccias may be concordant with bedding or highly discordant. Discordant breccias occur as irregular masses, tabular dikes (Fig. 3F), or cylindrical pipes. In some cases, coherent beds progressively break up into concordant and then discordant breccias (Fig. 4A). Contacts with unbrecciated strata can be sharp (Fig. 3, B and F) or gradational (Figs. 3A and 4A). No preferred orientation or imbrication of the clasts was observed.

Folds and faults

Where the Bass Islands and Tymochtee dolomites are laminated, folds and faults are commonly observed within individual horizons that are bounded top and bottom by undeformed strata. Folds vary in wavelength from centimeters to meters, most being highly disharmonic. Some regular wave trains are present, but most folds are isolated or are rootless (Fig. 3E). Overturning directions vary inconsistently, reversals in the space of a few meters being common.

Faults occur on scales from centimeters to meters and exhibit both normal and reverse senses of displacement. Faults with opposite offset sense occur side by side (Fig. 4B). In some outcrops, larger listric faults with small displacements extend tens of meters (Fig. 4C). The listric faults generally dip to the north, indicating a general transport of material to the north. In some cases, discordant breccias are developed along or are truncated by the listric faults (Fig. 4D).

Sediment intrusions

Discordant and concordant masses of sediment, interpreted to be intrusive into the surrounding beds, are common in the Bass

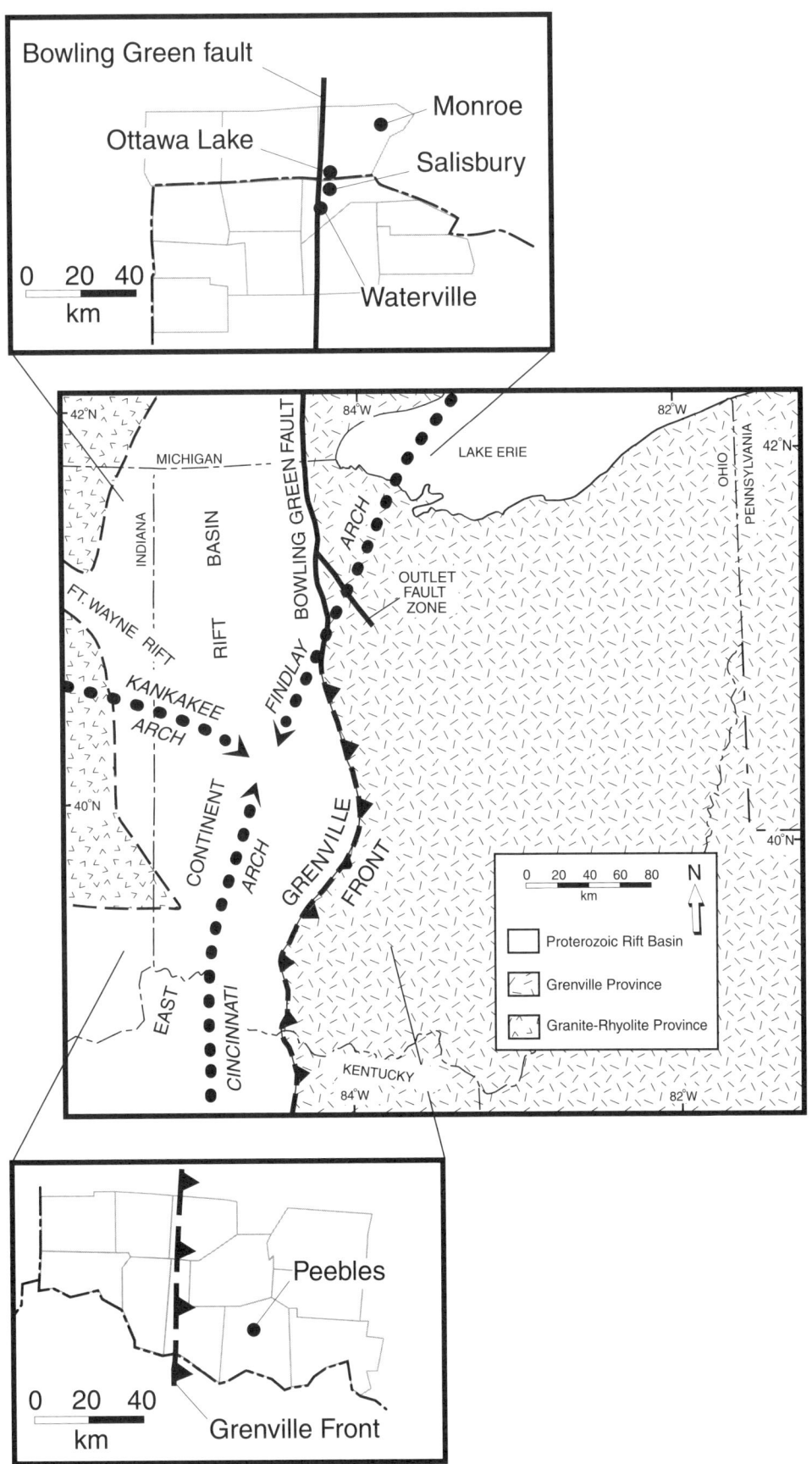

Figure 1. Tectonic setting of western Ohio and southeastern Michigan, and locations of quarries described in this study and their proximity to nearby structural features.

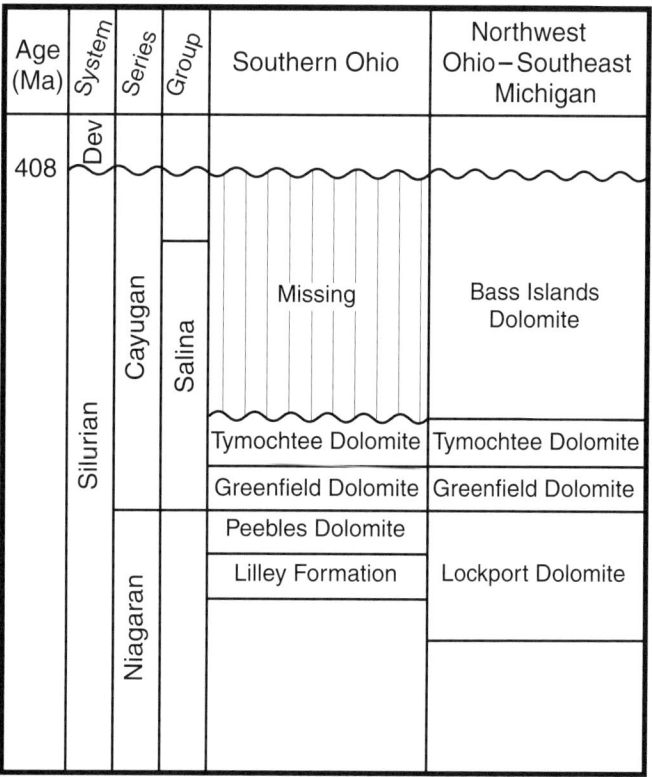

Figure 2. Stratigraphic column for Silurian rocks in northwestern Ohio–southeastern Michigan and southern Ohio. Modified from Hansen (1998).

Islands Dolomite in all locations examined. The intrusive nature is shown by sediment fill exotic to the enclosing strata fill that can, in some cases, be traced to the nearby beds. The fill ranges from clast- or matrix-supported breccia to moderately well sorted, fine-grained carbonate mud. Discordant intrusions are found as irregular masses, dikes, and pipes. Concordant intrusions (sills) occur in association with the discordant intrusions where they extend away from discordant masses into the surrounding bedding (Fig. 4E). In the discordant intrusions, the general direction of sediment transport was upward, as indicated by upturned bedding along the margins of the intrusions (Fig. 3F) and by tracing the source of clasts. Some intrusions have downturned bedding along the margins, indicating localized downward movement of the sediment.

Homogenized zones

Many beds consist of structureless dolomite. Although this texture may be primary and depositional in some beds, spatial variations in the development of this texture suggest that homogenization occurred after deposition. Single beds grade laterally and/or vertically from well laminated to disrupted laminations to structureless (Fig. 4F).

Fault-graded beds

In his original description of seismites, Seilacher (1969) described a soft-sediment deformation structure that he termed fault-graded bedding. It consisted, from top to bottom, of a homogenized zone, a rubble zone where beds were progressively disrupted, a segmented zone with faults whose displacement dies out with depth, and undisturbed sediment. This progressive change from brittle to ductile deformation is common in the Bass Islands Dolomite (Fig. 5A).

DISCUSSION

Physical state of sediment at the time of deformation

Previous studies of the structures described here, in particular the breccias, attributed the deformation to processes other than seismic activity. Sparling (1970) and Carman (1927) believed that most of the breccias in the Bass Islands Dolomite of northwest Ohio and southeast Michigan were a product of karst formation. In the same area, Johnson (1974) also argued for karst collapse, but added that solution collapse of underlying evaporites contributed to formation of some breccias. A key difference between the interpretations of these authors and the seismite hypothesis being evaluated here is the physical state of the material at the time of deformation. In karst or solution-collapse models, deformation would have taken place after lithification whereas in a seismite model, it would have occurred prior to lithification. Many of the features described in the preceding sections argue for deformation of incompletely lithified sediments.

The bent shape of many breccia clasts, along with the tails extending into the matrix (Fig. 3, D and E) show that many clasts were incompletely lithified at the time of brecciation. The progressive disruption of single beds into breccias and then into homogenized sediment (Fig. 4F) argues the same. Sediment intrusions could be collapse features, but sediment transport directions, as indicated by upturned margins (Fig. 3F) and clast provenance, show that sediment moved up, not down, in most cases. Finally, the association of the breccias with a wide range of other soft-sediment structures such as the folds and faults and with fault-graded bedding indicates that brecciation was accomplished prior to lithification.

Although the sediment at the time of deformation was generally unlithified, there was significant variation in the degree of lithification. The shape of breccia clasts varies from blocky and angular (Fig. 3C) to rounded with wispy tails (Fig. 3D), indicating that lithification was nearly complete in some beds, whereas adjacent beds were completely unlithified. The variation in the degree of lithification at the time of deformation appears to be related to grain size. Examination of thin sections shows that the grain size of the dolomite in the angular clasts is finer (4–8 μm) than that in the rounded or plastically deformed clasts (5–20 μm). Although the present grain size of the dolomite is likely not the same as that of the original carbonate sediment, it

TABLE 1. SUMMARY OF CHARACTERISTICS OF MIDDLE AND UPPER SILURIAN UNITS EXAMINED

Unit	Thickness	Lithologies	Sedimentary structures	Depositional environment	References
Bass Islands Dolomite	Up to 150 m	Bedded dolomite, massive, brecciated dolomite, and argillaceous dolomite with carbonaceous partings	Overall lack of fauna, mottling, oolites, stratiform stromatolites, flat pebble conglomerate, cut and fill, amalgamated bedding, low-angle erosion surfaces, mud cracks, and bipolar cross-stratification	Tidal flat	Carman (1927), Sparling (1970), Johnson (1974), Carlson (1992)
Tymochtee Dolomite	20–35 m	Thinly bedded to laminated gray dolomite containing abundant argillaceous and bituminous parting laminae	Stromatolites, rare thrombolites, mud cracks, storm layers, flat pebble conglomerates, cross-bedding, molds and pseudomorphs of calcite after gypsum, gypsum molds, anhydrite molds, and raindrop impressions	Tidal flat	Kahle and Floyd (1971), Court and Kahle (1993), Kahle (1997)
Lilley Formation	15–24 m	Extremely argillaceous, thinly bedded, blue-gray dolomite to light gray, pure, thick-bedded dolomite that may contain abundant fossil fragments	Bipolar cross stratification	Intertidal to shallow subtidal	Schmidt et al. (1961), Horvath (1969), Swineford (1985), Kleffner (1990)

should be proportional. If so, then beds composed of fine-grained sediment behaved in a more brittle fashion than those with a coarser grain size, indicating that grain size may have been a primary control on the style of soft-sediment deformation.

Origin of soft-sediment deformation features in carbonates of the eastern U.S. Midcontinent

We evaluate the seismite hypothesis by first considering the mechanisms of soft-sediment deformation and then the nature of the deforming stresses. As reviewed in the Introduction, two prerequisites are necessary for soft-sediment deformation: reduction in sediment strength and a deviatoric stress. The reduction in strength is most likely accomplished by an increase in pore-fluid pressure, leading to liquefaction or fluidization. The triggers for liquefaction and/or fluidization include (1) rapid sediment loading, (2) artesian groundwater flow, (3) wave action (Dalrymple, 1979, 1980), (4) slope failure, (5) flood surges, and (6) seismic activity (Lowe, 1975; Owen, 1987; Moss and Howells, 1996; Pope et al., 1997). Of these, rapid sediment loading, groundwater flow, slope failure, and flood surges can be ruled out in the rocks examined. Rates of carbonate sedimentation on carbonate tidal flats can range from slow to fast (Wright, 1984). Rapid rates of sedimentation on carbonate tidal flats can take place during storms, especially hurricanes (Hardie, 1977; Wright, 1984; Wanless et al., 1988), but the thickness of single storm deposits is typically only 5 mm to 2 cm (Hardie, 1977; Wanless et al., 1988). In modern environments, such a thickness of sediment is not known to create soft-sediment deformation structures in underlying sediments (Hardie, 1977; Wanless et al., 1988). Analogy suggests that the same processes and thicknesses are applicable to Silurian carbonate tidal flats represented by the Bass Islands and Tymochtee Dolomites reported herein, and that rapid sediment loading can be ruled out as a cause of soft-sediment deformation in these units. The low relief and lack of nearby highlands would preclude any significant topographically driven groundwater flow. Except for tidal channels, there would be no significant topographic relief to cause slope failure. Therefore, we are left with wave action and seismic activity as possible triggers. Of these, seismic activity is better able to produce widespread horizons with soft-sediment deformation.

The second prerequisite for soft-sediment deformation is a deviatoric stress sufficient to deform the sediment. Common sources of stress include (1) gravitational instabilities deriving from topographic slopes or density inversions, (2) uneven loading from topographic relief on the sediment surface, (3) drag from currents, debris flows, or glaciers, and (4) shear stress from the propagation of seismic waves (Lowe, 1975; Owen, 1987). Of these, currents, debris flow, or glacial drag can be ruled out, in view of the tidal-flat environment of deposition. In considering topography, only a very low slope may be necessary for failure in liquefied or fluidized sediments. Field et al. (1982) found that slopes as low as 0.25° may be subject to failure. The Bass Islands and Tymochtee dolomite exposures in northwest Ohio and southeast Michigan are located on the southern margin of the Michigan basin. The center of this basin is known to have been deeper than its margins (Nurmi and Friedman, 1977), so the slope may have been sufficient to allow weakened sediments to slide northward. The listric normal faults observed in the Monroe and Ottawa Lake quarries could be a manifestation of this process. Locally, the banks of tidal channels could fail and produce soft-sediment deformation, but

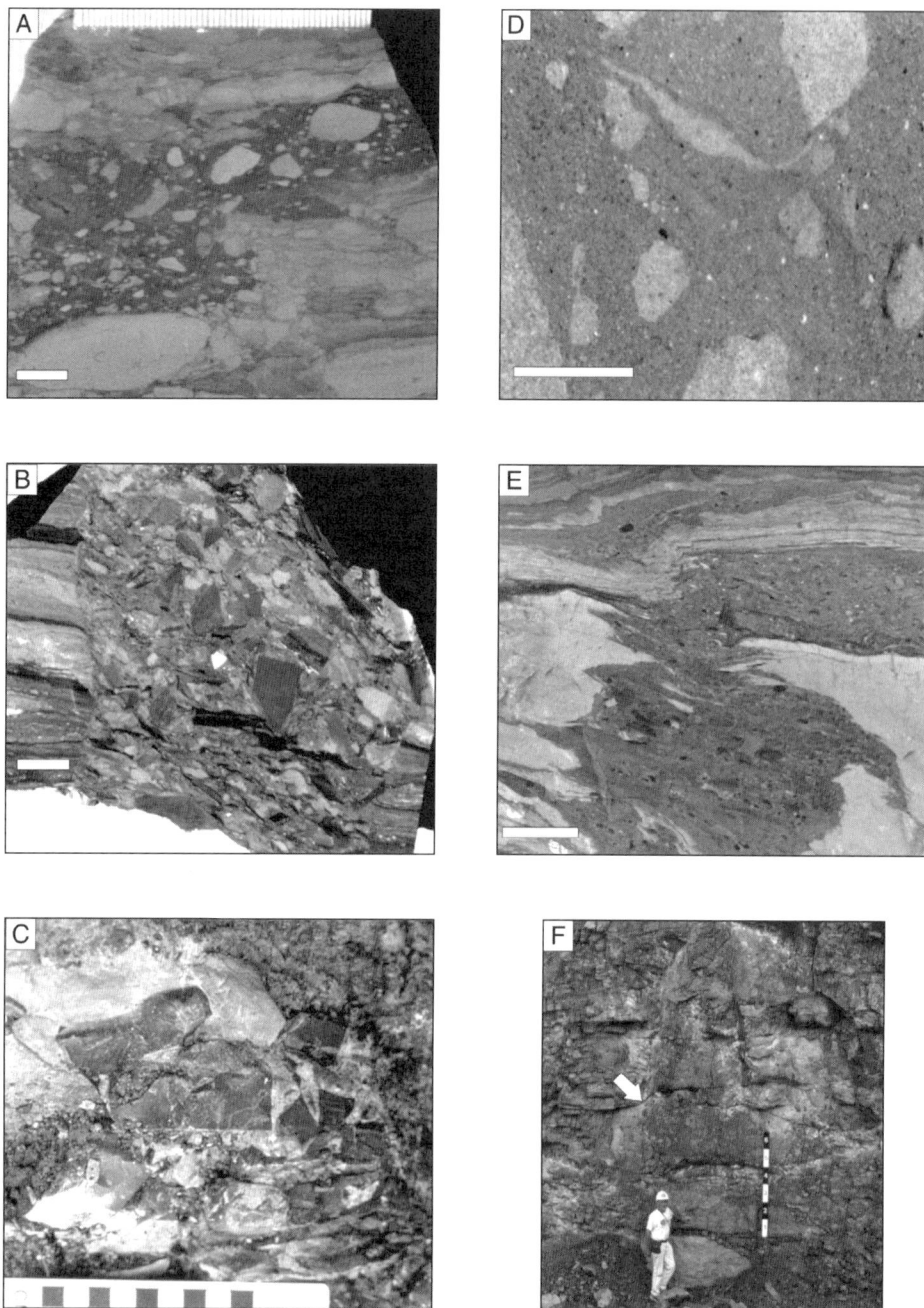

Figure 3. Soft-sediment deformation features. A: Monomict breccia with moderately rounded clasts ranging from clast-supported on the right to matrix-supported on the left. Clasts are derived from adjacent beds. Bass Islands Dolomite. Scale bar = 1 cm. B: Polymict breccia dike cutting bedding (seen on extreme left) at high angle in Bass Islands Dolomite. Scale bar = 1 cm. C: Angular breccia clasts from Lilley Formation. D: Photomicrograph of breccia in Bass Islands Dolomite. Note highly elongate clast with curved tail. Scale bar = 100 μm. E: Separation of beds in early stage of breccia formation in Bass Islands Dolomite. Note highly ductile nature of light-colored bed. Soft-sediment folds are visible near top of photo. Scale bar = 10 cm. F: Discordant breccia mass in Bass Islands Dolomite. Subvertical contact to left of person separates horizontally bedded Bass Islands Dolomite on left from discordant breccia to right. Note upturn of bedding along contact (arrow).

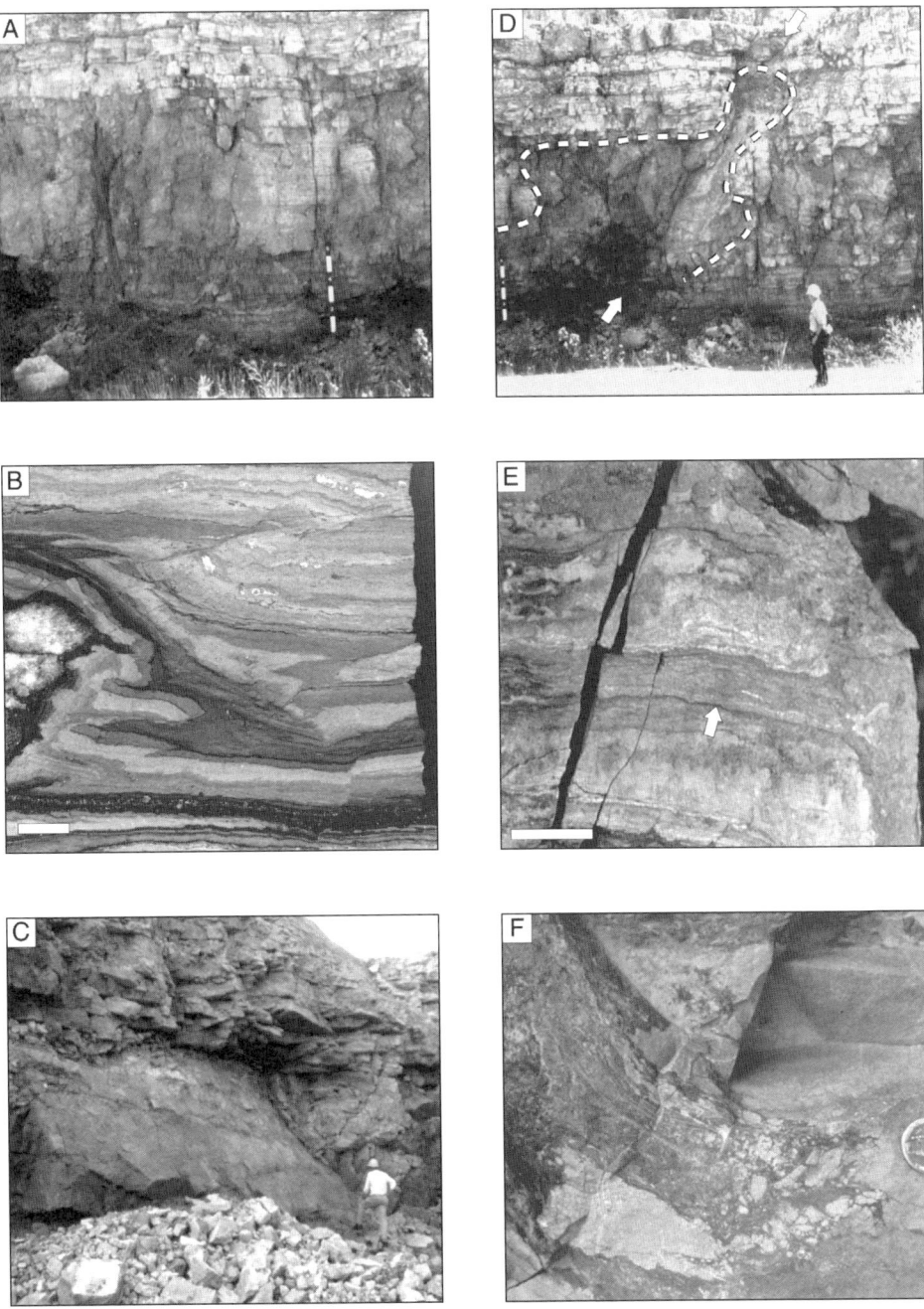

Figure 4. Soft-sediment deformation features. A: Thick layer of concordant breccia in Bass Islands Dolomite. Upper contact is at base of bedded unit near top of photo. Unbrecciated, bedded zone occurs within brecciated horizon above and just to left of pole. Contacts with surrounding breccia are gradational. Pole is scaled in feet. B: Soft-sediment faults in Bass Islands Dolomite. Vergence on thrusts in left center indicate horizontal shortening, and normal fault near lower right corner indicates horizontal extension. Scale bar = 2 cm. C: Listric normal fault surface in Bass Islands Dolomite. Dip is toward the north. D: Breccia mass injected along listric normal fault (arrows) in Bass Islands Dolomite. Dashed line outlines contact of breccia with bedded dolomite. Note spatial relationship between breccia and fault. E: Sediment intrusion of sill (arrow) extending off of vertical dike. Scale bar = 10 cm. F: Progressive brecciation of Bass Islands dolomite bed. From right to left: unbroken bed, incipient brecciation, matrix-supported breccia. Clasts are bent and have indistinct margins. Coin is 2 cm in diameter.

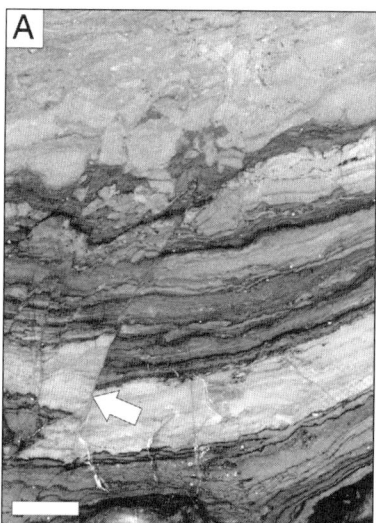 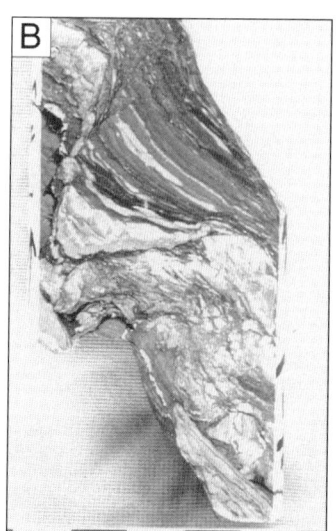 

Figure 5. Soft-sediment deformation features. A: Fault graded bed (after Seilacher, 1969) in Bass Islands Dolomite. Fault indicated by arrow dies out downward. Upward, it changes from a sharp contact to a gradational one as light-colored bed is progressively broken up. Uppermost light layer is homogenized near top of photo. Scale bar = 1 cm. B: Faults and breccia in vertical core of Tymochtee Dolomite from Bowling Green fault zone. Younging direction is toward top of core. Note steep dip of bedding, which is characteristic of bedding in the fault zone. Scale at bottom is in centimeters. C: Bowling Green fault zone. Tymochtee Dolomite is on left; Bass Islands Dolomite on the right. Stratigraphic separation is about 70 m, west side (right) down. Note apparent ductile deformation of light-colored bed indicated by arrow. Steeply dipping light streaks are calcite veins. Scale bar = 1 m.

this would be very limited spatially. The action of seismic waves therefore appears to be the best choice to trigger liquefaction and/or fluidization.

From the preceding sections, the hypothesis that the soft-sediment deformation in the carbonate units studied is a result of seismic activity is consistent with the observations; however, many authors have pointed out that it is difficult or impossible to unequivocally prove that any particular structure is a product of seismic activity (e.g., Leeder, 1987; Guiraud and Seguret, 1987). Several authors have developed criteria for attributing soft-sediment deformation to seismic activity (e.g., Sims, 1975; Leeder, 1987). These criteria include (1) lateral continuity of disturbed strata over large areas, (2) bounding top and bottom by undeformed strata, (3) similarity to structures formed experimentally or in proximity to modern seismic activity, and (4) spatial association to a fault active at the time of sedimentation. The seismites described here meet all of these criteria.

Although the true extent of any deformed horizon in these rocks cannot be determined, because of sparse outcrop, the horizons are exposed over extensive areas within the large quarries, as well as in adjacent quarries. Additionally, if exposures of Bass Islands Dolomite in northwest Ohio described by other authors (Carman, 1927; Sparling, 1970; Johnson, 1974) are included, then the deformation includes at least 10 000 km^2 in northwest Ohio and southeast Michigan. Except for the discordant sediment intrusions and breccia bodies, the deformed horizons in all locations studied are bounded top and bottom by undeformed strata. The nature of the structures is similar to those described in modern seismically active areas. Obermeier (1995) described structures that resulted from liquefaction and fluidization resulting from recent seismic activity. These include breccias, homogenized zones, sediment dikes and pipes, sediment sills, convolute laminations, and faults. All of these resemble those present in the Silurian carbonate rocks described here. Experimental studies in siliciclastic (Amini and Sama, 1999; Moretti et al., 1999) and carbonate (Weaver and Jeffcoat, 1978) sediments also produced similar structures. Finally, perhaps the most convincing evidence in support of seismic origin is the proximity of the rocks with soft-sediment deformation to faults known to have been active during Silurian time (Onasch and Kahle, 1991; Onasch, 1995).

All of the exposures examined lie close the Grenville Front, the most important crustal boundary in the basement of the eastern Midcontinent (Fig. 1). This boundary is thought to be a late Precambrian (ca. 1.0 Ga) suture separating the 1.2–1.4 Ga age anorogenic rocks of the Granite-Rhyolite province on the west from the 1.0–1.3 Ga age metamorphic rocks of the Grenville province to the east (Hinze et al., 1980; Lucius and Von Frese, 1988; Pratt et al., 1989). Faults in Paleozoic rocks overlying the boundary are thought to result from reactivation of basement structures associated with the front (Wickstrom and Gray, 1988; Onasch and Kahle, 1991; Stark, 1997). One such fault in northwest Ohio and southeast Michigan is the Bowling Green fault (Fig. 1). Stratigraphic and structural relations indicate a complex history with as many as seven episodes of displacement, beginning in the Late Cambrian and continuing to the end of the Paleozoic (Onasch and Kahle, 1991; Onasch, 1995). Most of these episodes involved dip-slip displacement, but some may have a significant strike-slip component (Wickstrom, 1990). Thickness variations across the fault and disruption of sedimentary facies in nearby shallow-water carbonate rocks show that some of the episodes were synchronous with sedimentation of several units, including those in the Silurian (Onasch and Kahle, 1991).

At its closest, the surface trace of the Bowling Green fault is less than 10 km from the Ottawa Lake, Waterville, and Salisbury quarries and 30 km from the Monroe quarry (Fig. 1). Production of liquefaction structures is known to occur only within limited distances from the epicenter for an earthquake of given magnitude (Leeder, 1987). For example, with a magnitude 7 earthquake, liquefaction should occur within approximately 50 km of the epicenter (Ambraseys, 1988). For even a small magnitude earthquake, the rocks examined are well within the zone of liquefaction. Several authors have stated that a minimum magnitude of 5 is required to produce liquefaction (Youd, 1977; Obermeier et al., 1991); however, Musson (1998) found abundant soft-sediment deformation structures, including sand volcanoes indicative of liquefaction and fluidization, that resulted from an earthquake with a local magnitude of only 2.5–3.5. The proximity of the Bowling Green fault to the exposures studied, along with evidence for displacement during deposition of Silurian units, clearly satisfies the criteria of association with syndepositional faulting.

In southern Ohio, the Peebles quarry also is located close to the Grenville front (Fig. 1). Numerous faults are present in and around the quarry (Schmidt et al., 1961); however, it is not known at this time whether any of these were active during deposition of the Lilley Formation.

Evidence from the Waterville quarry, where the Bowling Green fault is well exposed, suggests that unlithified units within the fault zone itself may have been disrupted during some displacement episodes. Two very different styles of deformation are found in the Tymochtee and Bass Islands dolomites in the Waterville quarry. Brittle faults with cataclasite and sparry calcite veins indicate brittle deformation at shallow levels. Other features, however, suggest an earlier deformation, in which the rocks behaved in a ductile fashion. Breccias and soft-sediment folds and faults (Fig. 5B) similar to those described in the other quarries appear to be spatially related to the fault zone. In the fault zone, beds pinch and swell, and some beds are deformed plastically where they abut fault surfaces (Fig. 5C). The ductile deformation cannot be the result of deformation of rock at higher temperatures and pressure, because the rocks exposed were never buried more than 1 km (Ramsey and Onasch, 1998) and there is no evidence of metamorphism. Therefore, we believe that recurring displacements deformed variably lithified sediments in the fault zone. Younger, post-Silurian displacement episodes occurred when the rocks were fully lithified and behaved in a brittle fashion.

Liquefaction and fluidization in carbonate rocks

Descriptions of soft-sediment deformation in carbonate rocks are not as numerous as for siliciclastic rocks. This may be a reflection of the relative abundances of the two rock types or a fundamental difference in mechanical behavior. The factors affecting liquefaction and fluidization of carbonate sediments appear to differ somewhat from those for siliciclastic sediments. It has been shown that liquefaction will be most effective in siliciclastic sediment in which the grain size is fine sand or silt (Lowe, 1975; Owen, 1987; Moretti et al., 1999). Finer or coarser grain sizes should not undergo significant liquefaction (Lowe, 1975). In carbonates, this generalization does not appear to hold true (e.g., Pope et al., 1997). For example, Plaziat et al. (1990) reported numerous liquefaction- and fluidization-related structures that formed in carbonate rocks whose grain size at the time of deformation was silt or finer. The grain size of the Silurian rocks described here was probably <10 μm, clearly less than the sand favored for siliciclastics.

Another difference in behavior is in the type of structures that result from liquefaction. Load structures, such as ball-and-pillow, are a common type of soft-sediment deformation in siliciclastics (e.g., Moss and Howells, 1996). This structure is uncommon in carbonates (see Weaver and Jeffcoat, 1978) and was not found in the rocks examined in this study. Conversely, breccias are a common structure in carbonate seismites (Plaziat et al., 1990; Pope et al., 1997; this study), but they are less abundant in siliciclastics.

The difference in behavior between carbonate and siliciclastic sediment may also affect the relationship between the epicenter

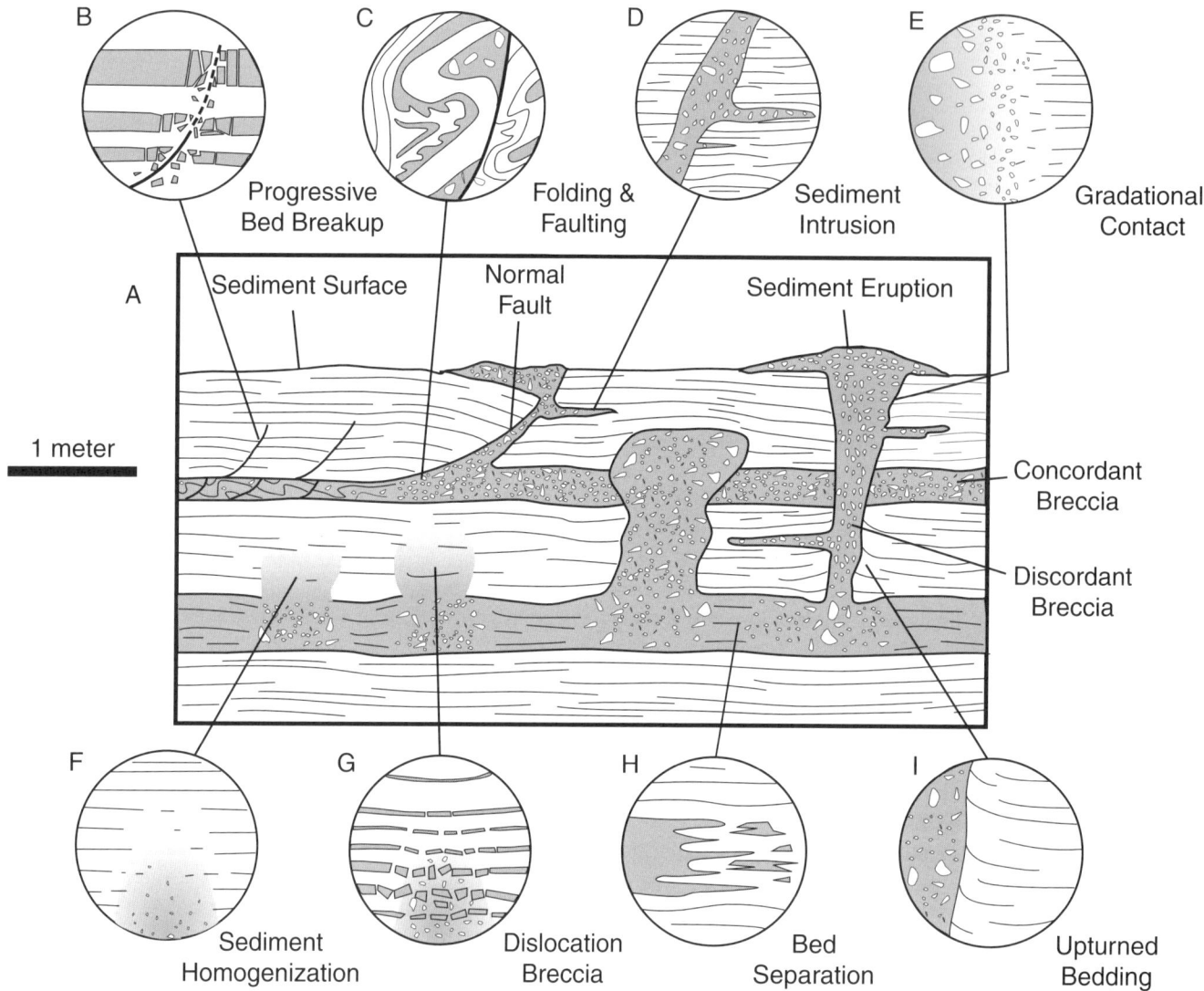

Figure 6. Summary diagram showing types of features in the rocks examined. A: Overview of outcrop, showing spatial relationships between features. B: Progressive breakup of beds adjacent to small fault. C: Soft-sediment folding and faulting, showing juxtaposition of compressional and extensional structures. D: Intrusion of sediment in form of dike and sill. E: Gradational contact at margin of sediment intrusion. F: Progressive homogenization of sediment above liquefied layer. G: Brittle failure of more lithified layer during extension, forming dislocation breccia (Plaziat et al., 1990). H: Progressive disintegration of bed by ductile extension. I: Upturned bedding at margin of sediment intrusion indicating upward movement of breccias.

location and the farthest liquefaction features. Specifically, carbonate sediments may be susceptible to liquefaction at distances greater than siliciclastic sediment. In interpreting widespread presence of seismites in carbonates, Pope et al. (1997) questioned whether the relationship between the epicenter location and farthest liquefaction feature for siliciclastics applies to carbonates.

A key difference between carbonate and siliciclastic sediment that may account for these behavioral differences during liquefaction and fluidization is the rate and degree of cementation. Carbonates can undergo cementation shortly after deposition without significant burial (Bathurst, 1971). This process can form a crust that caps less well cemented sediments below. As cementation progresses, it may do so in a nonuniform fashion, some layers becoming cemented while adjacent ones remain uncemented (Halley and Harris, 1979). Lateral variations are also likely. This nonuniform cementation will provide permeability barriers that are important in preventing fluid escape during pore-pressure buildup. They may also be important for causing fluidization. When the unlithified sediment below a cemented cap moves unevenly, the cap may fail brittlely in certain areas, allowing pore pressures to escape rapidly, thereby providing sufficient pore-fluid velocity to cause fluidization. Partial cementation may also provide the cohesion necessary for the hydroplastic behavior of carbonate muds. In contrast, electrostatic attraction of clay minerals provides the cohesive forces in clay minerals in siliciclastic sediments.

CONCLUSIONS

We believe that the exposures described herein contain the best developed and most extensive assemblage of soft-sediment deformation structures yet described in carbonate rocks. Although other causes cannot be ruled out, the evidence is strongly in support of these structures being a consequence of seismic activity on synsedimentary faults associated with and located along the Grenville Front. Seismogenic faulting on the Bowling Green and other faults in the region (Fig. 1) resulted in hydroplastic deformation, liquefaction, and fluidization in tidal-flat carbonate sediments in the surrounding areas. This in turn resulted in extensive development of soft-sediment structures (Fig. 6A). Settling of variably lithified sediments during liquefaction disrupted bedding, resulting in formation of dislocation breccias (Fig. 6G) (Plaziat et al., 1990) and local homogenized zones (Fig. 6F). Denser, more lithified beds broke up and sank into underlying less lithified beds, producing a variety of folds, faults, and breccias (Fig. 6, C and G). Pore pressures were released locally by semibrittle failure of overlying, more lithified beds. The escape of pore fluids was locally fast enough to fluidize the sediment, producing sediment intrusions (Fig. 6, D and I) and discordant breccia bodies (Fig. 6A). The regional slope into the Michigan basin may have been sufficiently large to induce large-scale sliding of sediment masses to the north on overpressured horizons. As these sheets moved down very low slopes, pore-fluid pressures were released suddenly, causing localized fluidization of the sediment and injection of breccias along the faults (Fig. 6A). Further movement produced internal deformation in liquefied horizons in the form of folds and faults (Fig. 6C). Vertical and lateral variations in the degree of cementation led to variations in the mechanical behavior of deforming sediment, ranging from brittle (Fig. 6B) to highly ductile (Fig. 6H).

This study has shown that carbonates may contain a wealth of information useful in accessing the paleoseismicity of an area. Those we examined have been shown to be useful in obtaining a better understanding of the tectonic history of the eastern U.S. Midcontinent during the Silurian.

ACKNOWLEDGEMENTS

The The authors would like to thank the quarry owners for their hospitality during our many visits. Matthew Duty provided valuable assistance in the field and was instrumental in gaining access to important exposures. The manuscript benefited considerably from the comments of reviewers Robert V. Demicco and Dennis Kolata.

REFERENCES CITED

Allen, J.R.L., 1982, Sedimentary structures: Their character and physical basis: Amsterdam, Elsevier, 643 p.

Alvarez, W., Colacicchi, R., and Montanari, A., 1985, Synsedimentary slides and bedding formation in Apennine pelagic limestones: Journal of Sedimentary Petrology, v. 55, p. 720–734.

Ambraseys, N.N., 1988, Engineering Seismology: Journal of the International Association of Earthquake Engineering, v. 17, p. 1–105.

Amini, F., and Sama, K.M., 1999, Behavior of stratified sand-silt-gravel composites under seismic liquefaction conditions: Soil Dynamics and Earthquake Engineering, v. 18, p. 445–455.

Bathurst, R.G.C., 1971, Carbonate sediments and their diagenesis: Amsterdam, Elsevier, Developments in Sedimentology, v. 12, 620 p.

Carlson, E.H., 1992, Reactivated interstratal karst: Example from the Late Silurian rocks of western Lake Erie (U.S.A.): Sedimentary Geology, v. 76, p. 273–283.

Carman, J.E., 1927, The Monroe division of rocks in Ohio: Journal of Geology, v. 35, p. 481–506.

Carman, J.E., 1946, The geologic interpretation of scenic features in Ohio: The Ohio Journal of Science, v. 46, p. 241–283.

Court, P.R., and Kahle, C.F., 1993, Stratigraphy and sedimentology of some Silurian rocks in southwestern Ohio: Great Lakes Section Society for Sedimentary Geology Fall Field Trip, 30 p.

Dalrymple, R.W., 1979, Wave-induced liquefaction: A modern example from the Bay of Fundy: Sedimentology, v. 26, p. 835–844.

Dalrymple, R.W., 1980, Wave-induced liquefaction: An addendum: Sedimentology, v. 27, p. 461.

Dugue, O., 1995, Seismites dans le Jurassique supérieur du Bassin anglo-parisien (Normandie, Oxfordien supérieur, Calcaire greseux de Hennequeville): Sedimentary Geology, v. 99, p. 73–93.

Field, M.E., Gardner, J.V., Jennings, A.E., and Edwards, B.D., 1982, Earthquake induced sediment failures on a 0.25° slope, Klamath River delta, California: Geology, v. 10, p. 542–546.

Guiraud, M., and Seguret, M., 1987, Soft-sediment microfaulting related to compaction within the fluvio-deltaic infill of the Soria strike-slip basin (northern Spain), in Jones, M.E., and Preston, R.M.F., eds., Deformation of sediments and sedimentary rocks: Geological Society [London] Special Publication 29, p. 123–136.

Halley, R.B., and Harris, P.M., 1979, Fresh-water cementation of a 1,000 year-old oolite: Journal of Sedimentary Petrology, v. 49, p. 969–988.

Hansen, M.C., 1998, Geology of Ohio: The Silurian: Ohio Geology, Spring, p. 1–7.

Hardie, L.A., 1977, Sedimentation on the modern carbonate tidal flats of northwest Andros Island, Bahamas: Baltimore, Maryland, The Johns Hopkins University Press, 202 p.

Hempton, M., and Dewey, J.F., 1983, Earthquake-induced deformational structures in young lacustrine sediments, east Anatolian fault, southeast Turkey: Tectonophysics, v. 98, p. T7–T14.

Hinze, W., Braile, L.W., Keller, G.R., and Lidiak, E.G., 1980, Models for midcontinental tectonism, in Continental tectonics: National Academy of Science Studies in Geophysics, p. 73–83.

Horvath, A.L., 1969, Relationships of Middle Silurian strata in Ohio and West Virginia: The Ohio Journal of Science, v. 69, p. 321–342.

Jeyapalan, J.K., Duncan, J.M., and Seed, H.B., 1983, Investigation of flow failures of tailings dams: Journal of Geotechnical Engineering, v. 109, p. 172–189.

Johnson, R.L., 1974, Geology and environmental interpretation of the Upper Cayugan Bass Islands dolomite, Southeastern Michigan [M.S. thesis]: Ann Arbor, University of Michigan, 89 p.

Kahle, C.F., 1997, Geology of some Silurian rocks in Ohio: The view from stone quarries: Fifth Annual Technical Symposium, Ohio Geological Survey and the Oil and Gas Association, p. 48–51.

Kahle, C.F., and Floyd, J.C., 1971, Stratigraphic and environmental significance of sedimentary structures in Cayugan (Silurian) tidal flat carbonates, northwestern Ohio: Geological Society of America Bulletin, v. 82, p. 2071–2096.

Kleffner, M.A., 1990, Wenlockian (Silurian) conodont biostratigraphy, depositional environments, and depositional history along the eastern flank of the Cincinnati arch in southern Ohio: Journal of Paleontology, v. 64, p. 319–328.

Laird, M.G., 1968, Rotational slumps and slump scars in Silurian rocks, western Ireland: Sedimentology, v. 10, p. 111–120.

Leeder, M., 1987, Sediment deformation structures and the paleotectonic analysis of sedimentary basins, with a case-study from the Carboniferous of northern England, *in* Jones, M.E, and Preston, R.M.F., eds., Deformation of sediments and sedimentary rocks: Geological Society [London] Special Publication 29, p. 137–146.

Lowe, D.R., 1975, Water escape structures in coarse-grained sediments: Sedimentology, v. 22, p. 157–204.

Lowe, D.R., 1976, Subaqueous liquefied and fluidized sediment flows and their deposits: Sedimentology, v. 23, p. 285–308.

Lucius, J.E., and Von Frese, R.R.B., 1988, Aeromagnetic and gravity anomaly constraints on the crustal geology of Ohio: Geological Society of America Bulletin, v. 100, p. 104–116.

McKee, E.D., Reynolds, M.A., and Baker, C.H., 1962, Experiments on intraformational recumbent folds in crossbedded sand, *in* Short papers in geology and hydrology and topography: U.S. Geological Survey Professional Paper 450-D, p. D155–D160.

Mohindra, R., and Bagati, T.N., 1996, Seismically induced soft-sediment deformation structures (seismites) around Sumdo in the lower Spiti Valley (Tethys Himalaya): Sedimentary Geology, v. 101, p. 69–83.

Moretti, M., Alfaro, P., Caselles, O., and Canas, J.A., 1999, Modeling seismites with a digital shaking table: Tectonophysics, v. 304, p. 369–383.

Moss, S.J., and Howells, C.G., 1996, An anomalously large liquefaction structure, Oligocene Ombilin basin, west Sumatra, Indonesia: Journal of Southeast Asian Earth Sciences, v. 14, p. 71–78.

Musson, R.M.W., 1998, The Barrow-in-Furness earthquake of 15 February 1865: Liquefaction from a very small magnitude event: Pure and Applied Geophysics, v. 152, p. 733–745.

Nataraja, M.S., and Gill, H.S., 1983, Ocean wave-induced liquefaction analysis: Journal of Geotechnical Engineering, v. 109, p. 573–590.

Nurmi, R.D., and Friedman, G.M., 1977, Sedimentology and depositional environments of basin-center evaporites, lower Salina Group (Upper Silurian), Michigan Basin Studies, *in* Fisher, J.H., ed., Studies in Geology, Volume 5: American Association of Petroleum Geologists, p. 23–52.

Obermeier, S.F., 1995, Using liquefaction-induced features for paleoseismic analysis, *in* Obermeier, S.G., and Jibson, R.W., eds., Using ground-failure features for paleoseismic analysis: U.S. Geological Survey Open-File Report 94-663, p. 1–50.

Obermeier, S.F., and nine others, 1991, Evidence of strong earthquake shaking in the lower Wabash Valley from prehistoric liquefaction features: Science, v. 251, p. 1061–1063.

Onasch, C.M., 1995. Structural evolution of the Bowling Green fault, *in* Wickstrom, L.H., and Berg, T., eds., Proceedings of the Canton Symposium: Ohio Geological Survey No. 3, p. 7–17.

Onasch, C.M., and Kahle, C.F., 1991, Recurrent tectonics in a cratonic setting: An example from northwestern Ohio: Geological Society of America Bulletin, v. 103, p. 1259–1269.

Owen, G., 1987, Deformation processes in unconsolidated sands, *in* Jones, M.E., and Preston, R.M.F., eds., Deformation of sediments and sedimentary rocks: Geological Society [London] Special Publication 29, p. 11–24.

Plaziat, J.C., and Ahmamou, M., 1998, Les differents mecanismes a l'origine de la diversité des seismites, leur identification dans le Pliocene du Saiss de Fes et de Meknes (Maroc) et leur signification tectonique: Geodinamica Acta, v. 11, p. 183–203.

Plaziat, J.C., Purser, B.H., and Philobbos, E., 1990, Seismic deformation structures (seismites) in the syn-rift sediments of the NW Red Sea (Egypt): Bulletin de la Société Géologique de France, v. 3, p. 419–434.

Pope, M.C., Read, J.F., and Hofmann, H.J., 1997, Late Middle to Late Ordovician seismites of Kentucky, southwest Ohio and Virginia: Sedimentary recorders of earthquakes in the Appalachian basin: Geological Society of America Bulletin, v. 109, p. 489–503.

Pratt, B.R., 1994, Seismites in the Mesoproterozoic Altyn Formation (Belt Supergroup), Montana: A test for tectonic control of peritidal carbonate cyclicity: Geology, v. 22, p. 1091–1094.

Pratt, T., Culotta, R., Hauser, E., Nelson, D., Brown, L., Kaufman, S., and Oliver, J., 1989, Major Proterozoic basement features of the eastern midcontinent of North America revealed by recent COCORP profiling: Geology, v. 17, p. 505–509.

Ramsey, D.W., and Onasch, C.M., 1998, Fluid migration in a cratonic setting: The fluid history of two fault zones in the eastern midcontinent: Tectonophysics, v. 305, p. 307–323.

Schmidt, R.G., McFralan, A.C., Nosow, E., Bowman, R.S., and Alberts, R., 1961, Examination of Ordovician through Devonian stratigraphy and the Serpent Mount Chaotic structure area: Geological Society of America Guidebook for Field Trips, Cincinnati Meeting, p. 261–293.

Seed, H.B., 1968, Landslides during earthquakes due to liquefaction: Journal of Soil Mechanics and Foundations, v. 94, p. 1053–1122.

Seilacher, A., 1969, Fault-graded beds interpreted as seismites: Sedimentology, v. 13, p. 155–159.

Seilacher, A., 1984, Sedimentary structures tentatively attributed to seismic events: Marine Geology, v. 55, p. 1–12.

Sims, J.D., 1975, Determining earthquake recurrence intervals from deformational structures in young lacustrine sediments: Tectonophysics, v. 29, p. 141–152.

Sparling, D.R., 1970, The Bass Islands Formation in its type region: Ohio Journal of Science, v. 70, p. 1–33.

Stark, T.J., 1997, The east continent rift complex: Evidence and conclusions, *in* Ojakangas, R.W., Dickas, A.B., and Green, J.C., Middle Proterozoic to Cambrian Rifting, Central North America: Geological Society of America Special Paper 312, p. 253–266.

Swineford, E.M., 1985, Geology of the Peebles quadrangle, Adams County, Ohio: Ohio Journal of Science, v. 85, p. 218–230.

Terzaghi, J., 1947, Shear characteristics of quicksand and soft clay: Proceedings of 7th Texas Conference on Soil Mechanics and Foundation Engineering, 10 p.

Vanneste, K., Meghraoui, M., and Camelbeeck, T., 1999, Late Quaternary earthquake-related soft-sediment deformation along the Belgian portion of the Feldbiss fault, Lower Rhine graben system: Tectonophysics, v. 309, p. 67–79.

Wanless, H.R., Tyrrell, K.M., Tedesco, L.P., and Dravis, J.J., 1988, Tidal flat sedimentation from Hurricane Kate, Caicos Platform, British West Indies: Journal of Sedimentary Petrology, v. 58, p. 724–738.

Weaver, J.D. and Jeffcoat, R.E., 1978, Carbonate ball and pillow structures: Geological Magazine, v. 115, p. 245–253.

Wickstrom, L.H., 1990, A new look at Trenton (Ordovician) structure in northwestern Ohio: Northeastern Geology, v. 12, p. 103–113.

Wickstrom, L.H., and Gray, J.D., 1988, Geology of the Trenton Limestone in northwestern Ohio, *in* Keith, B.D., ed., The Trenton Group (Upper Ordovician Series) of Eastern North America: American Association of Petroleum Geologists, Studies in Geology No. 29, p. 159–172.

Wright, V.P., 1984, Peritidal carbonate facies models: A review: Geological Journal, v. 19, p. 309–325.

Youd, T.L., 1977, Brief review of liquefaction during earthquakes in Japan: Discussion: Soils Foundation, v. 17, p. 82–85.

Youd, T.L., 1978, Major cause of earthquake damage is ground failure: Civil Engineering, April, p. 47–51.

MANUSCRIPT ACCEPTED BY THE SOCIETY MAY 11, 2001

Interpreting ancient marine seismites and apparent epicentral areas for paleo-earthquakes, Middle Ordovician Lexington Limestone, central Kentucky

Frank R. Ettensohn
Mark A. Kulp
Nicholas Rast
Department of Geological Sciences, University of Kentucky, Lexington, Kentucky 40506-0053, USA

ABSTRACT

Three horizons of soft-sediment deformation in the Middle Ordovician Brannon Member of the Lexington Limestone, a platform carbonate sequence in central Kentucky, are in stratigraphic, structural, and temporal circumstances that suggest seismogenic origins. However, because such deformation may not be a unique response to seismicity, application of four concurrent criteria to such horizons can reduce the ambiguity. For the Brannon horizons, concurrence of these criteria, which include deformation consistent with a seismogenic origin, deformation in widespread, temporally and stratigraphically constrained horizons, deformation that shows systematic increases in frequency or intensity toward a likely epicentral area, and the ability to exclude other possible causes, strongly supports a seismogenic origin; other criteria reinforce this interpretation.

Because deformation intensity, as indicated by preserved sedimentary structures, can reflect original energy input, mapping distribution of deformation types in the Brannon horizons has effectively generated isoseismal maps that were used to pinpoint apparent epicentral areas. Although use of these maps with other stratigraphic data has suggested interpretations regarding ancient epicenters, earthquake directivity, earthquake clustering, magnitude, and recurrence intervals, such interpretations must remain tentative until better data and modern, marine analogues can be found to illustrate the influence of site effects in similar marine facies.

The example of Brannon seismites shows that seismicity was an important process affecting epicontinental sedimentation and that well-constrained, seismically deformed beds are potentially important chronostratigraphic event horizons, which can provide information about the nature and occurrence of ancient Phanerozoic earthquakes.

INTRODUCTION

Horizons of soft-sediment deformation, which may include pseudonodules, flow rolls, ball-and-pillow structures, and sedimentary dikes, abound throughout the Middle and Upper Ordovician Lexington Limestone in central Kentucky (Figs. 1 and 2). Although such features are common and well known from siliciclastic sequences, they have not commonly been reported from carbonates. The origin of the deformation, moreover, has been largely ascribed to dewatering or liquefaction, which is thought to have been induced by slope-related creep or slump, artesian pressure, waves, or rapid loading during burial. The possibility of

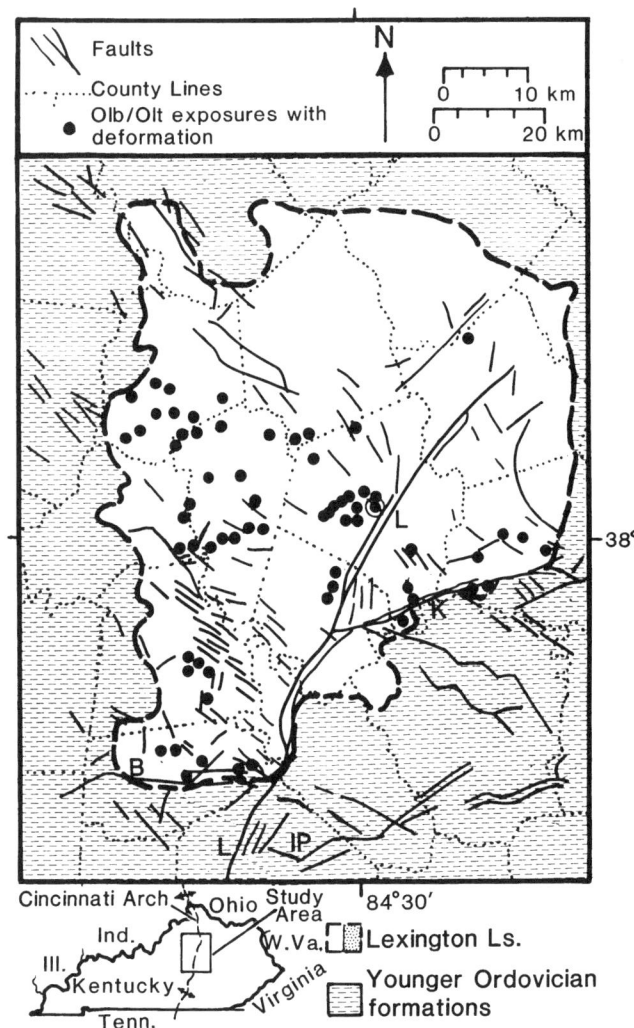

Figure 1. Location map for central Kentucky area, showing approximate distribution of Lexington Limestone, location of exposures with soft-sediment deformation in the Brannon Member (Olb) and equivalent parts of the Tanglewood Member (Olt), and major extant faults crossing the region. Many of these faults apparently have basement precursors that were periodically reactivated during Lexington deposition. L—Lexington fault zone; K—Kentucky River fault zone; B—Brumfield fault zone; IP—Irvine–Paint Creek fault zone; bull's-eye near center of map represents the city of Lexington.

seismically induced soft-sediment deformation is also widely recognized, but it is an interpretation that has largely been restricted to terrestrial or marginal marine, Tertiary to Quaternary, siliciclastic sequences on active margins. The possibility of seismic influence on older, intracratonic, marine sediments, especially epicontinental carbonates, has seldom been seriously considered until recently (e.g., Rast and Ettensohn, 1995; Pope et al., 1997; Rast et al., 1999). Yet, an abundance of recent, intracratonic seismicity and resulting soft-sediment deformation has been documented in the U.S. Midcontinent (New Madrid and Wabash Valley areas), a tectonically quiescent part of the North American continent (e.g., Russ, 1979, 1982; Obermeier et al., 1990, 1993, 1996; Wesnousky and Leffler, 1994; Pond and Martin, 1996; Tuttle et al., 1996a, 1996b). If this evidence gives any indication of expected frequency, then seismicity and resulting soft-sediment deformation should also be anticipated in former intracratonic marine settings like the central Kentucky area, located only 400 km from the active Taconian margin in Middle and Late Ordovician time. In fact, soft-sediment deformation structures in these Ordovician rocks, once interpreted to represent slope-induced deformation, are now being reinterpreted as "seismites," or seismically deformed sediments (Seilacher, 1969, 1984; Cita and Ricci Lucchi, 1984), largely on the basis of general characteristics of timing and location and comparison with Holocene seismites (e.g., Pope et al., 1997; Rast et al., 1999).

In this chapter, we examine in detail three temporally and stratigraphically constrained horizons of soft-sediment deformation in the Middle Ordovician (Shermanian; mid-Caradocian) Brannon Member of the Lexington Limestone (Fig. 2). We show that a combination of criteria and circumstances support a seismite interpretation for the Brannon structures, and we indicate how these interpretations might be used in well-constrained situations to locate apparent epicentral areas and determine possible recurrence intervals and magnitudes for ancient (pre-Cenozoic) Phanerozoic earthquakes.

GEOLOGIC FRAMEWORK

The Lexington Limestone is a complex, generally shallow water, carbonate unit (Fig. 2) deposited across the Lexington platform during late Middle to early Late Ordovician time (Rocklandian-Edenian; mid-Caradocian) (Cressman, 1973; Keith, 1989; Ettensohn, 1992a; Pope and Read, 1997; Pope et al., 1997). Although the unit was originally deposited at about lat 20°–25° S on the southeast margin of Laurentia, temperate carbonates seem to predominate (Pope and Read, 1997), and most of the units are composed of mid- and upper-ramp, bioclastic calcarenites that reflect storm reworking. The Brannon Member is a prominent exception, being composed largely of even-bedded calcisiltites and fine-grained calcarenites interbedded with shales (Fig. 3), which represent deep-ramp, distal storm deposits and background suspension sedimentation (Kulp, 1995). Commonly, the Brannon sharply overlies and is sharply overlain by coarser, shallow-ramp calcarenites and calcirudites of the Tanglewood Member; the contacts are typically marked by hardgrounds. At the northeastern limit of its distribution, the Brannon grades laterally into Tanglewood facies (Fig. 2), and the hardground at the base of the Brannon can be traced locally into the Tanglewood.

At depth, these carbonates overlie the mid-Proterozoic East Continental Rift basin, a late Proterozoic–Middle Cambrian Iapetan rift, and the Grenville Front (Black, 1986; Shumaker, 1986; Drahovzal et al., 1992; Ettensohn and Pashin, 1992; Rast and Goodmann, 1995), all of which are reflected in a series of basement faults that are prone to reactivation right up to the present (e.g., Herrmann et al., 1982; Van Arsdale, 1986; Street et al.,

1993). In fact, many of the extant fault systems in central Kentucky (Fig. 1) have basement precursors (Black, 1986; Drahovzal et al., 1992) and in the Lexington Limestone are associated with abrupt facies changes, nonuniform variations in thickness, and local unconformities, which suggest repeated synsedimentary structural reactivation by growth faulting at depth (Ettensohn et al., 1986a; Ettensohn, 1992b; Kulp, 1995; Ettensohn and Kulp, 1995). Moreover, growth faulting has been reported from individual outcrops and based on stratigraphic relationships (Black and Haney, 1975; Pope and Read, 1995; Pope et al., 1997). Mapping shows that some areas of soft-sediment deformation are close to or distinctly parallel to major fault zones in the area, although deformation is also present more broadly. However, the association of deformation with nearby structures has been used as evidence of seismogenic origin (Kulp, 1995; Rast and Ettensohn, 1995; Pope et al., 1997; Rast et al., 1999), on the basis of their similarity to soft-sediment deformation evidence found in other areas near reactivated faults (e.g., Sims, 1973; Leeder, 1987; Anand and Jain, 1987; Guiraud and Plaziat, 1993; Obermeier, 1996). Similar features are now widely cited as evidence for prehistoric seismicity along reactivated faults in the New Madrid and Wabash Valley regions (Obermeier et al., 1990, 1993, 1996; Wesnousky and Leffler, 1994; Pond and Martin, 1996; Tuttle et al., 1996a, 1996b). Although the tectonic regime of the New Madrid and Wabash Valley regions differs from that of central Kentucky during Ordovician time, a similar reactivation of Precambrian to Cambrian faults has been suggested (Hamilton and Zoback, 1982; Hildebrand, 1985).

Horizons of soft-sediment deformation and examples of structural control of facies are present throughout the upper Middle and Upper Ordovician rocks of central Kentucky; these coincide with the final phase of the Taconian orogeny, only 400 km to the east (e.g., Ettensohn, 1991). Although soft-sediment deformation of the scale and distal location observed in the Brannon Member has not previously been related to far-field transmission of tectonic stresses, foreland stratigraphic responses such as the structural control of facies are typical of orogenic complexes and have been interpreted to reflect distal transmission of stresses across hundreds of kilometers, commonly focused on preexisting zones of foreland basement weakness (e.g., Ziegler, 1987).

Horizons of soft-sediment deformation are common throughout the Lexington Limestone (Fig. 2), but we focus here on three horizons in the Brannon Member (Figs. 2 and 4), because they can be temporally and stratigraphically constrained by a bentonite at the base of the unit (Black et al., 1965; Kulp, 1995). This bentonite, however, has not been chemically fingerprinted throughout its distribution.

Much of the upper part of the Lexington Limestone is a complex facies mosaic of shoal-related calcarenites and calcirudites (Tanglewood Member) that intertongue with mid-ramp nodular carbonates and shales (Millersburg Member) in the Tanglewood buildup (Fig. 2). The Tanglewood buildup appears to represent an anomalous area of regressive, shallow-water sedimentation (Tanglewood and Millersburg Members; Fig. 2) in the

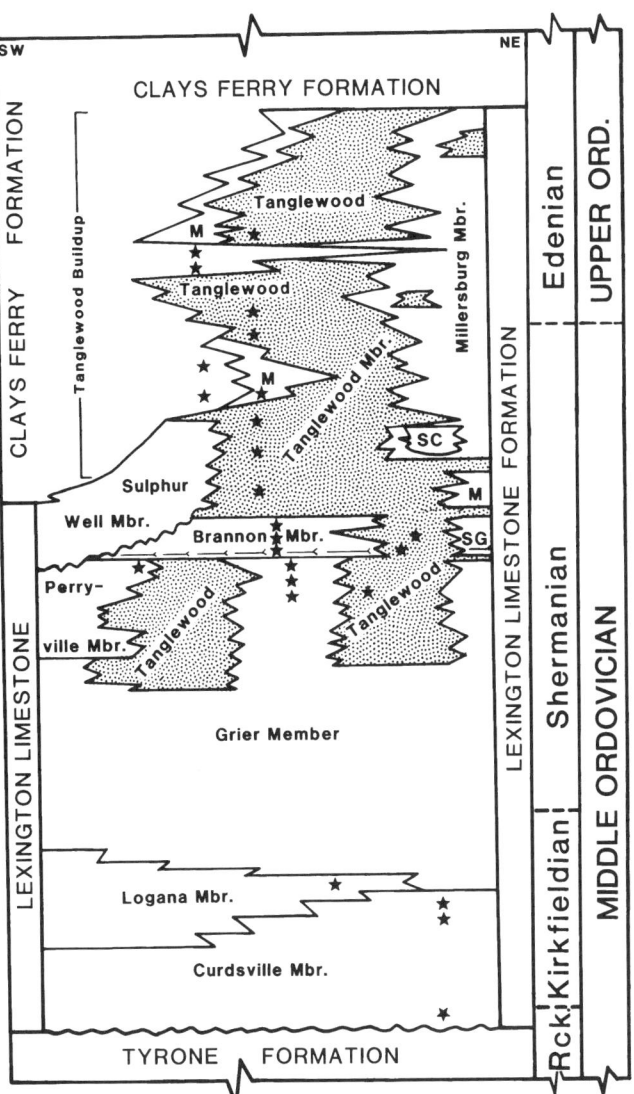

Figure 2. Generalized stratigraphic column for the Lexington Limestone, showing position of the Brannon Member, location of bentonite at the base of the unit (fletched lines), and relative positions of known horizons of soft-sediment deformation (stars) in the Lexington Limestone and related parts of the Clays Ferry Formation. The Brannon is eroded to the southwest and tongues out into the Tanglewood Member to the northeast. Lower parts of the Lexington reflect regional transgression through the Brannon Member. Upper parts of the Lexington represent a locally regressive shoal complex (Tanglewood and Millersburg Members), which comprises the Tanglewood buildup, a series of stacked shoals that intertongue in all directions with deeper water shales and micrograined limestones of the Clays Ferry Formation. M—Millersburg Member; SC—Strodes Creek Member; SG—Stamping Ground Member; Rck.—Rocklandian.

central Kentucky area, which developed on apparently reactivated structures at a time when deeper water sediments represented by the shale and micrograined limestone of the Clays Ferry Formation and its equivalents were being deposited beyond the buildup (Ettensohn, 1992b; Ettensohn and Kulp, 1995; Fig. 2).

Figure 3. A 2.9-m-thick, undeformed exposure of the upper Brannon Member, showing the even-bedded, micrograined limestones and interbedded shales that characterize the unit. The limestone layers are interpreted to represent distal tempestites (Kulp, 1995). Outcrop near junction of Bluegrass Parkway and Kentucky Route 33.

Figure 4. Brannon exposure showing three deformed horizons (dashed lines at base of each horizon) separated by undeformed shale beds. Reentrant at the very base of the exposure is the basal bentonite. Liquefaction was the predominant deforming process in all three beds, and examples of vertical piping and floating clasts are visible within the dashed-line circles in the second deformed horizon. Hammer is 28 cm long.

Amid the complex facies development, the Brannon is one of the most distinctive, well-constrained, and widespread units in the buildup (Figs. 2 and 3) and thus represents an ideal unit in which to study possible seismogenic influence.

PROCEDURES AND RATIONALE

Eighty-six exposures and nine cores that contained the Brannon Member were measured and described in the central Kentucky area; descriptions and locations are provided in Kulp (1995). Of these, the Brannon in 66 sections and two cores contained one to three horizons of soft-sediment deformation that were examined for this study (Figs. 5–7). In addition to standard descriptions, the orientations of fold axes in the deformation were measured, where possible, and plotted on rose diagrams to determine the possibility of preferential orientation; the results of 199 measurements are also presented in Kulp (1995).

Each horizon of soft-sediment deformation was located relative to the bentonite at the base of the unit and was characterized for the predominance of deformation type. Using the classification of Lowe (1975), we have characterized the predominant deformation as hydroplastic, liquefied, or fluidized. Deformation is controlled by the presence of susceptible sediments (water-saturated silt and fine sand; e.g., Allen, 1977, 1984) confined below impermeable layers (e.g., Obermeier et al., 1990) and a triggering input of energy sufficient to generate enhanced pore pressure (e.g., Sieh, 1978; Scott and Price, 1988; Holzer et al., 1989; Guiraud and Plaziat, 1993). If thixotropic muds are involved, deformation may occur in unconfined circumstances, especially if the triggering event is abrupt and rapid. However, if sediments are partially consolidated and cohesive, as the coherent layer segments in most Brannon deformation indicate, burial below impermeable layers may have been essential to generate the overpressuring necessary for deformation; otherwise, deforming pore fluids would seep out at the surface and stress would be dissipated. Triggering mechanisms for the deformation may include rapid loading during burial, gravity-induced mass wasting, dewatering during rapid stage changes in tidal and fluvial channels, storm waves, and seismicity.

Deformation processes reflect energy input via the minimum fluidization velocity (U_o), which can be recorded in the resulting deformation (Lowe, 1975), and similar relationships have been noted between the severity of earthquakes and resulting soft-sediment deformation (e.g., McAlpin and Nelson, 1996). At lower energies and resulting pore-fluid velocities well below U_o, hydroplastic deformation is typical and is characterized by simple contortion or folding of beds and preservation of primary lamination or bedding (Fig. 8); elutriation, or the separation of finer grains in

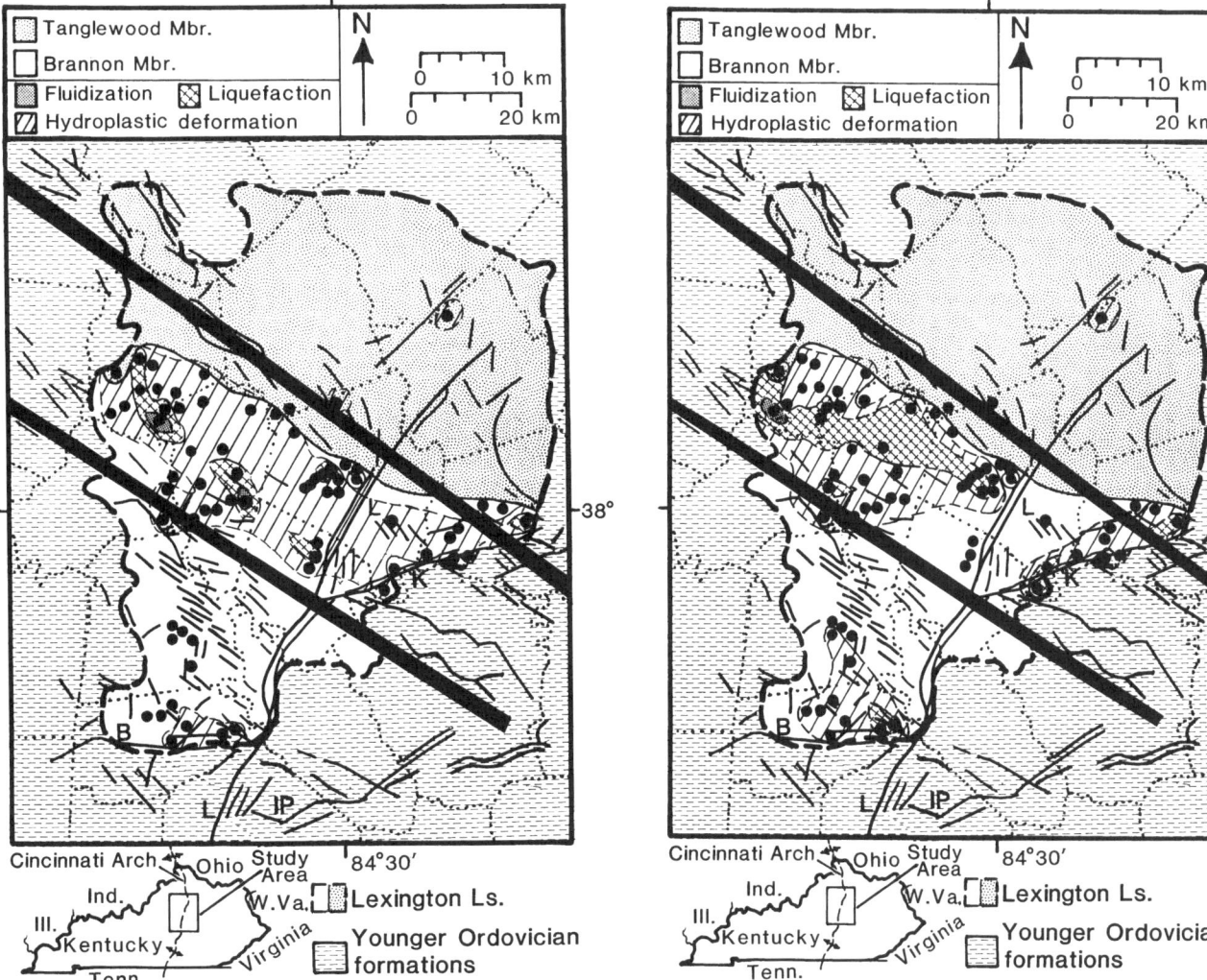

Figure 5. Distribution of soft-sediment structure mapped by deformation processes (different patterns) for the lower Brannon horizon of soft-sediment deformation. Aside from a small area of deformation near the Brumfield fault (B) in the south, deformation is visible in a broad, northwest-southeast–trending belt between two structural trends (diagonal heavy lines). Stippled area to the north represents shallow-water Tanglewood equivalents of the deeper water Brannon Member, and the northwest-southeast–oriented line separating the Brannon (no pattern) from the Tanglewood is an approximate facies boundary (Fig. 2) that roughly coincides with an upper structural trend or lineament. Note outliers of deformation in Tanglewood equivalents. The near coincidence of the facies boundary and lineament suggests synsedimentary downdrop to the southwest along growth faults; the similarly oriented southern trend reflects downdrop to the north (Kulp, 1995). The concentration of deformation between the two trends may reflect influence of reactivation along the trends or presence of site effects in the troughlike basin defined by the trends.

Figure 6. Distribution of deformation mapped by deformation process (different patterns) for the middle Brannon horizon of soft-sediment deformation. Most deformation is concentrated in the northwest parts of the area defined by the two structural trends; outliers are near the Kentucky River and Brumfield fault zones and in the Tanglewood equivalent. Other aspects of the map are similar to Figure 5.

a rising column of water, is negligible. At higher energies and a U_o approaching the minimum fluidization velocity, liquefaction becomes common and is characterized by vertical piping, sedimentary dikes, or sand volcanoes that initiate destruction of primary lamination and bedding (Fig. 4); the flow is laminar and some elutriation of finer grains occurs. At very high energies and resulting velocities at or greater than U_o, fluidization predominates and is characterized by complete destruction of primary lamination and bedding and homogenization of the unit (Fig. 9); the flow is turbulent and elutriation of finer grains is ubiquitous.

Individual deformation types reflect differing energy inputs. As a result, if individual deformed horizons can be characterized in exposure by the predominance of one of the three types of deformation, and if the types of deformation can be mapped in well-constrained horizons, then point sources of energy input of the type encountered in seismicity can be recognized by an area of intense deformation (complete homogenization) and a roughly concentric pattern of decreasing-energy deformation bands surrounding it (Ettensohn et al., 2000). Such isoseismal bands of

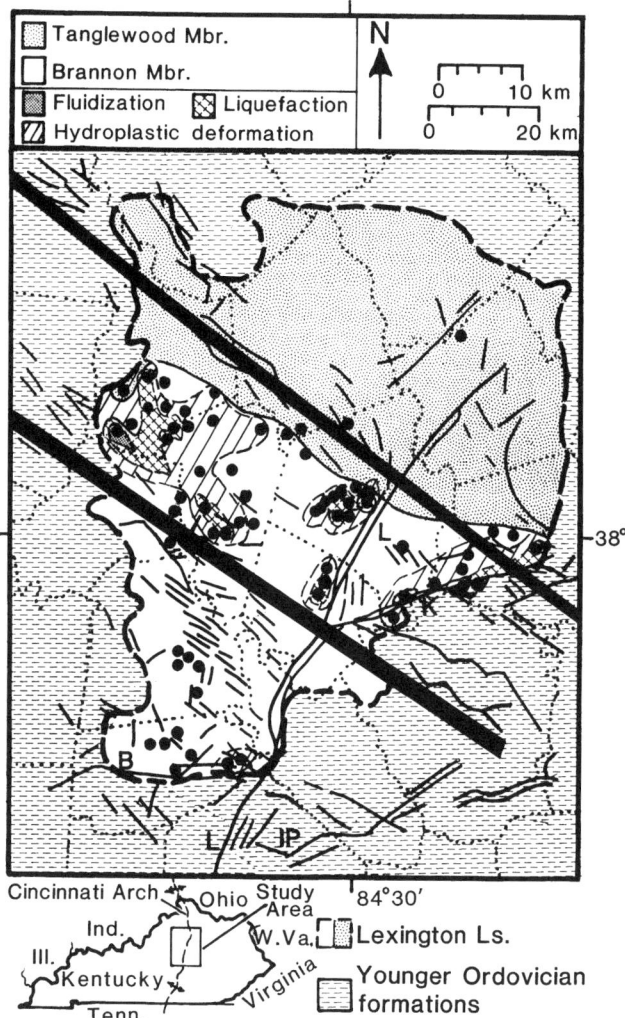

Figure 7. Distribution of deformation mapped by deformation process (different patterns) for the upper horizon of Brannon soft-sediment deformation. Most deformation is concentrated in the far northwestern parts of the area defined by the two structural trends; outliers are near the Lexington, Kentucky River, and Brumfield fault zones. Other aspects of the map are similar to Figure 5.

intensity are generally concentric about the epicenters of modern quakes, with local variations caused by site effects, and the area of most intense deformation—because it likely represents the area of greatest energy input—is inferred to be the apparent epicentral area (e.g., Reiter, 1990). In contrast, the distribution of deformation types resulting from other causes is expected to be random or have peculiar geometries related to channels or slopes. Consequently, mapping deformation types in well-constrained horizons has the potential to support interpretation of seismogenic origin and to identify apparent epicentral areas for very ancient earthquakes, just as seismologists used intensity evaluations to locate likely epicentral areas before the development of seismographs (e.g., Reiter, 1990; Yeats et al., 1997). Although other geological methods, such as measuring sedimentary-dike widths, are available for locating epicentral areas (Youd and Perkins, 1978), these techniques are useful only in our middle range of deformation (liquefaction), and they were developed for use primarily in terrestrial settings. Although Lowe's (1975) scheme is a very general one, it is the only one appropriate for use in marine rocks, and we use it here to characterize deformation intensity in the Brannon horizons.

ARE THE BRANNON HORIZONS SEISMITES?

Seilacher (1969, 1984) used the term "seismite" as a genetic or interpretive term for postdepositional, soft-sediment deformation inferred to have been earthquake-generated, whereas Cita and Ricci Lucchi (1984) expanded the term to include any coseismic deposition or deformation induced by earthquakes or tsunami waves. Seilacher (1991) realized that the characteristics of seismites were not unique and suggested that they had been underestimated in the geologic record due to misinterpretation and deletion by subsequent events such as storms. Deformed horizons in the Brannon have been interpreted variously as resulting from mass movement (McFarlan, 1943), rapid loading (Cressman, 1973), dewatering (Noger and Kepferle, 1985), and, most recently, seismicity (Ettensohn, 1992c; Kulp, 1995; Pope et al., 1997; Rast et al., 1999).

Criteria supporting seismogenic origin

Many criteria have been suggested for differentiating the effects of seismicity from those of other causes in soft-sediment deformation, but we differentiate deformation of seismogenic origin on the basis of the concurrence of four criteria outlined by Obermeier (1996, 1998), McAlpin and Nelson (1996), and Obermeier and Pond (1999), because they can be reasonably determined from the ancient, marine geologic record. The first of these is the presence of deformation consistent with a seismogenic origin, probably the easiest to assess. Features like those in the Brannon Member (Figs. 3, 8, and 9) have been produced experimentally in siliciclastic sediments (e.g., Selly and Shearman, 1962) and are widely reported from terrestrial and subaqueous sediments in the literature (e.g., Sieh, 1978; Scott and Price, 1988; Holzer et al., 1989; Audemard and de Santis, 1991; Guiraud and Plaziat, 1993; Lignier et al., 1998; Obermeier, 1996, 1998). Although multiple deformed horizons in a given section may support presence in a former seismically active region, it is important that analyzed horizons be temporally constrained so that horizons can be correlated regionally and so that the changes in deformation type can be convincingly traced from section to section. Several tens of deformed intervals are present in the Lexington Limestone, but few have been traced far in the complex Lexington facies mosaic (Fig. 2). The Brannon Member was specifically chosen because each of the three deformed horizons can be correlated regionally relative to a bentonite near the base of the unit (Black et al., 1965; Cressman, 1973) and a common basal hardground, which represents a regional flooding surface (Kulp, 1995). Both the basal Brannon bentonite and hardground can be traced locally across facies boundaries into the Tanglewood Member.

The widespread distribution of well-constrained, deformed horizons is a second criterion, inasmuch as seismic shaking is generally widely propagated. Thus, in relatively uniform units like the Brannon, widespread distribution of deformation is important because it largely precludes the likelihood of more locally expressed causes like mass movement and artesian effects. The areal extent of individual deformed horizons in the Brannon ranges from 800 to 2400 km^2 (Figs. 5–7), distributions well in excess of those commonly expected for more local causes.

A third criterion is a continuum of deformational features that show a regional pattern of increased frequency or increasing scale of deformation toward a possible epicentral area. Using the rationale and criteria of Lowe (1975), the mapped bands of deformation in Figures 5–7 reflect the effects of outwardly decreasing energy, exhibiting an irregular, but uniform pattern in all three deformed Brannon horizons.

The final criterion is the ability to exclude other possible nonseismic causes, which in this case, include artesian-induced deformation, rapid loading during burial, slope-induced deformation, and wave- or storm-induced deformation. Credible arguments against most of these alternatives have already been advanced by Pope et al. (1997) for deformed horizons in the Lexington Limestone. Therefore, we merely point out the most salient arguments as they apply to the Brannon Member.

Excluding other causes

Artesian-induced deformation most commonly occurs near fluvial or tidal channels due to rapid stage changes (Wunderlich, 1967; Kolb, 1976; Li et al., 1996; Obermeier, 1998), but the open-marine setting rules out fluvial causes, and the deep-ramp nature of Brannon deposition and absence of tidal structures suggest that there was little or no tidal influence. For similar reasons, the influence of tidal shear drag can also be excluded (Pope et al., 1997).

Depositional loading of coarser, denser sands over finer, water-saturated silts and clays may also cause soft-sediment deformation (e.g., McKee and Goldberg, 1969; Obermeier, 1996), but in the Brannon deep-ramp setting, the only major source of coarser materials would have been storms. Although distal storm sediments are common in the Brannon Member (Kulp, 1995), the predominant silt- and fine-sand–size sediments in these deposits seem incapable of generating the necessary load and density differences to produce deformation; otherwise, deformation would be found nearly everywhere in the Brannon, but it is simply not present beyond the three noted horizons. Pope et al. (1997) also argued that the deep-ramp setting, slow sedimentation rates, and absence of depositional point sources largely preclude load-induced deformation.

Slope-induced mass movement may also generate substantial soft-sediment deformation in marine settings (e.g., Martinsen,

Figure 8. Simple folding or contortion of beds due to hydroplastic deformation in the upper deformed horizon of the Brannon Member. Note truncation by the overlying Tanglewood Member. Outcrop near junction of Kentucky Route 52 and Balls Branch Run. Hammer is 28 cm long.

Figure 9. Close-up view of "homogenized" Brannon Member and remnants of a single bed in the process of "homogenization" due to fluidization processes. Clasts of partially homogenized Brannon float in a uniform matrix. Crude, dipping laminae result from internal flow lamination during fluidization. Outcrop near junction of U.S. 60 and Benson Creek. Hammer is 28 cm long.

1994). However, it is thought that the Ordovician ramp slope in Kentucky was, for the most part, less than the 2° threshold (Lowe, 1975) necessary for most nonseismic, slope-induced movement (Pope et al., 1997). However, major earthquakes may initiate mass movement on sea floors with slopes as gentle as 0.25° (Field et al., 1982), and it is possible that some Brannon deformation could represent seismically induced slumping. The widespread, uniform, generally fine grained nature of the Brannon Member and its planar, commonly hardground-encrusted basal contact indicate that the Brannon was deposited during a major flooding event that further inundated an already featureless, corroded, and sediment-starved surface with little or no inherited relief. The one exception would have been at the northeastern limit of its distribution, where Brannon calcisiltites and shales and Tanglewood calcarenites intertongue (Cressman, 1973; Kulp, 1995) (Figs. 2 and 5). Along this nearly linear boundary, channel-form bodies of skeletal calcarenites within the Brannon (Cressman and Karklins, 1970; Noger and Kepferle, 1985; Pope et al., 1997) suggest a slope down which Tanglewood shoal sands were transported into deeper water Brannon environments. Although these sand bodies are commonly deformed in ways that could suggest loading or slumping, the facts that they are everywhere part of a widespread Brannon deformed horizon and do not truncate that horizon indicate that their deformation, even if it was by local loading or slumping, was ultimately induced by the same larger cause that deformed other parts of the horizon; loading or slumping on this scale are otherwise just too localized to explain the widespread deformation in the three horizons. Moreover, the absence of the glide planes that would be expected in slumping (Kulp, 1995; Pope et al., 1997) and the random fold axes that characterize Brannon deformation (Kulp, 1995) militate against slumping, wherein most fold axes are preferentially oriented (Rupke, 1976; Woodcock, 1976).

Major storms can also induce soft-sediment deformation through overpressuring by high-amplitude storm waves (Kraft et al., 1985; Okusa, 1985) or by the drag force of bottom storm currents (Lowe, 1976; Orange and Breen, 1992). The Lexington area apparently sat astride major storm pathways in Ordovician time (Marsaglia and Klein, 1983; Duke, 1985; Ettensohn et al., 1986b), and storm-deposited or storm-reworked sediments are ubiquitous throughout the Lexington Limestone (e.g., Ettensohn, 1992b). In fact, the many micrograined limestone beds in the Brannon Member (Fig. 3) have been interpreted to represent distal tempestites formed by storm-generated back flow that interrupted muddy, background, suspension sedimentation in deep-ramp Brannon environments (Kulp, 1995), similar to the distal tempestites of Aigner (1982). Despite the abundance and widespread distribution of storm-related sedimentation in the Brannon, except for the three deformed horizons, each of which involves deformation of several smaller beds, other deformation of storm-related beds is absent. Moreover, there is an almost complete absence of sedimentary structures that reflect oscillatory water movements of the sort essential for generating the pressure differences that cause deformation, even though oscillation ripples and hummocky cross-strata are common in both underlying and overlying units. Finally, given the centimeter-size scale of most Brannon tempestites and the deep-ramp setting of the Brannon, the scale of most Brannon deformation (0.5–2.0 m) is too large to attribute to storm loading; deformation trends are nearly perpendicular to predicted storm pathways (see Marsaglia and Klein, 1983); and patterns of deformation intensity (Figs. 5–7) are difficult to explain using storm waves. Overall, it appears that most Brannon deposition occurred below storm wave base, and that storms, though recorded in the area, were rarely if ever strong enough to directly cause bottom overpressuring.

Other criteria

Many other individual criteria for differentiating seismogenic from other causes have been proposed, but we view none of them as strongly as we do positive results from the concomitant application of the above four criteria. However, we mention here three other criteria that support the Brannon seismite argument. The first of these is presence in a currently or formerly active seismic region and association with a potentially originating fault (e.g., Sims, 1975). On the basis of the presence of synsedimentary faulting in the Ordovician sequence (Black and Haney, 1975) and the coincidence of many facies boundaries in the Lexington Limestone with structural lineaments (Ettensohn et al., 1986a; Ettensohn, 1992b; Kulp, 1995; Rast and Ettensohn, 1995; Ettensohn and Kulp, 1995), synsedimentary reactivation of structures must have been common. Moreover, the thickest Brannon sequence (Fig. 10) and the most extensive areas of Brannon deformation (Figs. 5–7) are between two structural lineaments, suggesting active subsidence along the structures at depth and that seismicity was at least periodically part of this activity. Although no evidence indicates direct seismicity along the two structural trends, major areas of deformation and trends of deformation intensity approximately parallel the two lineaments (Figs. 5–7).

A second important criterion is an unusual crosscutting relationship between deformed and undeformed rocks in individual Brannon Member exposures. Deformed horizons are commonly found interlayered with undeformed rocks, but, more important, the deformed horizons crosscut both overlying and underlying undeformed layers at very subtle angles (Fig. 11A). Although crosscutting of lower rocks could be explained by erosion during emplacement, the crosscutting of overlying rocks from below is more difficult to explain. Moreover, continuation of bedding and lithologic characteristics from one side of the deformed horizon to the other precludes the possibility of onlap as an explanation for the "upper truncation." Additionally, in a few other places, isolated, lens-shaped bodies of deformed rock several meters long are wholly enclosed by undeformed rock (Fig. 11B). Both relationships reflect the fact that most marine, seismogenic, soft-sediment deformation probably takes place below the sediment-water interface and below a more impermeable overlying layer that acts to confine and enhance the effects of cyclic seismic loading on ambient pore pressure (e.g., Obermeier et al., 1990; Obermeier, 1996, 1998). As a result, seismically augmented pore pressure will deform susceptible sediments both within or obliquely truncating normally bedded, confining sediments, some of which will not deform because of local variations in conditions such as cementation or clay content. With the possible exception of storms, no other causes are known to generate such features; thus, these unusual relationships have also been used to support a seismogenic origin in other deposits (Davenport and Ringrose, 1987).

Perhaps more important is the fact that all horizons of deformation extend beyond the area of most active Brannon subsid-

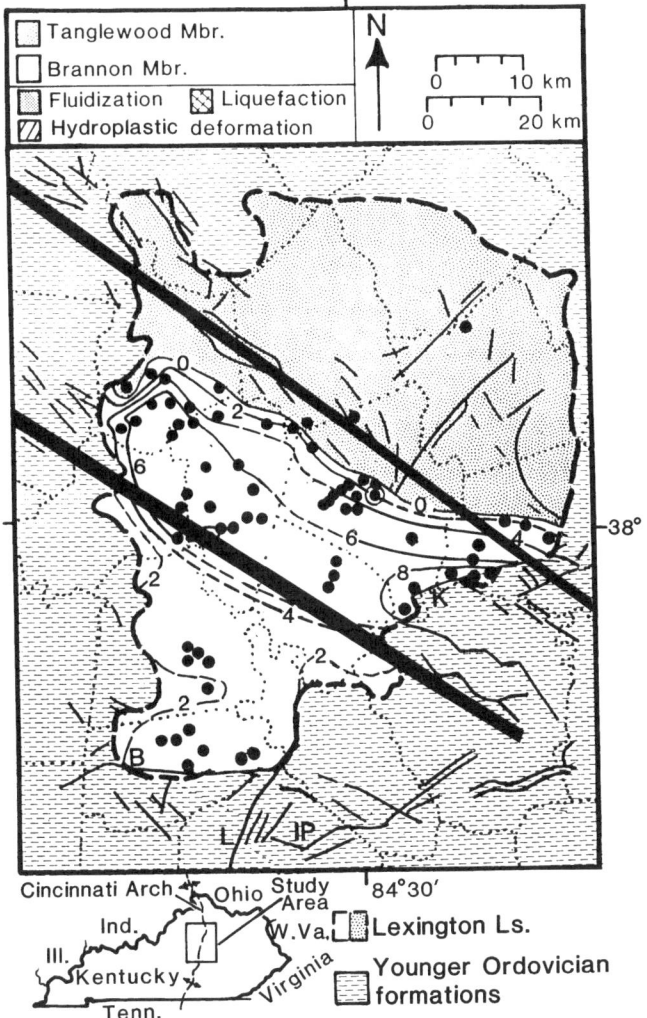

Figure 10. Isopach map (in meters) of the Brannon Member, showing relationships between thickness and structural trends noted in Figures 5–7. The fact that the thickest part of the Brannon is between these structural trends suggests synsedimentary growth faulting and graben formation along the trends and development of a small, troughlike Brannon basin in the northern part of its distribution. Concentration of Brannon seismites in this basin could reflect movement on faults associated with the trends or amplification due to increased sediment thickness. The basin is very similar in size and apparent origin to the Quaternary Seattle basin, where amplification factors of 6–10 have been noted (Brocher et al., 2000).

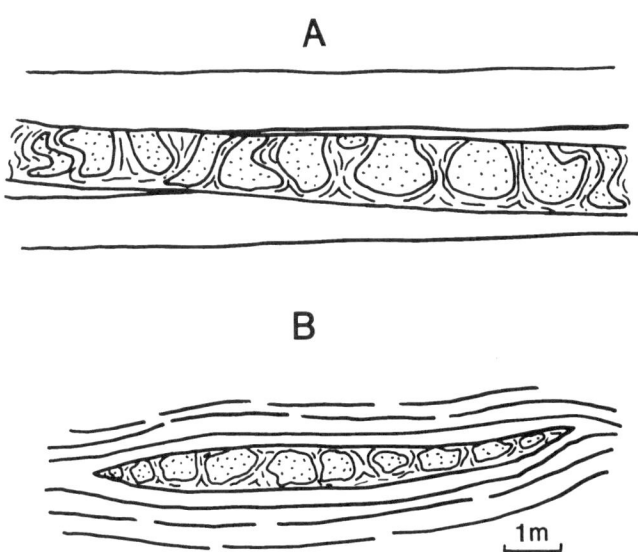

Figure 11. Schematic diagrams based on exposures illustrating A, a deformed horizon that crosscuts three beds, and B, a wholly enclosed, deformed horizon of limited continuity. Stippling represents flow rolls and contortion in calcisiltites or fine-grained calcarenites; dashed lines represent shaly sediments; and parallel lines represent undeformed beds.

ence between the two lineaments (Figs. 5–7) and cross facies boundaries into Tanglewood calcarenites along and just north of the Brannon-Tanglewood transition (Figs. 5–7). This means that deformation reflects a more basic cause that was not facies or subsidence specific—again, in combination with above criteria, excluding all but seismic causes.

DISCUSSION

The preponderance of evidence based on the above criteria points to seismicity as the most likely explanation for the three horizons of deformation in the Brannon Member. Although the ability to exclude other likely causes is important, more important is the widespread nature of the deformation, the ability to constrain each deformed horizon, and patterns of increasing intensity of deformation toward one or more possible sources of energy input (Figs. 5–7). In fact, mapping types of deformation is like mapping earthquake intensity, the violence of earthquake shaking based on the amount of damage done to structures, perceptions by humans, and secondary effects such as landslides and soil deformation (Richter, 1958). Although each earthquake varies in intensity from one place to another, the intensity is usually highest near the epicentral area and decreases away from it (e.g., Reiter, 1990; Yeats et al., 1997). The central area of greatest shaking and resultant damage and deformation is called the meizoseismal region, and it is commonly assumed to coincide with the epicentral area in analyses of prehistoric and preinstrumental earthquakes (Reiter, 1990). However, site effects, such as the type and thickness of sediments involved, basin configuration, and topography, can influence intensity distribution and lead to situations in which the meizoseismal area does not coincide with the epicentral area (Reiter, 1990).

Finding apparent epicentral areas

In the only other attempt to locate epicentral areas for ancient marine seismites, Pope et al. (1997) used the relative abundance of deformation and thickness trends in deformed horizons across the Appalachian basin to suggest possible broad epicentral locations. Ours is the first attempt to use intensity of deformation to locate specific epicentral areas in ancient rocks. If we assume relatively uniform site conditions and that lines demarcating deformation types are effectively isoseismals, or contour lines of equal intensity, then each map (Figs. 5–7) is an isoseismal map. Because deformation involving fluidization requires the greatest energy input (Lowe, 1975), we assume that areas of sediment homogenization (Fig. 9), which indicate fluidization, were at or near epicentral areas where energy output would have been greatest. With the exception of the lowest deformed horizon (Fig. 5), each horizon shows one area of homogenized sediments and concentric bands of deformation of decreasing intensity surrounding it (Figs. 5–7). The lowest horizon, in contrast, exhibits two areas of similar orientation where fluidization was effective, suggesting that nonuniform site conditions contributed to fluidization beyond the epicentral area or that the earthquake triggered an aftershock or another earthquake on a nearby fault segment (e.g., Stein et al., 1994; Harris et al., 1995; Kilb et al., 2000), which generated penecontemporaneous deformation. Inasmuch as there is a tendency for some earthquakes to cluster in time and space (Kagan and Jackson, 1991), and the most intense and broadest area of Brannon deformation is concentrated in the northwestern parts of each horizon (Figs. 5–7), it is likely that this represents the probable Ordovician epicentral area for the Brannon earthquakes.

Site effects

Site effects encompass several local conditions, such as basin shape, sediment thickness, and sediment type, which can affect earthquake expression in a given area (Reiter, 1990). A potentially important site effect relative to Brannon deformation is the presence of a local subsiding "trough" or possible graben of thickened Brannon sediments between the two structural lineaments (Fig. 10); in the trough, deformation is everywhere greater than 0.5 m thick, whereas beyond it, deformed horizons are generally thinner. In each deformed horizon, moreover, the area of most extensive and most intense deformation is in the northwestern parts of the trough and exhibits a prominent northwest-southeast orientation, suggesting a relationship to the northwest-southeast–oriented trough in which it occurs. Although there is no definite explanation for these patterns, the greater thicknesses of the deformed horizons in the trough may reflect proximity to epicentral areas (e.g., Obermeier, 1996; Obermeier and Pond, 1999), whereas orientation and concentration of deformed horizons could represent site effects or proximity to movement along another northwest-southeast–oriented fault paralleling the two boundary faults. This pronounced northwest-southeast orientation could also reflect earthquake directivity, the preferential orientation of ground motion due to propagation of seismic waves from a moving fault rupture rather than a point rupture (e.g., Reiter, 1990; Yeats et al., 1997).

Site effects, in particular, are difficult to analyze in such ancient situations, and in this case relevant effects could include presence in a structural trough, the bedrock configuration of that trough, increased sediment thickness, liquefaction susceptibility of that sediment, and seismological factors, such as focal depth and shaking frequency of the earthquake (e.g., Reiter, 1990; Obermeier, 1996). Such effects may work to amplify (i.e., increase the amplitude of earthquake waves), or attenuate (i.e., decrease the amplitude of earthquake waves) the earthquake energy released with time and distance traveled. Increased thicknesses of unconsolidated sediments may give rise to amplification factors as high as 2–2.5 (Idriss, 1990), and in the Quaternary Seattle basin, a structural basin of about the same size as the structurally defined area of thickened Brannon (Fig. 10), site effects related to basin geometry and sediment infill generated amplification factors between 6 and 10 (Brocher et al., 2000). Overall, the occurrence and patterns of deformation in the Brannon Member suggest that site effects were present, and their influence seems to have been one of overall amplification, especially in the trough.

Finally, the possible influence of other nearby structures must be considered. The distribution of deformation shown in Figures 5–7 indicates that the extent of deformation in successive ascending horizons decreases, perhaps reflecting a decreasing magnitude for each successive earthquake. Nonetheless, in each of the horizons, areas of equivalent deformation beyond the larger central deformed areas around apparent epicenters are largely associated with and seem to parallel extant structures in the area (Figs. 5–7). Clearly, Brannon deformation near the structures is not related to extant structures, because they crosscut the Brannon and younger units; however, the deformation could be related to subsurface precursors of extant structures, which stratigraphic evidence indicates were being episodically reactivated at the time (Ettensohn et al., 1986a; Ettensohn, 1992b; Kulp, 1995; Ettensohn and Kulp, 1995). Because there is a growing realization that earthquakes on one structure may trigger others on the same or nearby structures at nearly the same or later times through transfer of stress (e.g., Stein et al., 1994; Harris et al., 1995; Yeats et al., 1997; Reilinger et al., 2000; Kilb et al., 2000), it is tempting to suggest that similar processes of stress transfer were operative during Brannon deformation. Moreover, strong directivity of the major seismic events—in the Brannon horizons, possibly toward the southeast—can trigger penecontemporaneous and subsequent events on other structures in the same direction (e.g., Yeats et al., 1997). However, because the effects of erosion in intervening areas and other site effects cannot be ruled out, it must remain just that, a tempting suggestion.

Estimating earthquake magnitude

If specific soft-sediment deformation can be linked to a seismogenic origin, the minimum-magnitude earthquake required for liquefaction-produced deformation is 5, and more commonly 5.5 or larger (Carter and Seed, 1988; Ambraseys, 1988; Obermeier et al., 1990). This number provides a working minimum for the Brannon horizons, but there are other methods for estimating the strength of paleo-earthquakes (e.g., Obermeier, 1998). Of these, only the magnitude-bound method of Ambraseys (1988) is applicable to the Brannon horizons. This method requires knowledge of the epicentral area and relates distance from the epicenter to the farthest observed deformation to a given moment magnitude (M_w). This method is based on the fact that for shallow-focus earthquakes (<50 km), a greater input of energy, and hence a greater earthquake magnitude, is necessary to deform sediments increasingly distant from an epicenter, assuming uniform conditions (Youd and Perkins, 1978; Ambraseys, 1988). The epicentral distance to the farthest deformation is then compared to a curve developed from historic observations (e.g., Ambraseys, 1988; Obermeier, 1996, 1998). Epicentral distances were measured from Figures 5–7 for those parts of the deformed Brannon horizons immediately surrounding the apparent epicenters; comparison to the curve (e.g., Obermeier, 1996) gives the following ranges of possible moment magnitudes: for the lower horizon 6.7–7.7, for the middle horizon 6.4–7.4, and for the upper horizon 5.8–6.6. Although these estimates show decreasing magnitude with time, which is reflected in the actual nature and patterns of deformation, they may be far too precise for the data that generated them. In fact, it is possible that various site effects may have increased or decreased the degree of deformation, thereby skewing the results.

The magnitude-bound method and resulting curve, however, were developed for siliciclastic sediments in terrestrial settings, and exactly how the method and curve translate to carbonate-rich sediments in marine settings is uncertain; certainly, the method does not consider the effects of the overlying water column or the greater density of carbonates. Because fluidization velocity (U_0) is directly proportional to the density of the phases involved (Lowe, 1975), it is possible that more energy is required for the same deformation in carbonates than it is in siliciclastics. Thus, the magnitudes necessary to deform the Brannon horizons were at least in the range of 6–8, but they could have been even greater.

Possible recurrence intervals

Pope et al. (1997) calculated approximate recurrence intervals of 2.3–3.5 m.y. for magnitude 6 earthquakes, on the basis of the presence of four major seismite horizons throughout the Ordovician sequence in parts of the Appalachian basin. What we offer here, in contrast, is a closer look at a part of one of those intervals, which includes the Brannon Member. The Brannon and overlying parts of the Tanglewood Member represent a shallowing-upward, fourth-order sequence in the scheme of Pope and Read (1995); on the basis of sequence stratigraphy and absolute dates, they estimated the duration of such cycles to have been 50–130 k.y. If the Brannon is assumed to represent a half-cycle and the three deformed horizons reflect more or less regularly episodic earthquakes, then possible recurrence intervals of 8.3–21.7 k.y. can be calculated. As noted by Pope et al. (1997), even these recurrence intervals are much longer than those on many Holocene and historic faults, although the recurrence intervals are within the 10^3–10^5 yr range that characterizes most normal faults (Yeats et al., 1997). Such long-term recurrence intervals, however, must also be viewed with caution because of the likelihood of temporal and spatial clustering of earthquakes (Kagan and Jackson, 1991; Yeats et al., 1997). For example, it is possible that each Brannon horizon represents a cluster of related earthquakes, and that only one of those was of sufficient magnitude to manifest the intense deformation used here to characterize apparent epicentral zones. The horizons may also reflect only those times with earthquakes of magnitudes greater than 6, while the intervening intervals may have had no fewer earthquakes, but only ones with magnitudes that were not great enough to deform the sediments. The effects of erosion, intense bioturbation, and superimposed deformation offer yet other possibilities. Although there are many possibilities, there is some suggestion that clustering may be a factor in the Brannon horizons, because there are two apparent epicentral areas in the lower horizon, and each horizon exhibits one or more outlying loci of deformation that are separated from the larger central areas (Figs. 5–7). Nonetheless, if the large central deformed areas in each horizon are not the products of individual high-magnitude earthquakes, then the generally concentric arrangement of deformation around each apparent epicentral area calls for the very specific, and perhaps unlikely, spatial clustering of earthquakes by magnitude.

CONCLUSIONS

Because of the many uncertainties about the influence of seismicity in marine settings, the study of marine seismites is in its infancy. In fact, there are currently no modern marine analogues and very few studies of ancient marine seismites involving carbonates, especially in epicontinental settings. Of necessity, most of the applied analogues and models are based on examples from active margins that involve siliciclastic sediments in relatively young, terrestrial settings. In contrast, the Brannon seismites formed on a Middle Ordovician, epicontinental, carbonate platform in stratigraphic and structural settings that suggest frequent reactivation of basement structures by far-field processes generated by the coeval Taconian orogeny.

Soft-sediment deformation of the sort observed in the Brannon Member of the Lexington Limestone is not a unique indicator of seismicity. However, use of four concurrent criteria for deformation, which include deformation that is consistent with a seismogenic origin, is widespread in temporally and stratigraphically constrained horizons, shows a regional pattern of increasing frequency or intensity toward a likely epicentral area, and allows exclusion of other likely causes, can strongly reduce the ambiguity about seismic origin. Not only does deformation in the three Brannon horizons meet all the criteria above, but it also crosses facies boundaries, exhibits folds with random axes, demonstrates unusual crosscutting relationships with undeformed beds below it, and occurs in association with structures that were being reactivated at the time, on the basis of their concurrence with facies boundaries. Together, this association of criteria clearly suggests that the three horizons of Brannon soft-sediment deformation warrant interpretation as seismites.

Mapping the intensity of Brannon deformation on the basis of the types of deformation produced possible isoseismal maps that show at least one center of intense deformation for each deformed horizon. The mapped centers of intense deformation reflect former meizoseismal areas, or areas of severe shaking, that on many isoseismal maps would be interpreted as epicentral areas. However, because of likely site effects, or local conditions that amplify or attenuate the shaking, these areas in each Brannon horizon are best interpreted as "apparent" epicentral areas. Similarly, these maps, along with other stratigraphic information, suggest magnitudes ranging from 6 to 8 and recurrence intervals from 8.3 to 21.7 k.y. This study shows that the combined use of stratigraphy and intensity mapping for well-constrained and well-exposed seismite horizons can provide information about ancient seismic parameters such as possible epicenters, site effects, earthquake directivity, earthquake clustering, magnitude, and recurrence intervals; however, resulting interpretations must necessarily be very tentative until better data and definite modern, marine analogues are available.

Seismogenic horizons, like those in the Brannon, must abound in other cratonic sequences, and recognition of them has potential not only for use as local and regional chronostratigraphic marker horizons, but also for altering current ideas about the influence of seismicity on shallow-water, epicontinental sedimentation. The example of the Brannon seismites and the abundance of probable seismites in other parts of the Lexington Limestone and its equivalents suggest that earthquakes were characteristic of the craton interior at certain times in geologic history and that further interpretation of seismites from the old, Phanerozoic, marine sedimentary record may allow us to observe and quantify many of their effects.

ACKNOWLEDGMENTS

We thank C.E. Brett, S.F. Greb, and an anonymous reviewer for their constructive criticism of earlier versions of the manuscript; their comments have greatly improved the paper.

REFERENCES CITED

Aigner, T., 1982, Calcareous tempestites: Storm-dominated stratification in Upper Muschelkalk limestones (Middle Trias, SW-Germany), in Einsele, G., and Seilacher, A., eds., Cyclic and event stratification: Berlin, Springer-Verlag, p. 180–207.

Allen, J.R.L., 1977, The possible mechanics of convolute lamination in graded sandstone beds: Journal of the Geological Society of London, v. 134, p. 19–31.

Allen, J.R.L., 1984, Sedimentary structures: Their character and physical basis: Amsterdam, Elsevier, Developments in Sedimentology, v. 30, 1256 p.

Ambraseys, N.N., 1988, Engineering seismology: Earthquake Engineering & Structural Dynamics, v. 17, p. 1–105.

Anand, A., and Jain, A.K., 1987, Earthquakes and deformation structures (seismites) in Holocene sediments from the Himalayan-Andaman Arc, India: Tectonophysics, v. 133, p. 105–120.

Audemard, F.A., and de Santis, F., 1991, Survey of liquefaction structures induced by recent modern earthquakes: Bulletin of the International Association of Engineering Geology, v. 44, p. 5–16.

Black, D.B.F., 1986, Basement faulting in Kentucky, in Aldrich, M.J., and Laughlin, A.W., eds., Proceedings of the Sixth International Conference on Basement Tectonics: Salt Lake City, International Basement Tectonics Association, p. 125–139.

Black, D.B.F., and Haney, D.C., 1975, Selected structural features and associated dolostone occurrences in the vicinity of the Kentucky River fault system, Field Guide, Annual Field Conference of the Geological Society of Kentucky: Lexington, Kentucky Geological Survey, ser. 10, 27 p.

Black, D.B.F., Cressman, E.R., and MacQuown, W.C., Jr., 1965, The Lexington Limestone (Middle Ordovician) of central Kentucky: U.S. Geological Survey Bulletin 1224-C, 29 p.

Brocher, T.M., Pratt, T.L., Creager, K.C., Crosson, R.S., Steele, W.P., Weaver, C.S., Frankel, A.D., Tréhu, A.M., Snelson, C.M., Miller, K.C., Harder, S.H., and ten Brink, U.S., 2000, Urban seismic experiments investigate Seattle fault and basin: Eos (Transactions, American Geophysical Union), v. 81, p. 545, 551–552.

Carter, D.P., and Seed, H.B., 1988, Liquefaction potential of sand deposits under low levels of excitation, Report No. UCB/EERC-81/11: Berkeley, University of California, College of Engineering, 119 p.

Cita, M.B., and Ricci Lucchi, F., 1984, Preface: Seismicity and sedimentation: Marine Geology, v. 55, p. 1–4.

Cressman, E.R., 1973, Lithostratigraphy and depositional environments of the Lexington Limestone (Ordovician) of central Kentucky: U.S. Geological Survey Professional Paper 768, 61 p.

Cressman, E.R., and Karklins, D.L., 1970, Lithology and fauna of the Lexington Limestone (Ordovician) of central Kentucky, in Guidebook for field trips, Geological Society of America, Southeastern Section: Lexington, Kentucky Geological Survey, p. 17–28.

Davenport, C.A., and Ringrose, P.S., 1987, Deformation of Scottish Quaternary sediment sequences by strong earthquake motion, in Jones, M.E., and Preston, R.M.F., eds., Deformation of sediments and sedimentary rocks: Geological Society [London] Special Publication 29, p. 299–314.

Drahovzal, J.A., Harris, D.C., Wickstrom, L.H., Walker, D., Keith, B., and Furer, L.C., 1992, The East Continental Rift basin: A new discovery: Kentucky Geological Survey Special Publication 18, ser. 11, 25 p.

Duke, W.L., 1985, Hummocky cross-stratification, tropical storms, and intense winter storms: Sedimentology, v. 32, p. 167–194.

Ettensohn, F.R., 1991, Flexural interpretation of relationships between Ordovician tectonism and stratigraphic sequences, central and southern Appalachians, in Barnes, C.R., and Williams, S.H., eds., Advances in Ordovician geology: Geological Survey of Canada Paper 90-9, p. 213–224.

Ettensohn, F.R., 1992a, General Ordovician paleogeographic and tectonic framework for Kentucky, in Ettensohn, F.R., ed., Changing interpretations of Kentucky geology: Layer-cake, facies, flexure, and eustacy: Ohio Division of Geological Survey Miscellaneous Report No. 5, p. 19–22.

Ettensohn, F.R., 1992b, Regressive facies in the upper Lexington Limestone: Tanglewood-Millersburg relationships, in Ettensohn, F.R., ed., Changing interpretations of Kentucky geology: Layer-cake, facies, flexure, and eustacy: Ohio Division of Geological Survey Miscellaneous Report No. 5, p. 62–66.

Ettensohn, F.R., 1992c, Kentucky River fault zone, the Lexington Limestone and the Clays Ferry Formation near the southeastern margin of the Tanglewood buildup, in Ettensohn, F.R., ed., Changing interpretations of Kentucky geology: Layer-cake, facies, flexure, and eustacy: Ohio Division of Geological Survey Miscellaneous Report No. 5, p. 66–68.

Ettensohn, F.R., and Kulp, M.A., 1995, Structural-tectonic control on Middle-Late Ordovician deposition of the Lexington Limestone, central Kentucky, in Cooper, J.D., Droser, M.L., and Finney, S.C., eds., Ordovician odyssey: Short papers for the Seventh International Symposium on the Ordovician System: Fullerton, Pacific Section, SEPM (Society for Sedimentary Geology), p. 261–264.

Ettensohn, F.R., and Pashin, J.C., 1992, A brief structural framework of Kentucky, *in* Ettensohn, F.R., ed., Changing interpretations of Kentucky geology: Layer-cake, facies, flexure, and eustacy: Ohio Division of Geological Survey Miscellaneous Report No. 5, p. 6–9.

Ettensohn, F.R., Amig, B.C., Pashin, J.C., Greb, S.F., Harris, M.Q., Black, J.C., Cantrell, D.J., Smith, C.A., McMahan, T.M., Axon, A.G., and McHargue, C.J., 1986a, Paleoecology and paleoenvironments of the bryozoan-rich Sulphur Well member, Lexington Limestone (Middle Ordovician), central Kentucky: Southeastern Geology, v. 26, p. 199–219.

Ettensohn, F.R., Pashin, J.C., and Jacobs, G.W., 1986b, Characteristics of shallow-water, marine, shelf silts and sands: Two Paleozoic examples from eastern Kentucky, *in* Shumaker, R.C., ed., Appalachian Basin Industrial Associates Program, Spring Meeting: Morgantown, West Virginia University, v. 10, p. 197–211.

Ettensohn, F.R., Rast, N., and Kulp, M.A., 2000, Locating possible epicentral areas for paleoearthquakes, Middle Ordovician Lexington Limestone, central Kentucky: Geological Society of America Abstracts with Programs, v. 32, p. A215.

Field, M.E., Gardner, J.V., Jennings, A.E., and Edwards, B.D., 1982, Earthquake induced sediment failures on a 0.25° slope, Klamath River delta, California: Geology, v. 10, p. 542–546.

Guiraud, M., and Plaziat, J.-C., 1993, Seismites in the fluviatile Bima sandstones: Identification of paleoseisms and discussion of their magnitudes in a Cretaceous synsedimentary strike-slip basin (Upper Benue, Nigeria): Tectonophysics, v. 225, p. 493–522.

Hamilton, R.M., and Zoback, M.D., 1982, Tectonic features of the New Madrid seismic zone from seismic-reflection profiles, *in* McKeon, F.A., and Pakiser, L.C., eds., Investigations of the New Madrid, Missouri, earthquake region: U.S. Geological Survey Professional Paper 1236-I, p. 55–82.

Harris, R.A., Simpson, R.W., and Reasenberg, P.A., 1995, Influence of static stress changes on earthquake locations in southern California: Nature, v. 375, p. 221–224.

Herrmann, R.B., Langston, C.A., and Zollweg, J.E., 1982, The Sharpsburg, Kentucky, earthquake of 27 July 1980: Bulletin of the Seismological Society of America, v. 72, p. 1219–1239.

Hildebrand, T.G., 1985, Rift structure of the northern Mississippi embayment from analysis of gravity and magnetic data: Journal of Geophysical Research, v. 90, p. 12607–12622.

Holzer, T.L., Youd, T.L., and Hanks, T.C., 1989, Dynamics of liquefaction during the 1987 Superstition Hills California, earthquake: Science, v. 244, p. 59–69.

Idriss, I.M., 1990, Response of soft soil sites during earthquakes, *in* Proceedings, H. Bolton Seed Memorial Symposium: Vancouver, BiTech Publishers, Ltd., v. 2, p. 273–289.

Kagan, Y.Y., and Jackson, D.D., 1991, Long-term earthquake clustering: Geophysical Journal International, v. 104, p. 117–133.

Keith, B.D., 1989, Regional facies of Upper Ordovician Series of eastern North America, *in* Keith, B.D., ed., The Trenton Group (Upper Ordovician Series) of eastern North America, deposition, diagenesis, and petroleum: American Association of Petroleum Geologists, Studies in Geology No. 29, p. 1–16.

Kilb, D., Gomberg, J., and Bodin, P., 2000, Triggering of earthquake aftershocks by dynamic stresses: Nature, v. 408, p. 570–574.

Kolb, C.R., 1976, Geologic control of sand boils along Mississippi River levees, *in* Coates, D.R., ed., Geomorphology and engineering: Stroudsburg, Pennsylvania, Dowden, Hutchinson, and Ross, Inc., p. 99–114.

Kraft, L.M., Jr., Helfrich, S.C., Suhayda, J.N., and Marin, J.E., 1985, Soil responses to ocean waves: Marine Geotechnology, v. 6, p. 173–203.

Kulp, M.A., 1995, Paleoenvironmental interpretation of the Brannon Member, Middle-Upper Ordovician Lexington Limestone, central Bluegrass region of Kentucky [M.S. thesis]: Lexington, University of Kentucky, 222 p.

Leeder, M., 1987, Sediment deformation structures and the paleotectonic analysis of sedimentary basins, with a case study from the Carboniferous of northern England, *in* Jones, M.E., and Preston, R.M.F., eds., Deformation of sediments and sedimentary rocks: Geological Society [London] Special Publication 29, p. 137–146.

Li, Y., Craven, J., Sweig, E.S., and Obermeier, S.F., 1996, Sand boils induced by the 1993 Mississippi river flood: Could they be one day be misinterpreted as earthquake-induced liquefaction?: Geology, v. 24, p. 171–174.

Lignier, V., Beck, C., and Chapron, E., 1998, Caractérisation géometrique et texturale de perturbations synsédimentaires attribuées à des séismes, dans une formation quaternaire glaciolacustre des Alpes ("les Argiles du Trièves"): Comptes Rendus de l'Académie des Sciences, Serie 2, Sciences de la Terre et des Planètes, Earth and Planetary Sciences, v. 327, p. 645–652.

Lowe, D.R., 1975, Water escape structures in coarse-grained deposits: Sedimentology, v. 22, p. 157–204.

Lowe, D.R., 1976, Subaqueous liquefied and fluidized sediment flows and their deposits: Sedimentology, v. 23, p. 285–308.

Marsaglia, K.M., and Klein, G., deV., 1983, The paleogeography of Paleozoic and Mesozoic storm depositional systems: Journal of Geology, v. 91, p. 117–142.

Martinsen, O., 1994, Mass movements, *in* Maltman, A., The geological deformation of sediments: London, Chapman & Hall, p. 127–165.

McAlpin, J.P., and Nelson, A.R, 1996, Introduction to paleoseismology, *in* McAlpin, J.P., ed., Paleoseismology: San Diego, Academic Press, p. 1–32.

McFarlan, A.C., 1943, Geology of Kentucky: Lexington, University of Kentucky, 531 p.

McKee, E.D., and Goldberg, M., 1969, Experiments on formation of contorted structures in mud: Geological Society of America Bulletin, v. 80, p. 231–244.

Noger, M.C., and Kepferle, R.C., 1985, Stratigraphy along and adjacent to the Bluegrass Parkway, Geological Society of Kentucky Annual Field Conference: Lexington, Kentucky Geological Survey, 24 p.

Obermeier, S.F., 1996, Use of liquefaction-induced features for paleoseismic analysis: An overview of how seismic liquefaction features can be distinguished from other features and how their regional distribution and properties of source sediment can be used to infer the location and strength of Holocene paleo-earthquakes: Engineering Geology, v. 44, p. 1–76.

Obermeier, S.F., 1998, Liquefaction evidence for strong earthquakes of Holocene and latest Pleistocene ages in the states of Indiana and Illinois, USA: Engineering Geology, v. 50, p. 227–254.

Obermeier, S.F., and Pond, E.C., 1999, Issues in using liquefaction features for paleoseismic analysis: Seismological Research Letters, v. 70, p. 34–58.

Obermeier, S.F., Jacobson, R.B., Smoot, J.P., Weems, R.E., Gohn, G.S., Monroe, J.E., and Powars, D.S., 1990, Earthquake-induced liquefaction features in the coastal setting of South Carolina and in the fluvial setting of the New Madrid seismic zone: U.S. Geological Survey Professional Paper 1504, 44 p.

Obermeier, S.F., Martin, J.R., Frankel, A.D., Youd, T.L., Munson, C.A., and Pond, E.C., 1993, Liquefaction evidence for one or more strong Holocene earthquakes in the Wabash valley of southern Indiana and Illinois, with a preliminary estimate of magnitude: U.S. Geological Survey Professional Paper 543-D, 27 p.

Obermeier, S.F., Garniewicz, R.C., and Munson, P.J., 1996, Seismically induced paleoliquefaction features in southern half of Illinois: Seismological Research Letters, v. 67, p. 49.

Okusa, S., 1985, Measurements of wave-induced pore pressure in submarine sediments under various marine conditions: Marine Geotechnology, v. 6, p. 119–144.

Orange, D.L., and Breen, N.A., 1992, The effects of fluid escape on accretionary wedges, seepage force, slope failure, headless submarine canyons and vents: Journal of Geophysical Research, v. 97, p. 9277–9295.

Pond, E.C., and Martin, J.R., 1996, Estimates of prehistoric earthquake magnitude and regional attenuation characteristics based on geotechnical study of paleoliquefaction evidence in the Wabash valley: Seismological Research Letters, v. 67, p. 50.

Pope, M.C., and Read, J.F., 1995, Sequences and meter-scale cyclicity of Middle to Late Ordovician cool water carbonates and clastics of Kentucky, *in* Cooper, J.D., Droser, M.L., and Finney, S.C., eds., Ordovician odyssey: Short papers for the Seventh International Symposium on the Ordovician System: Fullerton, California, Pacific Section, SEPM (Society for Sedimentary Geology), p. 333–336.

Pope, M.C., and Read, J.F., 1997, High-resolution stratigraphy of the Lexington Limestone (Late Middle Ordovician), Kentucky, U.S.A.: A cool-water carbonate-clastic ramp in a tectonically active foreland basin, *in* James, N.P., and Clark, J.A.D., eds., Cool-water carbonates: SEPM (Society for Sedimentary Geology) Special Publication No. 56, p. 411–429.

Pope, M.C., Read, J.F., Bambach, R., and Hofmann, H.J., 1997, Late Middle to Late Ordovician seismites of Kentucky, southwest Ohio and Virginia: Sedimentary recorders of earthquakes in the Appalachian basin: Geological Society of America Bulletin, v. 109, p. 489–503.

Rast, N., and Ettensohn, F.R., 1995, Effects of seismic disturbance on epicontinental depositional systems in the Ordovician and Devonian rocks of central Kentucky: Geological Society of America Abstracts with Programs, v. 27, p. A381.

Rast, N., and Goodmann, P., 1995, Tectonic and sedimentary consequences of Late Proterozoic and early Middle Paleozoic overthrusting in Kentucky: Northeastern Geology, v. 37, p. 1–12.

Rast, N., Ettensohn, F.R., and Rast, D.E., 1999, Taconian seismogenic deformation in the Appalachian orogen and the North American craton, *in* MacNiocaill, C., and Ryan, P.D., eds., Continental tectonics: Geological Society [London] Special Publication 164, p. 127–137.

Reilinger, R., Tokosoz, N., McClusky, S., and Barka, A., 2000, 1999 Izmit, Turkey earthquake was no surprise: GSA Today, v. 10, p. 1–6.

Reiter, L., 1990, Earthquake hazard analysis, issues and insights: New York, Columbia University Press, 254 p.

Richter, C.F., 1958, Elementary seismology: San Francisco, W.H. Freeman, 768 p.

Rupke, N.A., 1976, Large-scale slumping in a flysch basin, southwestern Pyrenees: Journal of the Geological Society of London, v. 132, p. 121–130.

Russ, D.P., 1979, Late Holocene faulting and earthquake occurrence in the Reelfoot Lake area, northwestern Tennessee: Geological Society of America Bulletin, v. 90, p. 1013–1018.

Russ, D.P., 1982, Style and significance of surface deformation in the vicinity of New Madrid, Missouri: U.S. Geological Survey Professional Paper 1236-H, p. 95–114.

Scott, B., and Price, S., 1988, Earthquake-induced structures in young sediments: Tectonophysics, v. 147, p. 165–170.

Seilacher, A., 1969, Fault-graded beds interpreted as seismites: Sedimentology, v. 13, p. 155–159.

Seilacher, A., 1984, Sedimentary structures tentatively attributed to seismic events: Marine Geology, v. 55, p. 1–12.

Seilacher, A., 1991, Events and their signatures: An overview, *in* Einsele, G., Ricken, W., and Seilacher, A., eds., Cycles and events in stratigraphy: Berlin, Springer-Verlag, p. 222–226.

Shumaker, R.C., 1986, Structural development of Paleozoic continental basins in eastern North America, *in* Aldrich, M.J., and Laughlin, A.W., eds., Proceedings of the Sixth International Conference on Basement Tectonics: Salt Lake City, International Basement Tectonics Association, p. 82–95.

Sieh, K.E., 1978, Prehistoric large earthquakes produced by slip on the San Andreas Fault at Palett Creek, California: Journal of Geophysical Research, v. 83, p. 3907–3939.

Selly, R.C., and Shearman, D.J., 1962, The experimental production of sedimentary structures in quicksands: Proceedings of the Geological Society of London, v. 1599, p. 101–102.

Sims, J.D., 1973, Earthquake-induced sedimentary structures in Van Norman Lake, San Fernando, California: Geological Society of America Abstracts with Programs, v. 5, p. 107–108.

Sims, J.D., 1975, Determining earthquake recurrence intervals from deformational structures in young lacustrine sediments: Tectonophysics, v. 29, p. 141–152.

Stein, R.S., King, G.C., and Lin, J., 1994, Stress triggering of the 1994 M = 6.7 Northridge earthquake by its predecessors: Science, v. 265, p. 1432–1435.

Street, R., Taylor, K., Jones, D., Harris, J., Steiner, G., Zekulin, A., and Zhang, D., 1993, The 4.6 $m_{b,Lg}$ northeastern Kentucky earthquake of September 7, 1988: Seismological Research Letters, v. 64, p. 187–199.

Tuttle, M.P., Lafferty, R.H., Cande, R.F., Chester, J.S., and Haynes, M., 1996a, Evidence of earthquake-induced liquefaction north of the New Madrid seismic zone, central United States: Seismological Research Letters, v. 67, p. 58.

Tuttle, M.P., Schweig, E.S., Lafferty, R.H., and Guccione, M.J., 1996b, Update on paleoliquefaction study in the New Madrid seismic zone, central United States: Seismological Research Letters, v. 67, p. 58.

Van Arsdale, R.B., 1986, Quaternary displacement on faults within the Kentucky River fault system of east-central Kentucky: Geological Society of America Bulletin, v. 97, p. 1382–1392.

Wesnousky, S.G., and Leffler, L.M., 1994, A search for paleoliquefaction and evidence bearing on the recurrence behavior of the Great 1811–12 New Madrid earthquakes: U.S. Geological Survey Professional Paper 1538-H, 42 p.

Woodcock, N.H., 1976, Ludlow series slumps and turbidites and the form of the Montgomery Trough, Powys, Wales: Proceedings of the Geological Association of London, v. 87, p. 169–182.

Wunderlich, F., 1967, Die Enstehung von "convolute bedding" an Platenrändern: Senckenbergiana Lethaea, v. 48, p. 345–349.

Yeats, R.S., Sieh, K., and Allen, C.R., 1997, The geology of earthquakes: New York, Oxford University Press, 568 p.

Youd, T.L., and Perkins, D.M., 1978, Mapping liquefaction-induced ground failure potential: Journal of Geotechnical Engineering, v. 104, p. 433–446.

Ziegler, P.A., 1987, Late Cretaceous and Cenozoic intra-plate compressional deformations in the Alpine foreland: A geodynamic model: Tectonophysics, v. 137, p. 389–420.

MANUSCRIPT ACCEPTED BY THE SOCIETY MAY 11, 2001